Mobile-to-Mobile Wireless Channels

For a listing of recent titles in the
Artech House Mobile Communications Series, turn to the back of this book.

Mobile-to-Mobile Wireless Channels

Alenka Zajić

ARTECH HOUSE
BOSTON | LONDON
artechhouse.com

Library of Congress Cataloging-in-Publication Data
A catalog record for this book is available from the U.S. Library of Congress.

British Library Cataloguing in Publication Data
A catalog record for this book is available from the British Library.

ISBN-13: 978-1-60807-495-2

Cover design by Adam Renvoize

10 9 8 7 6 5 4 3 2 1

To
my husband, Miloš, and daughter, Mia
for their encouragement, support, and patience.

Contents

Preface

Mobile communications are rapidly becoming *the* communications in most people's minds, because they provide the ability to connect people anywhere and at any time, even on the move. While present-day mobile communications systems can be classified as "fixed-to-mobile" because they allow mobility only on one end (e.g., the mobile phone), while the other end (e.g., the base station) is immobile, emerging "mobile-to-mobile" communications systems allow mobile users or vehicles to directly communicate with each other.

The driving force behind mobile-to-mobile communications is consumer demand for better coverage and quality of service (e.g., in rural areas where base stations or access points are sparse or not present or in disaster-struck areas where the fixed infrastructure is absent), as well as increased mobility support, location-based services, and energy efficient communication (e.g., for cars travelling in opposite directions that exchange information about traffic conditions ahead or when mobile devices "gang together" to reach a far-away base station so that none of them expends a lot of powers). Although mobile-to-mobile is still a relatively young technology, it is already finding its way into wireless standards, such as IEEE 802.22 for cognitive radio, IEEE 802.11p for intelligent transportation systems, and IEEE 802.16 for WiMax systems.

Propagation in mobile-to-mobile (M-to-M) wireless channels is different from traditional fixed-to-mobile channels, and the quality of service, energy efficiency, mobility support, and other advantages of M-to-M communication all depend on having good models of the M-to-M propagation channels. This book provides a detailed introduction into the state of the art in M-to-M wireless propagation and models for such propagation. It is intended to serve as an educational tool for those just entering this exciting area and to serve as a reference for practitioners and designers of M-to-M communication systems.

The book begins with a brief introduction, review of radio propagation, and statistical characteristics used to describe wireless communication channels. The book continues with a description of fixed-to-mobile channel modeling to allow existing practitioners to contrast the M-to-M channel to the more familiar F-to-M channels. The book then covers typical mobile-to-mobile communication scenarios: vehicle-to-vehicle communications, air-to-ground communications, and underwater autonomous vehicle communications. Finally, the book provides a CD with MATLAB implementations of key channel models described in the book. More about the contents of the CD can be found in the README.TXT file. This CD is intended to facilitate use of these models in future systems and to provide a starting point for development of new models and algorithms. The author hopes that the book will be very useful to practitioners, researchers, and students at graduate and post-graduate level.

CHAPTER 1

Introduction

Mobile communications are one of the most active and rapidly growing areas of the telecommunications industry. The primary reason for this is that mobile communications offer connectivity to people "on the move." Over the past several decades, all mobile communications systems in widespread use offered mobility only on the user's end, while the other (service provider's) end is fixed (immobile). Examples of these fixed-to-mobile (F-to-M) communications include cellular telephony, WiFi, and aircraft-tower communication.

In contrast to this provider-to-user communication approach, several emerging wireless communication systems allow (or even depend on) direct communication between mobile users, who can be either people or vehicles. Such systems are often called mobile-to-mobile (M-to-M) communication systems, and the driving factors behind them are consumer demand for better quality of service, new applications, increased mobility support, the telecommunications industry's desire to expand its service offerings (and thus sources of revenue), and the governmental and military needs for more robust communications in disaster recovery and battlefield scenarios where fixed infrastructure may not be available or may not adapt quickly enough to the changing situation in the field. As a result, M-to-M systems are finding applications in mobile ad-hoc wireless networks, intelligent transportation systems, future cellular systems that employ cooperative diversity techniques, underwater communication systems, and in emerging disaster recovery and battlefield communication systems. Although M-to-M is a relatively young technology, it is already finding its way into wireless standards, such as IEEE 802.22 for cognitive radio, IEEE 802.11p for intelligent transportation systems, and IEEE 802.16 for WiMax systems.

Channel models and channel characterization are very important in traditional F-to-M communications, where the models are used for estimating channel capacity and other characteristics in the design phase, and for managing transmitted power (using channel sounding and feedback techniques) during actual communication. However, F-to-M communications systems can be designed with only basic channel models, using worst-case analysis and careful placement of fixed stations to provide good coverage to mobile users. In contrast, good channel models and channel characterization are absolutely essential for successful design of M-to-M systems, where channel capacity and link quality can vary greatly from one environment to another and change rapidly as one or both users move, and where energy for communication is often battery-supplied at both endpoints (in contrast to F-to-M communications where the fixed end of the communications is often far less energy-constrained).

The goal of this book is to describe wireless propagation, statistical characterization, and simulation models for M-to-M systems. To help understand this material, this book also compares and contrasts the propagation, characterization, and models for M-to-M channels to the relatively well-known modeling and simulation results for F-to-M cellular radio channels.

To introduce the reader to M-to-M channels, this chapter first presents an overview of M-to-M systems and introduces some key concepts in wireless propagation and statistical characterization of these channels. After that, the rest of this chapter consists of a brief overview of each remaining chapter in this book.

1.1 Mobile-to-Mobile Communication Systems

Wireless technology has become a part of our daily lives. Probably the most widespread use of wireless technology is mobile telephony, which is nearly ubiquitous throughout the world because it allows people to be connected to each other anywhere and at any time, even in places where little or no other high-technology infrastructure exists. In developed countries, wireless access to the Internet is also nearly ubiquitous, both through WiFi and through 3G/4G mobile access. It can be argued that any connectivity service, be it telephony, Internet access, music distribution, or social networking, cannot be called pervasive until it is available wirelessly to people on the go. This growing demand for new applications, better quality of service, and improved mobility support has been noticed by both service providers and consumers, and rapid development of M-to-M communications is a direct consequence of this trend. Although the field of M-to-M communications is still in its infancy, several specific technologies in this domain are already clearly defined: vehicle-to-vehicle communications, air-to-ground communications, and underwater vehicle-to-underwater vehicle communications. Here, we briefly overview these technologies as examples of M-to-M use scenarios and also as an illustration of the diversity of propagation environments and potential impact of future M-to-M communications.

1.1.1 Vehicle-to-Vehicle Communication Systems

Vehicle-to-vehicle (V-to-V) communications systems have recently drawn great attention because they have the potential to improve convenience, efficiency, and safety of automotive traffic (e.g., by allowing drivers or their vehicles to know in real time if a collision or a traffic jam is present further down the road, or by notifying police when the traffic situation demands their presence). Some of this may be possible using existing F-to-M technology, but it would become much more efficient and timely with a multi-hop network between vehicles equipped with communication devices. This mobile ad hoc network can then pass along relevant information about road conditions, accidents, and congestion without congesting the cellular network. Additional services that would be difficult to implement using F-to-M communications include making drivers aware of emergency braking by a preceding vehicle, efficient roadway utilization by platooning vehicles (strings of vehicles that communicate with each other so they can maintain tight intervehicle spacing) and enabling safe traffic through critical points like blind crossings (intersections without traffic lights) and entries to highways. Finally, future vehicles will allow for in-vehicle Internet access and intervehicle communication [1, 2] that can be provided much more efficiently and cost-effectively using a cooperative vehicle-to-vehicle network. Figure 1.1 illustrates one possible scenario where distributed vehicles cooperate to facilitate delivery of information.

There are currently many initiatives across the globe in V-to-V communications. An international standard, IEEE 802.11p, also known as Wireless Access in Vehicular Environments (WAVE), was recently published [3]. The simulation and performance evaluation of V-to-V systems, as well as the design of future, improved systems, requires a deep understanding of the underlying propagation channels. However, the time-frequency selective fading nature of such V-to-V channels is significantly different from the better-explored F-to-M cellular channel, and thus requires distinct measurement campaigns and models. Hence, Chapter 3 of this book presents the recent progress and current state of the art in V-to-V channel modeling and simulation. This will help communications system designers acquire an understanding of the pertinent channel characteristics and will help propagation researchers assess where the need for further work is the most pressing.

1.1.2 Air-to-Ground Communication Systems

Air-to-ground (A-to-G) communications systems have recently drawn great attention because they have potential to support a wide spectrum of applications ranging from two-way telecommunications (e.g., interactive video, and Internet access) to emergency response, intelligent transportation systems, remote

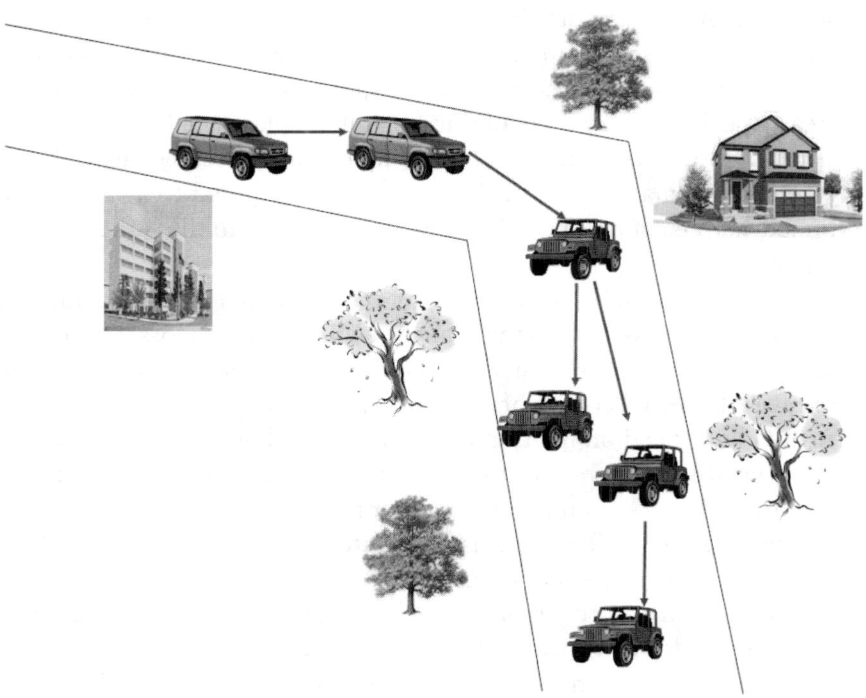

Figure 1.1 A distributed vehicle-to-vehicle communication scenario.

sensing, earth observation, navigation applications, pollution monitoring, meteorological measurements, real-time earth monitoring, land management and agriculture, and so on [4]. For example, emergency response A-to-G communications can have a significant advantage in emergency situations because they can be deployed very rapidly, can cover a large area, and are immune to many natural disasters such as floods and earthquakes. Rapid deployment of A-to-G systems, operating as airborne base stations capable of establishing "on-the-fly" cellular links (i.e., having rapid restoration capability) while allowing users to operate their existing mobile handsets also presents an attractive concept (e.g., when terrestrial networks are overloaded, perhaps due to a large concentration of users at a major event). A-to-G systems can also be deployed in military applications, where they can function as surrogates for satellites or as efficient and cost-effective relays for ground-based communications or for sensor surveillance to help battleships acquire ground targets [5]. Another example where A-to-G communication systems can play an important role is in environmental monitoring, such as on-board aerial imaging of forest fires, disaster monitoring of earthquake-stricken areas, volcanic eruptions, industrial accidents, land slides, avalanches, and oil spills.

An A-to-G communication system consists of sky and ground segments. The sky segment consists of a platform (i.e., an airplane or a lighter-than-air airship) positioned in the stratosphere. The aircraft is kept relatively stationary with respect to the ground by flying in a tight circle, while the airships employs stationary technology to counteract movements due to wind and remain stationary. The ground segments consist of mobile units, such as cellphones. An A-to-G communication system may provide mobile cellular coverage to regions ranging from high-density (urban) areas to low-density (rural) areas. The choice of the energy source is of fundamental importance. Fossil fuels are heavy, and therefore expensive to lift and maintain at altitude. The preferred form of energy source is solar power, since A-to-G systems are designed to have large surfaces (to maintain a steady position) that are suitable for lining solar panels. The A-to-G systems are situated at 17–22 km altitude (above flying civil aircraft) where the effect of winds is relatively small. The A-to-G systems can be linked to a satellite system, can be linked with each other, and can be linked with mobile nodes on the ground (see Figure 1.2).

One of the main advantages of A-to-G communication links, compared to terrestrial wireless links, is their line-of-sight (LoS) path loss profile. In terrestrial systems, the received power usually decays as a power-of-four function of the transmitter-receiver distance because of scattering along the propagation path. In contrast, in A-to-G links the dominant signal component is line of sight, whose power decays as a power-of-two function of the transmitter-receiver distance. Compared to satellite channels, A-to-G communication channels provide much lower propagation loss because high-altitude platforms are located in the stratosphere so the signals do not go through the (high-loss) ionosphere.

Simulation and performance evaluation of A-to-G systems requires understanding of the underlying propagation channels. Hence, Chapter 4 of this book presents recent progress and the current state of the art in A-to-G channel modeling and simulation.

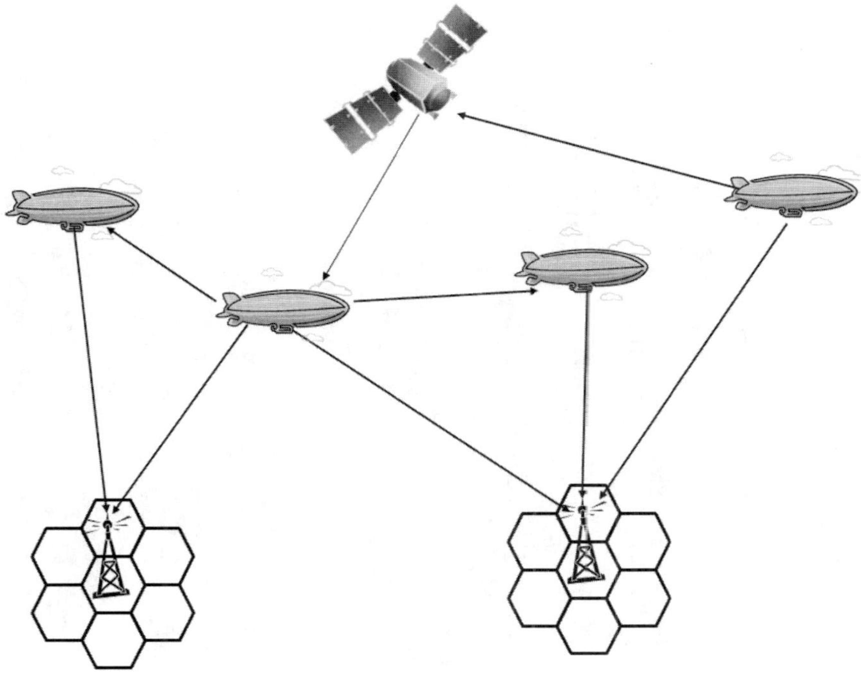

Figure 1.2 Integrated terrestrial, air-to-ground, and satellite systems.

1.1.3 Underwater Vehicle-to-Underwater Vehicle Systems

The need for underwater communications exists in applications such as data harvesting for environmental monitoring, communication between manned and/or unmanned underwater vehicles, and transmission of diver speech. An example of communications between fixed submerged devices are sensor networks deployed on the ocean bottom for seismic monitoring (e.g., to predict tsunamis), monitoring of deep-water oil wells, analysis of resource deposits, oceanographic studies, tactical surveillance, and pollution monitoring. Another example of underwater communications with fixed devices is the triggering of an event or action on a submerged device (e.g., a release device that allows an instrument package deployed on the ocean bottom to return to the surface) [6–8].

Underwater communication is also needed between underwater vehicles, such as unmanned or autonomous underwater vehicles (UUVs/AUVs) for deep-water construction, repairs of deep-water oil wells, scientific or resource exploration, or military UUVs (submarine drones), as illustrated in Figure 1.3.

Unfortunately, underwater communication is extremely challenging: use of wires/cables either severely limits or entirely precludes most of the important underwater communication scenarios, whereas existing wireless communications still provide extremely low data rates, work over very limited distances, and/or have low energy efficiency. The main reason for such poor performance is that the underwater environment is very harsh for traditional wireless communication approaches: radio signal range is only several meters because of severe losses, especially in seawater (which is a relatively good conductor), while optical communication has a very short range because of high dispersion due to murkiness and particles suspended

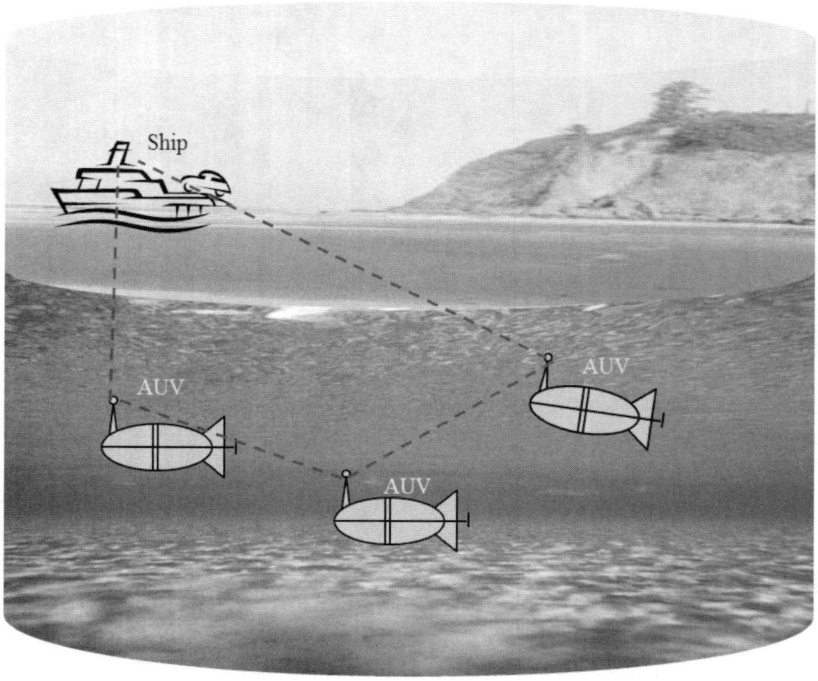

Figure 1.3 An acoustic network of autonomous underwater vehicles.

in the water. As a result, underwater wireless communication mostly uses acoustics (sound), which can travel long distances in water. Unfortunately, acoustic communication suffers from its own set of problems—the available spectrum is measured in kilohertz instead of gigahertz, propagation is orders of magnitude slower than with radio or optics, and there are many sources of signal distortion that further limit data rates that can be achieved using simple modulations and coding.

More advanced modulations and coding, especially those that exploit multipath propagation and provide bandwidth and energy efficiency, can dramatically improve data rates in acoustic underwater communication. However, the acoustic underwater propagation environment differs significantly from the air- and space-based radio propagation environment, so this environment must be studied in order to exploit advanced modulations and coding. Hence, Chapter 5 of this book presents recent progress and the current state of the art in UV-to-UV channel modeling and simulation.

1.2 The Wireless Channel

Understanding of wireless propagation is key to improving the performance of wireless communication systems. While each type of wireless channel has its own characteristics, many of the key concepts are common to all of the propagation environments presented in this book. We now briefly review these concepts, which are described in more detail in Chapter 2.

1.2.1 Fading

In general, *fading* refers to variation of received signal power over time, due to changes in the transmission medium or path. Fading is usually classified as *large-scale fading* or *small-scale fading*.

Large-scale fading effects are dominant when the transmitter or receiver moves over distances greater than several tens of the signal's wavelength, and they play an important role in determining the cell coverage area, outages, and handoffs. The main large-scale fading effects are *pathloss* and *shadowing*.

- *Pathloss*. The pathloss is calculated by averaging the received power at a particular distance over a sufficiently large area. Pathloss limits interference between sufficiently separated transmitters, but also rapidly diminishes the useful signal power. Any technique that improves pathloss is thus highly appreciated by network planners.
- *Shadowing*. Variation in the received power around the level predicted by pure pathloss is called shadowing. It is calculated by averaging the received power at a particular distance over an area whose radius is similar to the shadowing coherence distance. Shadowing is traditionally modeled as a random process, and is one of the most detrimental performance factors in modern communications systems—it cannot be absorbed by suitable channel codes, so it can result in link outages. As a result, any technique that improves shadowing outages is highly appreciated.

Small-scale fading is caused by multipath propagation. This effect plays an important role in determining link level performance in terms of bit error rates, average fade durations, and so on.

- *Multipath*. Not averaging the signal at all allows one to observe additional signal fluctuation beyond pathloss and shadowing. This fluctuation is caused by constructive and destructive addition of signal components traveling along multiple propagation paths. If channel fading changes from symbol to symbol due to fast variations of channel amplitude, then the channel is called *fast-fading*; otherwise, it is called a *slow-fading* channel. A fast-fading channel offers temporal diversity, which can be exploited by a suitable channel coder; a slow-fading channel does not offer temporal diversity, which can lead to significant outage periods. If multiple copies (path components) of the symbol can be resolved because their propagation delays are larger than the symbol duration, then the channel is called a *frequency-selective* fading channel; otherwise, it is called a *frequency-flat* or simply *flat* fading channel. A frequency-selective fading channel offers frequency diversity, which can be exploited by suitable signal processing coupled with a channel coder; a flat fading channel does not offer frequency diversity, which again can lead to prolonged outages. In terms of outages, the worst-case scenario is a slow and flat fading channel, while the best case is a fast and frequency-selective fading channel. Modern communications systems can be fast or slow fading, depending on the mobility they are subjected to; however, due to increasingly short symbol durations, they are typically frequency selective and hence inherently offer a healthy amount of diversity.

1.2.2 Statistical Properties of Channels

Modeling of the wireless channel has been one of the most difficult parts of the wireless system design. Propagation models are the base for channel modeling, as they try to describe the way a wireless signal changes during its travel from the transmitter to the receiver. Propagation models have traditionally focused on predicting large-scale and small-scale fading. Propagation modeling can be done using ray-tracing tools, which determine the field behavior for a deterministically predefined scenario. However, when the number of multipath components is large, or when the geometry and dielectric properties of the propagation environment are unknown, this approach is impractical. Similarly, if the number of reflectors in the environment is very large and/or reflector surfaces are not smooth, ray-tracing tools become less useful and we typically must use statistical approximations to characterize the propagation medium. Most mobile-to-mobile propagation environments are very complex, so this book focuses mainly on statistical characterization. For wireless communication systems, statistics of special importance are the correlation functions, Doppler spectrum, level crossing rate, and average fade duration.

- *Correlation functions*. The measure of how fast a wireless channel changes with time, movement, or frequency is very important for the design of wireless systems. This measure is provided by temporal, spatial, and frequency correlation functions.
  The *temporal autocorrelation* function describes how fast the wireless channel is changing over time. If the channel does not change at all, it does not fade and instead remains at some particular level (i.e., its autocorrelation is very high for a long time). However, separate instances of this channel (e.g., different paths) will be uncorrelated with one another, assuming that each of the scattered components fades independently. Once relative motion is introduced to the transmitter, receiver, and/or scatterers, the fading becomes both correlated and varying in time. Temporal autocorrelation is important for estimating the quality of communication, as it determines how long a measured property will be present in a wireless channel. For example, the temporal autocorrelation function determines how often training sequences should be sent to estimate the wireless channel.
  The *spatial correlation* function measures a correlation between the received average signal gain and the angle of arrival of a signal. Rich multipath propagation decreases spatial correlation by spreading the signal such that multipath components are received from many different spatial directions. Similarly, small antenna separations increase spatial correlation, as adjacent

antennas will receive similar signals. Spatial correlation is important for estimating the potential benefit from path diversity—whether multiple-input-multiple-output (MIMO) techniques can be effective—and/or for judging the need for techniques that reduce or eliminate intersymbol interference due to different delays along different paths.
The *frequency correlation* function measures a channel's frequency selectivity. As the channel's frequency response becomes more varied, an interacting signal will exhibit a broader envelope in the delay domain. In practice, delay spread is the result of different propagation delays of received signals from reflectors at large distances compared to the LoS path between the transmitter and the receiver.

- *Doppler power spectral density.* In wireless communications, the transmitter, the receiver, and/or scatterers are moving. The faster they are moving, the faster the wireless channel changes, and the more Doppler shift will be present. The Doppler power spectral density of a fading channel measures the Doppler spread (twice the Doppler shift); that is, it describes how much spectral broadening it causes.
- *Level crossing rate and average fade duration.* The level crossing rate is a measure of how often the wireless channel changes over time. It quantifies how often the fading crosses some threshold, usually in the positive-going direction. The average fade duration quantifies how long the signal spends below the threshold.

1.3 Overview of Remaining Chapters

The remainder of the book is organized as follows:

1.3.1 Chapter 2: Fixed-to-Mobile Cellular Radio Channels

Chapter 2 of this book reviews statistical characterization, modeling, and simulation of fixed-to-mobile (F-to-M) radio channels. The goal of this chapter is to define important wireless propagation concepts used throughout the book and to allow easy comparison between F-to-M and M-to-M channels. This chapter first describes the propagation principles that govern signal propagation in F-to-M cellular radio channels. The chapter proceeds to describe important concepts of propagation modeling such as large-scale and small-scale fading effects. Then, the statistical characterization of multipath propagation in F-to-M cellular radio channels is detailed. Several reference models and their statistics, such as correlation function, Doppler spectrum, power delay profile, and level crossing rate are detailed. Furthermore, this chapter describes the simulators for frequency-flat and frequency-selective multipath fading. Finally, because of the importance of the multipath structure in determining the small-scale fading effects, the rest of the chapter describes several wideband channel sounding techniques that are commonly used for channel measurements.

1.3.2 Chapter 3: Vehicle-to-Vehicle Radio Channels

Chapter 3 of this book details statistical characterization, modeling, and simulation of vehicle-to-vehicle (V-to-V) radio channels. It first details the propagation principles that govern signal propagation in V-to-V channels and describes large-scale and small-scale fading effects. Then, statistical characterization of multipath propagation in V-to-V channels is discussed and several two-dimensional (2-D) and three-dimensional (3-D) models are detailed. Finally, a validation of the 3-D wideband models against the measurements are presented.

1.3.3 Chapter 4: Air-to-Ground Radio Channels

Chapter 4 of this book details the propagation principles that govern signal propagation in air-to-ground (A-to-G) channels and describes large-scale and small-scale fading effects. Then, statistical characterization of multipath propagation in A-to-G channels is discussed and several 3-D models are detailed.

1.3.4 Chapter 5: Underwater Vehicle-to-Underwater Vehicle Acoustic Channels

Chapter 5 of this book details statistical characterization, modeling, and simulation of underwater vehicle–to–underwater vehicle (UV-to-UV) acoustic channels. It first briefly describes the propagation principles that govern signal propagation in acoustic underwater UV-to-UV channels and then proceeds to describe important concepts of propagation modeling such as large-scale and small-scale fading effects. Then, statistical characterization of multipath propagation in UV-to-UV channels is presented. Finally, several models for frequency-flat and frequency-selective channels and their validation against the measurements is presented.

1.3.5 Appendixes

Appendix A presents a brief review of the basic concepts of probability theory and random processes used in this book.

To help readers better understand, use, and implement the propagation theory presented in Chapters 2–5, this book also provides MATLAB-based simulators for fixed-to-mobile, vehicle-to-vehicle, air-to-ground, and underwater vehicle–to–underwater vehicle channels. Appendix B of this book explains how to use these simulators.

References

[1] FleetNet project Internet on the road, Available: www.et2.tu-harburg.de/fleetnet, 2000.

[2] Car2Car Communication Consortium, Available: www.car-to-car.org, 2005.

[3] 802.11p-2010-IEEE Standard for Information Technology, Available: ieeexplore.ieee.org.

[4] Elabdin Z., O. Elshaikh, R. Islam, A.P. Ismail, and O.O. Khalifa, "High altitude platform for wireless communications and other services," *International Conference on Electrical and Computer Engineering ICECE '06*, 19–21 Dec. 2006, pp. 432–438.

[5] Jamison L., G. S. Sommer, and I.R. Porche, "High-altitude airships for future force army," Technical Report, RAND Corporation, 2005.

[6] Lynn W., "3D, 4D, and beyond," *Journal of Petroleum Engineering*, Jan. 1998, p. 35.

[7] Proakis J. G., J. Rice, E. Sozer, and M. Stojanovic, "Shallow water acoustic networks," *Encyclopedia of Telecommunications*, J. G. Proakis, Ed. John Wiley and Sons, 2003.

[8] Akyildiz I. F., D. Pompili, and T. Melodia, "Underwater acoustic sensor networks: Research challenges," *Ad Hoc Networks (Elsevier)*, Vol. 3, March 2005, pp. 257–279.

CHAPTER 2

Fixed-to-Mobile Cellular Radio Channels

The design of spectrally and energy efficient cellular communication systems requires a detailed understanding of the fixed-to-mobile (F-to-M) cellular radio propagation environment. The characteristics of the radio channel vary greatly with the operating frequency and the propagation environment (i.e., rural, urban). A typical F-to-M cellular radio system consists of a collection of fixed base stations (BSs) that define the radio coverage areas, as shown in Figure 2.1. Typically, the BS antennas are usually well elevated above the local terrain and relatively free of local scatterers, as shown in Figure 2.2. In contrast, a mobile station (MS) moves, and its antennas have low elevation and are surrounded by buildings, mountains, and foliage. The BS communicates with the MS through either a line-of-sight (LoS) link or through a non-line-of-sight (NLoS) link, due to natural and man-made objects that are situated between the BS and MS. This chapter describes the propagation principles that govern signal propagation in F-to-M cellular radio channels and then explains how such channels are modeled, simulated, and measured.

2.1 Propagation Principles

The modulated signal at the BS is fed into an antenna of length l from which it is radiated in the form of electromagnetic (EM) waves with wavelength λ. The propagation of an EM wave through a medium is governed by Maxwell's equations. In the far field, where the distance between the BS and MS, d, is significantly larger than the wavelength, the EM waves can be approximated as the planar waves with electric and magnetic components $E = E_0 e^{j(\omega \cdot t - \mathbf{k} \cdot \mathbf{r})}$ and $H = H_0 e^{j(\omega \cdot t - \mathbf{k} \cdot \mathbf{r})}$, respectively. Here, $\mathbf{r} = d \cdot \mathbf{n}_r$ is the spatial translation vector, where $\mathbf{n}_r$ is a unit vector pointing from the BS to the observation point, $\mathbf{k} = k \cdot \mathbf{n}_k$ is the wave vector, where $\mathbf{n}_k$ is a unit vector pointing into the direction of wave propagation, $\omega = 2\pi f$ is the angular frequency, and t is time.

The waves propagate toward the MS via four main propagation mechanisms: *free-space propagation*, *reflection*, *diffraction*, and *scattering*. Free-space propagation occurs when energy from the BS makes its way directly to the MS via a LoS path. The other three propagation mechanisms are created by objects in the channel, such as buildings, trees, mountains, and other obstacles, which we collectively call the *channel scatterers*. The received waves are a superposition of the wave coming from the LoS path and all waves coming from different directions due to the scatterers in a channel. This effect is known as *multipath propagation*.

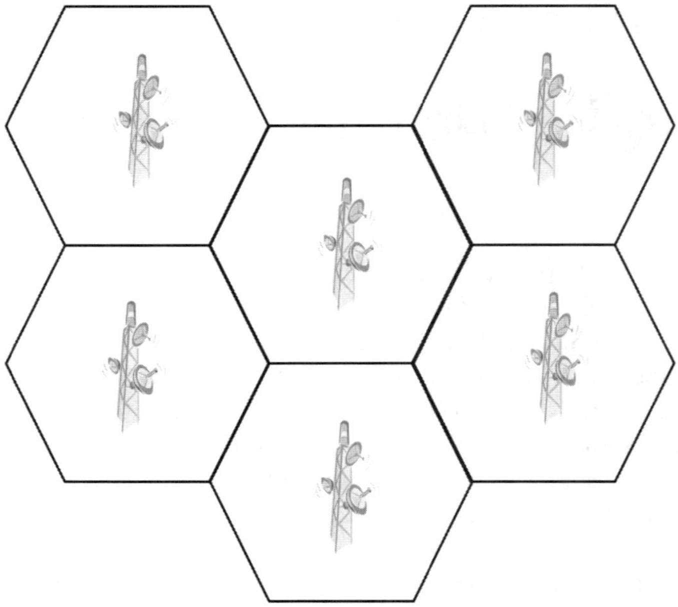

Figure 2.1 Illustration of a F-to-M cellular radio coverage.

2.1.1 Free Space Propagation

The EM waves undergo free space propagation when BS and MS antennas are in the far field of each other (i.e., $d \gg \lambda$) and there is an unobstructed LoS path between them. Under these assumptions, the received signal power is [1]

$$P_R = P_T G_T G_R \left(\frac{\lambda}{4\pi d}\right)^2 \tag{2.1}$$

where P_T is the transmitted power, G_T and G_R are the transmit and receive antenna gains, respectively, λ is the carrier wavelength, and d is the distance between the BS and MS. The free space equation in (2.1) (also known as the Friis equation) shows that the received power falls off as the square of the distance d. This implies that the received power decays with distance at a rate of 20 dB/decade.

2.1.2 Reflection

Reflection arises when an EM wave propagating in one medium impinges upon another medium having different electrical properties, as shown in Figure 2.2. The EM wave is then split into a reflected component and a transmitted (continuing) component. For example, the EM wave impinging on the building wall will be partially transmitted through the wall into the building (thereby providing indoor coverage), but part of it will be reflected on the building wall back into the surrounding air. Flat surfaces that are much larger than the wavelength may be modeled as reflective surfaces with *specular reflection*. The electric field intensity of the reflected and transmitted waves is related to the incident wave in the medium of origin through the Fresnel reflection coefficient Γ [2]. The reflection coefficient is a function of the material properties and generally depends on the wave polarization, angle of incidence, and frequency of the propagation wave.

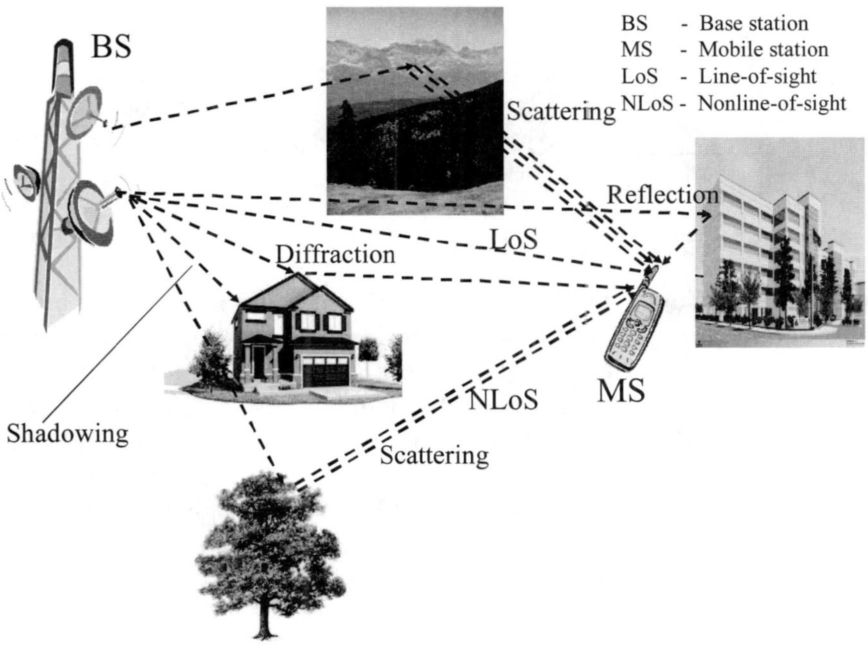

Figure 2.2 Typical F-to-M cellular radio propagation environment.

2.1.3 Diffraction

Diffraction occurs according to Huygen's principle, which states that when there is an obstruction between the BS and MS antennas, the secondary wavelets are created and combined to produce a new wavefront behind the obstructing body and in the direction of propagation, as shown in Figure 2.2. The field strength of a diffracted wave is the vector sum of the electric filed components of all the secondary wavelets in the space around the obstacle. Diffraction allows EM signals to propagate around the curved surface of the earth, beyond the horizon, and to propagate behind obstructions. Although the received field strength decreases rapidly as a receiver moves deeper into the obstructed region, the diffraction field still exists and often has sufficient strength to arrive at the MS.

2.1.4 Scattering

As mentioned in Section 2.1.2, flat surfaces that have much larger dimensions than a wavelength may be modeled as reflective surfaces. However, the texture (roughness) of such surfaces often also creates *scattering*—a propagation effect different from the specular reflection. Depending on the scatterer's surface characteristics, different scattering phenomena can be observed. Generally, the rougher the surface, the less the impact of the specular reflected component and the stronger the impact of the scattered component, as shown in Figure 2.3.

The effect of scattering is due to the constructive and destructive addition of waves reflected off different points on a rough surface. For carrier frequencies that are typically used in telecommunications, the well-established Rayleigh criterion [1] can be used to determine whether a surface is smooth (i.e., it exhibits only a specular

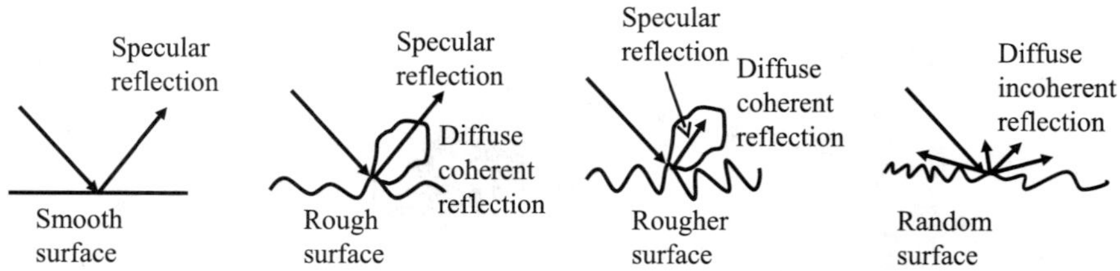

Figure 2.3 Reflection of a smooth surface, a rough surface, and a random surface.

component) or rough (i.e., it also exhibits a scattering components). The Rayleigh criterion states that a surface is considered smooth if its minimum to maximum protuberance h is less than $h_c = \lambda/(8\sin\theta_i)$, where θ_i denotes the angle of incidence, otherwise, a surface is considered rough. For an example, a GSM system operating at $f = 900$ MHz in an urban environment with houses having balconies with undulation $\Delta h = h/\sin\theta_i \approx 1$m reveals that the surfaces need to be considered rough since significant scattering occurs when $\Delta h/\lambda > 1/8$ (Rayleigh criterion).

2.2 Propagation Modeling

As pointed out in Section 1.2.1, fading refers to time variation of the received signal power due to changes in the transmission medium. Fading is usually classified into *large-scale fading* and *small-scale fading*. Large-scale fading is dominant when the receiver moves over distances greater than several tens of the carrier wavelength and includes *pathloss* and *shadow fading*. Large-scale fading plays an important role in determining cell coverage area, outage, and hand-offs. On the other hand, small-scale fading is caused by multipath propagation. This effect plays an important role in determining link level performance in terms of bit error rates, average fade durations, and so on.

2.2.1 Large-Scale Fading—Pathloss

Pathloss is the attenuation of the transmitted signal as it propagates from the BS to the MS. This attenuation may be caused by effects such as free-space loss, diffraction, reflection, and absorption. Pathloss is also influenced by terrain contours, environment (e.g., a city or a muddy path in the woods), propagation medium (e.g., dry or moist air), the distance between the BS and MS, and the height and location of antennae.

The simplest pathloss model assumes a LoS link between the BS and MS and propagation in free space. Under these assumptions, the received signal power is given as in (2.1), (i.e., it assumes a loss of 20 dB/decade). It can be formalized as $L(d) = L(d_0) \cdot (d/d_0)^2$, where d_0 is the reference distance and $L(d_0)$ is the power measured at the reference distance d_0. Note that the pathloss exponent (the power of the distance dependence) in this equation is 2 for free-space propagation. It is generally not applicable to terrestrial scenarios but finds some applicability in satellite and A-to-G channels as well as for short LoS channels.

The signals in wireless communications, however, do not experience free space propagation. Single-slope pathloss model assumes a loss of $10 \cdot n$ dB/dec, where n

is the pathloss exponent, which can range from $n = 1.5$ for waveguides, to $n = 2,\ldots, 4$ for LoS + clutter, to $n = 4,\ldots, 8$ for NLoS + clutter. This model can be formalized as $L(d) = L(d_0) \cdot (d/d_0)^n$. This model has applications in all forms of terrestrial communication systems including cellular radio systems.

However, the pathloss in terrestrial systems does not always follow the single-slope pathloss model. In the presence of two strong wave components (e.g., one direct and one reflected), the pathloss experiences a double-slope pathloss behavior [3]. The double-slope model assumes that for distances smaller than a given breakpoint d_{bp}, the decay follows a pathloss coefficient n_1 and thereafter a pathloss coefficient n_2. This can be formalized as $L(d) = L(d_0) \cdot (d/d_0)^{n_1}$ with typically $n_1 = 2$ for $d < d_{bp}$ and $L(d) = L(d_0) \cdot (d/d_0)^{n_2}$, with typically $n_2 = 2, \ldots, 6$ for $d \geq d_{bp}$. It is well applicable to long-range communication systems, such as cellular systems. The break point distance can be calculated approximately as $d_{bp} \approx 4h_r h_t / \lambda$, where h_r and h_t are the elevations of MS and BS, respectively.

More advanced modeling of pathloss can be done using ray-tracing and ray-launching tools, which determine the field behavior for a deterministically predefined scenario. It is generally a very complex modeling approach that finds applications in very specific propagation environments such as propagation behavior close to the head, within mobile phone, and so on. However, this approach is not necessarily better than the (much simpler) single- and double-slope modeling for other propagation environments.

Another very popular way to model pathloss is to use empirically fitted data. This approach is based on real measurements for a given environment. Once the measurements are taken, statistical fitting techniques are used to obtain the pathloss coefficients that match well-measured data. Examples of such a modeling approach are the Okumura-Hata, Lee, and Walfish-Ikegami, models [4]. The models are generally difficult to obtain, but once the statistical fitting is performed, they are very simple and fairly accurate models. They are typically used in academic simulators and in planning and optimization tools.

Finally, a very accurate approach to model the pathloss is to take real measurements for each point in the plane (or space) according to a predefined grid (e.g., every 20m in urban areas). While this approach is difficult and expensive, it is often used by large operators in order to optimize planning and rollout of their real-world cellular systems, especially in densely populated areas.

2.2.2 Large-Scale Fading—Shadowing

The pathloss models described in Section 2.2.1 assume that the pathloss is constant at a given distance. However, obstacles such as buildings and trees result in random variations of the received power at a given distance. This effect is called *shadow fading*.

Experimental results show that shadow fading can be modeled as a log-normal random variable with a distribution given as [4]

$$f_{\Omega_p}(x) = \frac{10}{x\sigma_\Omega\sqrt{2\pi}\ln 10}\exp\left[-\frac{(10\log_{10}x - u_{\Omega_p(\mathrm{dBm})})^2}{2\sigma_\Omega^2}\right] \tag{2.2}$$

where Ω_p denotes the mean squared envelope level, $\mu_{\Omega p}$ is the area mean expressed in dBm, and σ_Ω is the standard deviation of the shadow fading. Typical σ_Ω values range from 5 to 12 dB.

The random shadowing components are usually correlated in space. The degree of correlation mainly depends on the environment. We can distinguish two types of correlated shadowing: the auto-correlated shadowing and the site-to-site correlated shadowing, as illustrated in Figure 2.4.

The autocorrelated shadowing quantifies how shadowed power changes over time as the MS moves (i.e., over the MS spatial separation of Δd). This metric is very important for power control algorithms in general and for diversity gains in cooperative systems. The auto-correlation shadowing coefficient can be written as [6, 7]

$$R(\Delta d) = \mathrm{E}[S(d) \cdot S(d + \Delta d)] \propto e^{-\frac{\Delta d}{d_{\mathrm{corr}}}} \tag{2.3}$$

where E[·] denotes the expectation with respect to the random realizations of signal S, and d_{corr} is the correlation distance, which heavily depends on the distance d between the BS and MS. In cellular systems, the typical experimentally observed value for d_{corr} is 112 m for d = 4.8 m [3].

The site-to-site shadowing quantifies how shadowed power changes over angular separation between two BS stations $\Delta\phi$. This metric is used for quantifying interference power from adjacent cells in cellular systems and for macro-diversity gains in distributed antenna systems. The cross-correlation shadowing coefficient can be written as

$$R(\Delta\phi) = \mathrm{E}[S(d_1) \cdot S(d_2)] \propto \begin{cases} \sqrt{\dfrac{d_1}{d_2}} & \text{for } 0 \le \Delta\phi \le \phi_{\mathrm{thr}} \\ \dfrac{\phi_{\mathrm{thr}}}{\Delta\phi}\sqrt{\dfrac{d_1}{d_2}} & \text{for } \phi_{\mathrm{thr}} < \Delta\phi \le \pi \end{cases} \tag{2.4}$$

where d_1 and d_2 are the respective distances between the MS and the BSs, and $\phi_{\mathrm{thr}} = 2\arcsin(d_{\mathrm{corr}}/(2d_1))$.

Shadowing is one of the most detrimental performance factors of modern communication systems because it diminishes the useful signal power in a difficult-to-predict fashion. Since it cannot be compensated for by using suitable channel codes, shadowing causes *outages* (nonavailability of a link). However, shadowing can also be useful—it limits interference from nearby BSs.

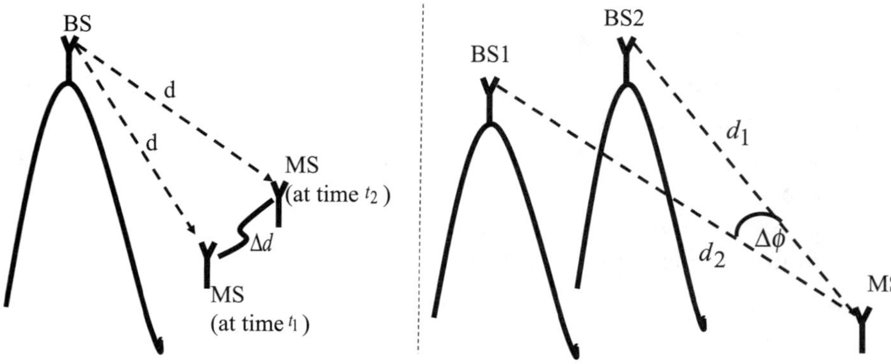

Figure 2.4 Auto-correlated shadowing (left) and site-to-site correlated shadowing (right).

2.2.3 Small-Scale Fading—Multipath Propagation

Multipath propagation is the propagation mechanism manifested when the transmitted signal reaches the receive antenna along two or more paths. The presence of local scattering objects, such as mountains and buildings, often obstructs a direct path between the BS and MS. This leaves only NLoS propagation paths between the BS and MS (i.e., the waves must propagate via reflection, diffraction, and scattering). Such waves typically arrive at the MS from many different directions and with different delays, and combine vectorially at the receiver antenna (a phenomenon called multipath fading) to produce a composite received signal.

Here, we consider the transmission of the pass-band signal

$$s(t) = \Re\left[\hat{s}(t)e^{j2\pi f_c t}\right] \tag{2.5}$$

where $\hat{s}(t)$ is the complex envelope of the transmitted signal, f_c is the carrier frequency, and $\Re[\cdot]$ denotes the real part operation. Due to multipath propagation, the received signal is a superposition of many attenuated, delayed, and phase-shifted replicas of the transmitted signal, each influencing each other. If the channel is comprised of L multipath components, then the noiseless[1] received pass-band waveform is

$$r(t) = \Re\left[\sum_{l=1}^{L} C_l e^{-j\phi_l(t)}\hat{s}(t-\tau_l)\right] \tag{2.6}$$

where C_l, $\phi_l(t)$, and τ_l are the amplitude, phase, and time delay, respectively, associated with the lth propagation path. From (2.6), it follows that the multipath channel can be modeled as a linear time-variant filter whose complex baseband impulse response is [4]

$$h(t,\tau) = \sum_{l=1}^{L} C_l e^{-j\phi_l(t)}\delta(\tau-\tau_l) \tag{2.7}$$

where $h(t, \tau)$ is the channel response at time t due to an impulse applied at time $t-\tau$, and $\delta(\cdot)$ is the Dirac delta function. The channel amplitude, phase, and time delay strongly depend on time and location. For example, for a fixed MS location, these values change with time due to mobility in the environment itself (scatterers move). Similarly, for a given time, these values differ from one location to another. Because of that, they are often modeled as statistical random variables (see Section 2.3 for more details).

2.2.4 Factors Influencing Small-Scale Fading

Many physical factors such as multipath propagation, speed of the MS, speed of the surrounding objects, and bandwidth of the transmitted signal influence small-scale fading. In this section, we will focus on the impact of multipath propagation, movement of the MS, and physical location of the MS on small-scale fading.

1. Each communication system is affected by the noise. However, in this book we are focused on understanding and modeling multipath propagation, and without loss of generality we can neglect the effect of noise because they are two independent mechanisms that affect communication systems.

2.2.4.1 Impact of Multipath Propagation on Small-Scale Fading

The movement of MS and the presence of reflecting objects and scatterers in the channel create multiple versions of the transmitted signal that arrive at the receiving antenna, displaced with respect to one another in time and frequency. The random phase and amplitudes of the different multipath components cause fluctuations in signal strength, thereby inducing small-scale fading, signal distortion, or both.

As indicated in (2.6), the received signal is a sum of many attenuated, delayed, and phase-shifted replicas of the transmitted signal, each influencing each other. A sum of these delayed signals creates *time dispersion* of the original signal. In the frequency domain, this manifests as *frequency selectivity* (i.e., the channel distorts the frequency response characteristic of the transmitted signal). Based on their degree of frequency selectivity, channels may be classified as *frequency-flat* or as *frequency-selective* channels. If all the transmitted frequencies undergo approximately identical amplitude and phase changes, the channel is called frequency flat. On the other hand, if all transmitted frequencies experience different amplitude and phase changes, the channel is called frequency selective. Frequency selectivity is measured in terms of the coherence bandwidth B_c (i.e., the bandwidth over which the channel's frequency response remains constant). The system is said to undergo flat fading if the coherence bandwidth is bigger than the symbols's bandwidth $B_s = 1/T_s$ (i.e., the transmission is narrowband), and frequency selective otherwise.

The selectivity in the frequency domain leads to intrasymbol and intersymbol interference. The first one refers to the overlap of symbol replicas within the symbol duration, while the second refers to the overlap of symbol replicas belonging to different symbols. While the intrasymbol interference is generally irreversible, the intersymbol interference can be reversed by means of suitable transceivers. One realization of such a transceiver includes use of an equalizer, which effectively performs the inverse filter function to the one given in (2.7). However, given the increasing number of multipath components that need equalization, this becomes more and more difficult. The second transceiver realization includes use of spread spectrum techniques, which is facilitated by the use of Rake receivers. However, the limitation might be the number of Rake fingers that can be implemented. The third transceiver realization includes the use of low-complexity orthogonal frequency division multiplexing (OFDM)–like techniques that essentially transform the frequency-selective time domain into a flat-frequency domain system. However, several drawbacks related to implementation prevail here also.

2.2.4.2 Impact of the MS Movement on Small-Scale Fading

A transmitted signal is impacted not only by multipath propagation, but also by other factors such as the speed of MS as well as its direction of movement. Due to the movement of the MS, the Doppler effect causes a frequency shift of each of the waves. The angle of arrival α, which is defined by the direction of arrival of the nth incident wave and the direction of motion of the MS, determines the Doppler frequency of the nth incident wave according to the relation $f = f_{\max} \cos \alpha$. This effect is called *time selectivity*. When viewed in the frequency domain, time selectivity manifests as Doppler shifts in the transmitted signal, causing a broadening of the transmitted signal spectrum, which is called *frequency dispersion*. A good measure of time selectivity is given by the channel coherence time, or equivalently, the Doppler spread (i.e., the time duration for which the channel can be considered as approximately time invariant). Based on the coherence time (rate with which the channel impulse response changes relative to the signal transmission rate), channels may be classified as *fast fading* or *slow fading*. Fast fading implies that the channel changes within the transmitted symbol duration, while slow fading implies that the channel is approximately constant within a symbol duration. The distinction between slow and fast fading has a profound impact on the choice of modulation, frequency of pilot symbols, and so on.

2.2.4.3 Impact of the MS Location on Small-Scale Fading

The channel also varies in the spatial domain due to the dependency of amplitudes, phases, and time delays on the physical location. This is important for multiple-input multiple-output (MIMO) systems, which deploy several transmit and receive antennas. A good measure of channel *spatial selectivity* is given by the channel coherence distance. The coherence distance depends on the environment where some heuristics suggest that it is inversely proportional to the angular spread of the incoming/outgoing signal. The system is said to undergo spatially nonselective fading if antenna spacing is smaller than the coherence distance and spatially selective otherwise. The distinction between selective and nonselective spatial fading has an important impact on the performance of MIMO systems, where a high correlation between antenna elements incurs significant performance losses.

2.3 Statistical Characterization of Multipath Propagation in F-to-M Cellular Radio Channels

Propagation modeling can be done using ray-tracing tools, which determine the field behavior for a deterministically predefined scenario. However, when the number of multipath components is large, or when the geometry and dielectric properties of the propagation environment are unknown, this approach is impractical. Similarly, if the number of reflectors is very large or the reflector surfaces are not smooth, ray-tracing tools become less useful and we typically must use statistical approximations to characterize the propagation medium. This section details statistical characterization of multipath propagation in F-to-M cellular radio channels.

2.3.1 Received Envelope Distribution

Under NLoS propagation conditions, the statistical time-varying nature of the received envelope of a flat fading signal, or the envelope of an individual multipath component, are commonly described by the Rayleigh distribution, whose probability density function (pdf) is given by

$$p(x) = \frac{x}{\sigma^2} \exp\left\{-\frac{x^2}{2\sigma^2}\right\} \tag{2.8}$$

where σ is the root mean square (rms) value of the received voltage signal and σ^2 is the time-average power of the received signal.

When a specular component (i.e., LoS or a strong reflected path) also arrives at the receiver, the fading is described by a Rician distribution, whose pdf is

$$p(x) = \frac{x}{\sigma^2} \exp\left\{-\frac{x^2 + A^2}{2\sigma^2} I_0\left(\frac{Ax}{\sigma^2}\right)\right\} \tag{2.9}$$

where parameter A denotes the peak amplitude of the dominant signal and $I_0(\cdot)$ is the modified Bessel function of the first kind and zero-order. Note that the Rayleigh distribution is a special case of the Ricean distribution, so the Ricean distribution is often described in terms of a parameter $K = A^2/(2\sigma^2)$ (i.e., K is the ratio between the deterministic signal power and the variance of the multipath). When $K = 0$, the channel exhibits Rayleigh fading; when $K = \infty$, the channel does not exhibit any fading at all; and when $0 < K < \infty$, the channel exhibits Ricean fading.

2.3.2 Clarke's Model for F-to-M Frequency Flat Multipath Fading Channels

Clarke [8] was the first to propose a channel model for NLoS F-to-M frequency flat multipath fading channels based on the statistical characteristics of the electromagnetic fields of the received signal. The model assumes a fixed transmitter with a vertically polarized antenna. The field incident on the mobile antenna is assumed to consist of N plane waves propagating in x-y plane. Each plane wave has random phase and random azimuth angle of arrival, and all waves have equal average amplitudes. It should be noted that the equal average amplitude assumption is based on the fact that, in the absence of a direct line-of-sight path, the scattered components arriving at a receiver will experience similar attenuation over small-scale distances.

Figure 2.5 illustrates a plane wave incident on a MS traveling with a velocity υ in the x direction. The angle of arrival is measured in the x-y plane with respect to the direction of motion. Every wave that is incident on the MS undergoes a Doppler shift due to the motion of the receiver and arrives at the receiver at the same time. Note that no variation in delay due to multipath is assumed in this model—this is a consequence of the frequency flat fading assumption. For the nth wave arriving at an angle α_n relative to the x-axis, the Doppler shift in Hertz is given by

$$f_n = \frac{\upsilon}{\lambda} \cos\alpha_n = f_d \cos\alpha_n \tag{2.10}$$

where f_d is the maximum Doppler frequency in Hertz, λ is the carrier wavelength, and υ is the speed of the MS. The angle of arrival, α_n, depends on the scattering environment and the antenna radiation pattern.

The vertically polarized plane waves arriving at the mobile have electric and magnetic field components given by

$$E_z(t) = E_0 \sum_{n=1}^{N} C_n \cos(2\pi f_c t + 2\pi f_n t + \phi_n) \tag{2.11}$$

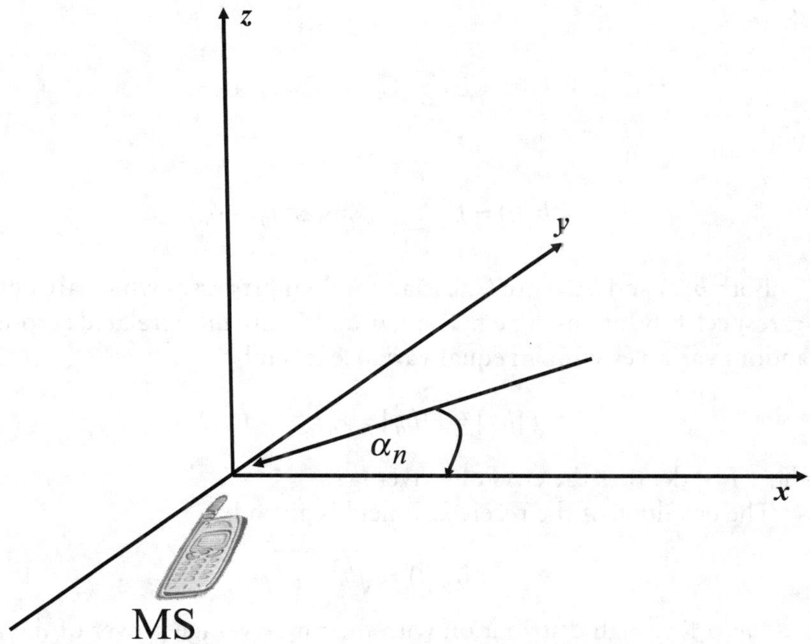

Figure 2.5 Illustration of plane wave arriving at the MS at random angle α_n.

$$H_x(t) = -\frac{E_0}{\eta}\sum_{n=1}^{N} C_n \sin\alpha_n \cos(2\pi f_c t + 2\pi f_n t + \phi_n) \tag{2.12}$$

$$H_y(t) = -\frac{E_0}{\eta}\sum_{n=1}^{N} C_n \cos\alpha_n \cos(2\pi f_c t + 2\pi f_n t + \phi_n) \tag{2.13}$$

where E_0 is the real amplitude of local average E-field (E_0 is assumed to be constant), C_n is a real random variable representing the amplitude of individual waves, η is the intrinsic impedance of free space (377Ω), and f_c is the carrier frequency. The random phase of the nth arriving component is denoted by ϕ_n.

The amplitudes of the E- and H-field are normalized such that the ensemble average of the C_n 's is given by

$$\sum_{n=1}^{N} E[C_n^2] = 1 \tag{2.14}$$

Since the Doppler shift is very small when compared to the carrier frequency, the three field components may be modeled as narrowband random processes with mutually independent random variables C_n, α_n, and ϕ_n. If N is sufficiently large, the three components E_z, H_x, and H_y can be approximated as Gaussian random variables (according to the central limit theorem [4]). The phase angles ϕ_n are assumed to have a uniform probability density function (pdf) on the interval (0, 2π]. Based on the analysis in [9], the E-field can be expressed in an in-phase and quadrature form as

$$E_z(t) = h_i(t)\cos(2\pi f_c t) - h_q(t)\sin(2\pi f_c t) \tag{2.15}$$

where

$$h_i(t) = E_0 \sum_{n=1}^{N} C_n \cos(2\pi f_n t + \phi_n) \tag{2.16}$$

and

$$h_q(t) = E_0 \sum_{n=1}^{N} C_n \sin(2\pi f_n t + \phi_n) \tag{2.17}$$

Both $h_i(t)$ and $h_q(t)$ are Gaussian random processes which are denoted as h_i and h_q, respectively, at any time t. Then, h_i and h_q are uncorrelated zero-mean Gaussian random variables with an equal variance given by

$$E[h_i^2] = E[h_q^2] = E[E_z^2] = E_0^2/2 \tag{2.18}$$

where $E[\cdot]$ denotes the ensemble average.

The envelope of the received E-field is given by

$$|E_z(t)| = \sqrt{h_i^2 + h_q^2} \tag{2.19}$$

and has a Rayleigh distribution with the time-average power of the received signal equal to $\sigma^2 = E_0^2/2$.

In Section 2.2.3, we noted that the multipath channel is generally modeled as a linear time-variant filter with low-pass impulse response in (2.7). The Clarke's model can be related to this general channel model by characterizing flat frequency fading as the transmission of an unmodulated carrier. Then, $\hat{s}(t)$ becomes 1 in (2.5), and the received pass-band signal in (2.6) obtains the quadrature form as in (2.15). Then, the multipath channel has the *complex baseband impulse response*

$$h(t,\tau) = (h_i(t) + jh_q(t))\delta(\tau - \hat{\tau}) = E_0 \sum_{n=1}^{N} C_n e^{j(2\pi f_n t + \phi_n)} \delta(\tau - \hat{\tau}) \tag{2.20}$$

where all time delays τ_n are equal to $\hat{\tau}$ due to flat frequency fading.

In the presence of a specular component, (i.e., LoS or a strong reflected path), the complex baseband impulse response becomes a superposition of a strong specular component and scattered components in (2.20). To account for a specular component, the Clarke's model needs to be modified as follows

$$h(t) = \frac{E_0\sqrt{K}}{\sqrt{K+1}} e^{j(2\pi f_d t \cos\alpha_0)} + \frac{E_0}{\sqrt{K+1}} \sum_{n=1}^{N} C_n e^{j(2\pi f_n t + \phi_n)} \tag{2.21}$$

where K is the ratio of the received specular to scattered power (i.e., Ricean factor) and α_0 is the angle of arrival of the specular component.

2.3.3 Correlation Functions and Doppler Power Spectra for Isotropic Scattering Channels

The correlation function and the Doppler spectrum are important statistics that characterize temporal selectivity of a communication system. As pointed out in Section 1.2.2, these statistics enable the system designer to make informed decisions when choosing modulation, interleaving, and coding schemes at the transmitting end and the type of channel estimator and decoder at the receiving end.

Assuming that the received pass-band signal $r(t)$ is wide sense stationary (WSS),[2] the autocorrelation of $r(t)$ is

$$\begin{aligned} R_{rr}(\tau) &= \mathrm{E}[r(t)r(t+\tau)] \\ &= \mathrm{E}\left[h_i(t)h_i(t+\tau)\right]\cos(2\pi f_c\tau)] - \mathrm{E}[h_q(t)h_i(t+\tau)]\sin(2\pi f_c\tau) \\ &= R_{h_ih_i}(\tau)\cos(2\pi f_c\tau) - R_{h_qh_i}(\tau)\sin(2\pi f_c\tau) \end{aligned} \tag{2.22}$$

where

$$R_{h_ih_i}(\tau) = R_{h_qh_q}(\tau) \tag{2.23}$$

$$R_{h_ih_q}(\tau) = -R_{h_qh_i}(\tau) \tag{2.24}$$

It is reasonable to assume that the phases ϕ_n and ϕ_m are independent for $m \neq n$ since their associated delays and Doppler shifts are independent. Furthermore, since the phases are uniformly distributed over $[0, 2\pi)$, the autocorrelation $R_{h_ih_i}(\tau)$ can be obtained as follows:

$$\begin{aligned} R_{h_ih_i}(\tau) &= \mathrm{E}_{\tau,\alpha_n}[h_i(t)h_i(t+\tau)] \\ &= \frac{E_0^2}{2}\mathrm{E}_{\alpha_n}[\cos(2\pi f_d\tau\cos\alpha_n)] \end{aligned} \tag{2.25}$$

Similarly, the cross-correlation $R_{h_ih_q}(\tau)$ is

$$\begin{aligned} R_{h_ih_q}(\tau) &= \mathrm{E}_{\tau,\alpha_n}[h_i(t)h_q(t+\tau)] \\ &= \frac{E_0^2}{2}\mathrm{E}_{\alpha_n}[\sin(2\pi f_d\tau\cos\alpha_n)] \end{aligned} \tag{2.26}$$

To evaluate the expectations in (2.25) and (2.26), statistical channel models assume a particular distribution of incident power on the receiver antenna $p(\alpha)$. Recall from Section 2.3.2 that Clarke's two-dimensional (2-D) *isotropic scattering* model assumes that the plane waves propagate in the x-y plane and arrive at the MS from all directions with equal probability—that is, $p(\alpha) = 1/(2\pi)$, for $\alpha \in [0, 2\pi)$ with an isotropic receiver antenna with gain $G(\alpha) = 1$. Then, the expectation in (2.25) can be obtained as follows:

$$\begin{aligned} R_{h_ih_i}(\tau) &= \frac{E_0^2}{2}\int_{-\pi}^{\pi}\cos(2\pi f_d\tau\cos\alpha)p(\alpha)G(\alpha)d\alpha \\ &= \frac{E_0^2}{2}\frac{1}{2\pi}\int_{-\pi}^{\pi}\cos(2\pi f_d\tau\cos\alpha)d\alpha \\ &= \frac{E_0^2}{2}\frac{1}{\pi}\int_{0}^{\pi}\cos(2\pi f_d\tau\sin\alpha)d\alpha \\ &= \frac{E_0^2}{2}J_0(2\pi f_d\tau) \end{aligned} \tag{2.27}$$

where $J_0(\cdot)$ is the zeroth-order Bessel function of the first kind.

2. See Section 2.3.7 for details on WSS.

Similarly, the cross-correlation function for 2-D isotropic scattering and an isotropic antenna with $G(\alpha) = 1$ becomes

$$\begin{aligned} R_{h_ih_q}(\tau) &= \frac{E_0^2}{2}\frac{1}{2\pi}\int_{-\pi}^{\pi} \sin(2\pi f_d\tau\cos\alpha)\,d\alpha \\ &= 0 \end{aligned} \tag{2.28}$$

This result implies that $h_i(t)$ and $h_q(t)$ are uncorrelated functions. Furthermore, because h_i and h_q are Gaussian random variables, it means that they are also independent random processes. However, note that the fact that $h_i(t)$ and $h_q(t)$ are independent is a direct result of the symmetry of the 2-D isotropic scattering environment and the isotropic antenna gain pattern. In case of an un-isotropic scattering environment, $h_i(t)$ and $h_q(t)$ may not be independent random processes.

The autocorrelation of the complex baseband impulse response $h(t) = h_i(t) + jh_q(t)$ can be written as

$$R_{hh}(\tau) = \frac{1}{2}\mathrm{E}[h^*(t)h(t+\tau)] = R_{h_ih_i}(\tau) + jR_{h_ih_q}(\tau) \tag{2.29}$$

The autocorrelation function of the complex baseband impulse response in the presence of both a strong specular component and a scatter component can be obtained by substituting (2.21) into (2.29) and evaluating the expectation, and the result is

$$\begin{aligned} R_{hh}(\tau) = {} & \frac{1}{K+1}\frac{E_0^2}{2}J_0(2\pi f_d\tau) + \frac{K}{K+1}\frac{E_0^2}{2}\cos(2\pi f_d\tau\cos\alpha_0) \\ & + j\frac{K}{K+1}\frac{E_0^2}{2}\sin(2\pi f_d\tau\cos\alpha_0) \end{aligned} \tag{2.30}$$

The power spectral density (psd) of $h_i(t)$ and $h_q(t)$ is the Fourier transform of $R_{h_ih_i}(\tau)$ or $R_{h_qh_q}(\tau)$. For autocorrelation in (2.27), the corresponding psd is ([10], eq. 6.6711.7)

$$\begin{aligned} S_{h_ih_i}(f) &= \mathcal{F}[R_{h_ih_i}(\tau)] \\ &= \begin{cases} \dfrac{E_0^2}{2\pi f_d}\dfrac{1}{\sqrt{1-(f/f_d)^2}} & |f| \le f_d \\ 0 & \text{otherwise} \end{cases} \end{aligned} \tag{2.31}$$

The power spectral density of the complex impulse response $h(t) = h_i(t) + jh_q(t)$ is

$$S_{hh}(f) = S_{h_ih_i}(f) + jS_{h_ih_q}(f) \tag{2.32}$$

Often, $S_{hh}(f)$ is called the *Doppler power spectrum*. For 2-D isotropic scattering and isotropic antennae, $R_{h_ih_q}(\tau) = 0$ and the Doppler power spectrum is $S_{hh}(f) = S_{h_ih_i}(f)$ (which is real and even). If both a specular component and a scattered component are present, the Doppler power spectrum becomes

$$S_{bb}(f)=\begin{cases}\frac{1}{K+1}\frac{E_0^2}{2\pi f_d}\frac{1}{\sqrt{1-(f/f_d)^2}}+\frac{K}{K+1}\frac{E_0^2}{2}\delta(f-f_d\cos\alpha_0) & |f|\le f_d\\ 0 & \text{otherwise}\end{cases} \tag{2.33}$$

The pass-band Doppler power spectrum of the received pass-band signal $r(t)$ can be obtained from the baseband Doppler power spectrum as follows

$$S_{rr}(f)=\frac{1}{2}[S_{bb}(f-f_c)+S_{bb}(-f-f_c)] \tag{2.34}$$

For 2-D isotropic scattering and isotropic antennae, the pass-band Doppler poer spectrum becomes

$$S_{rr}(f)=\begin{cases}\frac{E_0^2}{4\pi f_d}\frac{1}{\sqrt{1-((f-f_c)/f_d)^2}} & |f-f_c|\le f_d\\ 0 & \text{otherwise}\end{cases} \tag{2.35}$$

2.3.4 Correlation Functions and Doppler Power Spectrum for Nonisotropic Scattering Channels

One of the main assumptions in the Clarke's channel model is that scattering is isotropic (i.e., angles of arrival of multipath components at the MS are uniformly distributed). However, it has been argued [11–15] and experimentally demonstrated [16–22] that scattering encountered in many environments, such as suburban area, urban area between buildings, street canyons, and highways, is nonisotropic, resulting in a nonuniform probability density function (pdf) for angles of arrival (AoA) at the MS. As has been discussed in [21], the assumption of a uniform pdf for the AOAs introduces small errors on the first-order statistics of the received signal, but a significant error on the second-order statistics, like correlation functions and level crossing rates.

In the literature, several different scatterer distributions, such as quadratic pdf [23], cosine pdf [24], Laplace and Gaussian pdfs [25], and von Mises pdf [26], have been used to characterize a nonuniform scattering environment. In addition to showing good fit to measurements [27], the von Mises pdf approximates many of the other distributions (e.g., uniform, sinusoid, and Gaussian) and, in contrast to most other distributions, leads to closed-form solutions for many useful statistics. Hence, this distribution is often used to characterize nonisotropic scattering.

The von Mises pdf was introduced by R. von Mises in 1918 to study the deviations of measured atomic weights from integral values [28]. This pdf plays a prominent role in statistical modeling and analysis of angular variables [24].

The von Mises pdf is defined as [28]

$$p(\alpha)\triangleq\frac{1}{2\pi I_0(k)}\exp[k\cos(\alpha-\mu)] \tag{2.36}$$

where $\alpha\in[-\pi,\pi)$ represents AoAs, $I_0(\cdot)$ is the zeroth-order modified Bessel function of the first kind, $\mu\in[-\pi,\pi)$ is the mean angle at which the scatterers are distributed, and k controls the spread of scatterers around the mean. When $k=0$, $p(\alpha)=1/(2\pi)$ is a uniform distribution yielding 2-D isotropic scattering. As k increases, the scatterers become more clustered around angle μ and the scattering becomes increasingly nonisotropic. For small k, this function approximates the cardioid pdf [29], which is similar to the cosine pdf [24], while for large k it resembles a Gaussian pdf with mean μ and standard deviation $1/\sqrt{k}$ [29]. In general, the von Mises pdf can approximate the wrapped Gaussian pdf [29]. It is interesting to note that the von Mises pdf appears in a number of other applications in telecommunications. For example, this pdf is referred to as the Tikhonov pdf in partially coherent communications [30] and has been used in phase-locked loop–related problems [31]. It also has been shown that the phase of a sine wave with Gaussian noise for large signal to noise ratios has a von Mises pdf [32].

Assuming that the received band-pass signal $r(t)$ is wide sense stationary, the autocorrelation of $r(t)$ can be evaluated starting from (2.22), as in Section 2.3.3. By assuming that the phases ϕ_n and ϕ_m are independent for $m \neq n$ and uniformly distributed over $[0, 2\pi)$, the autocorrelation $R_{h_ih_i}(\tau)$ and the cross-correlation $R_{h_ih_q}(\tau)$ can be written as in (2.25) and (2.26), respectively. To evaluate the expectations in (2.25) and (2.26), the nonisotropic scattering model assumes that the plane waves propagate in the x-y plane and arrive at the MS from all directions with nonequal probability (i.e., the von Mises pdf describes the angles of arrival). To simplify calculations, the nonisotropic scattering model still assumes an isotropic receiver antenna with gain $G(\alpha)=1$. Then, the autocorrelation function in (2.25) can be calculated as follows:

$$\begin{aligned} R_{h_ih_i}(\tau) &= \frac{E_0^2}{2}\int_{-\pi}^{\pi}\cos(2\pi f_d \tau\cos\alpha)p(\alpha)G(\alpha)d\alpha \\ &= \frac{E_0^2}{2}\frac{1}{2\pi I_0(k)}\int_{-\pi}^{\pi}\cos(2\pi f_d\tau\cos\alpha)\exp(k\cos(\alpha-\mu))\,d\alpha \end{aligned} \tag{2.37}$$

Using the quality $\int_{-\pi}^{\pi} e^{a\sin(c)+b\cos(c)}dc = 2\pi I_0\left(\sqrt{a^2+b^2}\right)$ [10, eq. 3.338-4], the autocorrelation $R_{h_ih_i}(\tau)$ becomes

$$R_{h_ih_i}(\tau) = \frac{E_0^2}{2}\Re\left\{\frac{I_0\left(\sqrt{k^2-4\pi^2 f_d^2\tau^2+j4\pi k f_d\tau\cos\mu}\right)}{I_0(k)}\right\} \tag{2.38}$$

where $\Re\{\cdot\}$ denotes the real part operation. For $k = 0$, the von Mises pdf becomes the uniform pdf and the autocorrelation function in (2.38) becomes identical to the autocorrelation function in (2.25). To illustrate how the autocorrelation function $R_{h_ih_i}(\tau)$ changes as the spread of scatterers k changes, Figure 2.6 plots the normalized autocorrelation function $R_{h_ih_i}(\tau)/(E_0^2/2)$ against the normalized time delay $f_d\tau$ for several values of $k \in \{0,1,2,5,10\}$ and $\mu = 0°$.

Similarly, the cross-correlation function for 2-D nonisotropic scattering and an isotropic antenna with $G(\alpha) = 1$ is

$$R_{h_ih_q}(\tau) = \frac{E_0^2}{2}\Im\left\{\frac{I_0\left(\sqrt{k^2-4\pi^2 f_d^2\tau^2+j4\pi k f_d\tau\cos\mu}\right)}{I_0(k)}\right\} \tag{2.39}$$

where $\Im\{\cdot\}$ denotes the imaginary part operation. Note that the cross-correlation function is not zero in nonisotropic scattering channels.

The autocorrelation of the complex impulse response $h(t) = h_i(t) + jh_q(t)$ can be obtained as

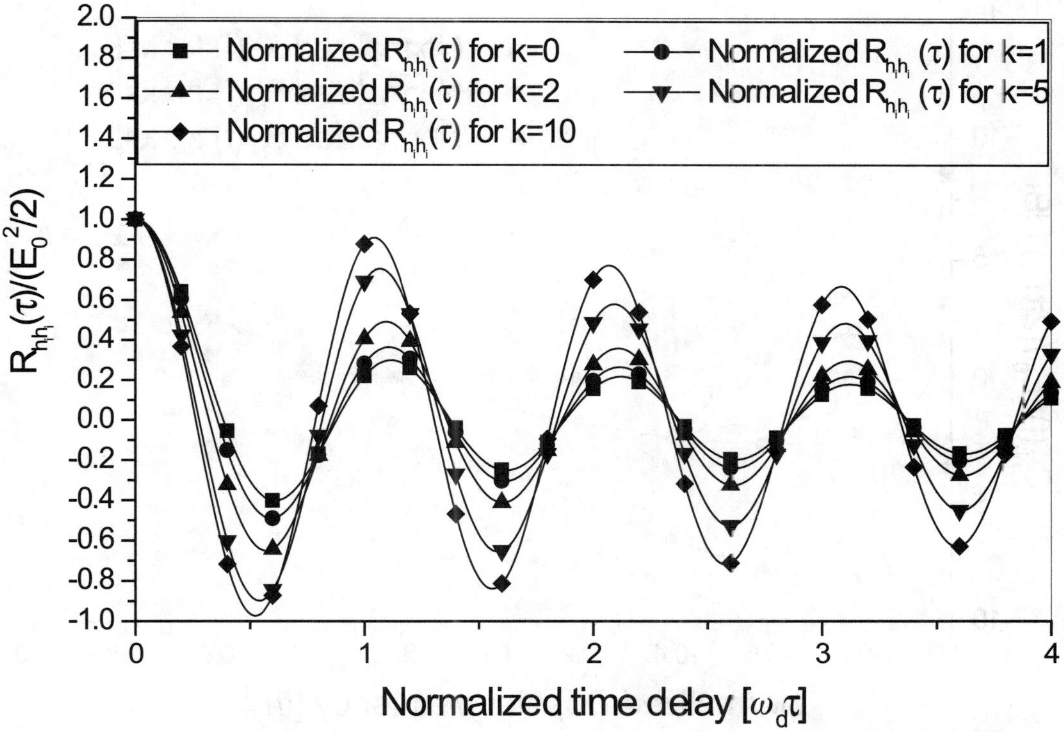

Figure 2.6 The normalized autocorrelation function $R_{h_ih_i}(\tau)/(E_0^2/2)$ as a function of the normalized time delay $\omega_d\tau$ for $k \in \{0,1,2,5,10\}$.

$$R_{hh}(\tau) = \frac{1}{2}\mathrm{E}[h^*(t)h(t+\tau)] = R_{h_ih_i}(\tau) + jR_{h_ih_q}(\tau)$$

$$= \frac{E_0^2}{2}\frac{I_0\left(\sqrt{k^2-4\pi^2 f_d^2\tau^2+j4\pi k f_d \tau\cos\mu}\right)}{I_0(k)} \tag{2.40}$$

The autocorrelation function of the complex impulse response in the presence of both specular and scattered components can be obtained by substituting (2.21) into (2.40) and evaluating the expectation, and the result is [26]

$$R_{hh}(\tau) = \frac{1}{K+1}\frac{E_0^2}{2}\frac{I_0\left(\sqrt{k^2-4\pi^2 f_d^2\tau^2+j4\pi k f_d \tau\cos\mu}\right)}{I_0(k)}$$

$$+\frac{K}{K+1}\frac{E_0^2}{2}\exp(j2\pi f_d\,\tau\cos(\alpha_0)) \tag{2.41}$$

The Doppler power spectrum of the complex impulse response $h(t) = h_i(t) + jh_q(t)$ can be obtained by calculating Fourier transform of the autocorrelation function of the complex impulse response in (2.41). The result of this Fourier transform is [26]

$$S_{hh}(f) = \frac{1}{K+1}\frac{E_0^2}{2\pi f_d\sqrt{1-(f/f_d)^2}}\frac{\exp(kf\cos\mu/f_d)\cosh\left(k\sin\mu\sqrt{1-(f/f_d)^2}\right)}{I_0(k)}$$

$$+\frac{K}{K+1}\frac{E_0^2}{2}\delta(f-f_d\cos\alpha_0), |f| \le f_d \tag{2.42}$$

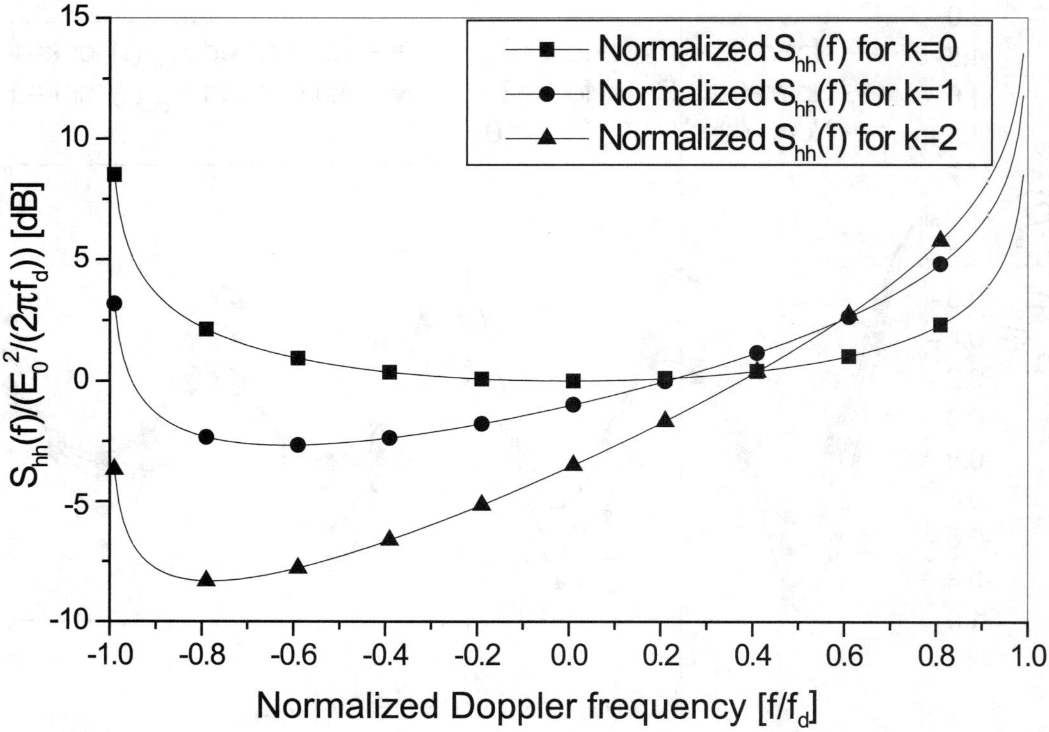

Figure 2.7 The normalized Doppler power spectrum as a function of the normalized Doppler frequency f/f_d for $k \in \{0,1,2\}$.

where cosh $(\cdot)$ is the hyperbolic cosine. For $k = 0$, this Doppler power spectrum reduces to (2.33) (i.e., the Doppler power spectrum for the Clarke's 2-D isotropic scattering). Figure 2.7 illustrates the effect spread of scatterers k has on the normalized Doppler power spectrum $S_{hh}(f)/(E_0^2/(2\pi f_d))$ for $\mu = 0°$ and $K = 0$.

2.3.5 Second-Order Statistics—Level Crossing Rates and Fade Durations

Two important second-order statistics associated with envelope fading are the *level crossing rate* (how often the envelope crosses a specified level) and the *average fade duration* (how long the envelope remains below a specified level). These quantities are second-order statistics because they are affected not only by the scattering environment but also by the velocity of the MS. These statistics are very useful for designing error control codes and diversity schemes in communication systems, because they relate the rate of change of the received signal over time to the signal level and velocity of the MS.

2.3.5.1 Envelope Level Crossing Rate

The level crossing rate (LCR), $L(R)$, is defined as the rate at which the signal envelope crosses level R in the positive (or negative) direction and can be written as [4]

$$L(R) = \int_0^\infty \dot{\beta} \cdot p_{\beta,\dot{\beta}}(R,\dot{\beta}) d\dot{\beta} \tag{2.43}$$

where $\beta = |h(t)|$ is the envelope level, $\dot{\beta} = |\dot{h}(t)|$ is the envelope slope, $p_{\beta,\dot{\beta}}(R,\dot{\beta})$ is the joint pdf of the envelope level and the envelope slope, and $|\cdot|$ denotes the absolute value operation. To evaluate the LCR, we need to know the pdf of the fading envelope $\beta = |h(t)|$, the pdf of the envelope slope $\dot{\beta} = |\dot{h}(t)|$, and the joint pdf $p_{\beta,\dot{\beta}}(R,\dot{\beta})$.

For 2-D isotropic scattering channels that have both specular and scattered components, the joint pdf $p(\beta,\dot{\beta})$ can be separated into a product of a Ricean pdf $p_{|h(t)|}(\beta)$ and a Gaussian pdf $p_{|\dot{h}(t)|}(\dot{\beta})$, because the in-phase and quadrature components of the complex faded envelope $h(t)$ are uncorrelated. Then, the joint pdf $p(\beta,\dot{\beta})$ can be written as [4]

$$p(\beta,\dot{\beta}) = \sqrt{\frac{1}{2\pi b_2}}\exp\left(-\frac{\dot{\beta}^2}{2b_2}\right)\cdot\frac{\beta}{b_0}\exp\left(-\frac{(\beta^2+K/(K+1))}{2b_0}\right)I_0\left(\frac{\beta\sqrt{\frac{K}{K+1}}}{b_0}\right)$$

$$= p(\beta)\cdot p(\dot{\beta}) \tag{2.44}$$

where K denotes the Rice factor (ratio of specular to scatter received power), the parameters b_0 and b_2 are defined as [9]

$$b_0 \triangleq \mathrm{E}_\alpha[h_i(t)^2] = \mathrm{E}_\alpha[h_q(t)^2] = R_{hh}(\tau)_{|\tau=0} \tag{2.45}$$

$$b_2 \triangleq \mathrm{E}_\alpha[\dot{h}_i(t)^2] = \mathrm{E}_\alpha[\dot{h}_q(t)^2] = -\frac{d^2R_{hh}(\tau)}{d\tau^2}_{|\tau=0} \tag{2.46}$$

$\dot{h}_i(t)$ and $\dot{h}_q(t)$ denote the first derivative of $h_i(t)$ and $h_q(t)$ with respect to time t, and $R_{hh}(\tau)$ denotes the autocorrelation function of the complex low-pass impulse response defined in (2.29). For 2-D isotropic scattering, the parameters b_0 and b_2 simplify to

$$b_0 = \frac{E_0^2}{2}\frac{1}{K+1} \tag{2.47}$$

$$b_2 = \frac{b_0(2\pi f_d)^2}{2} \tag{2.48}$$

Substituting the joint pdf in (2.44) into (2.43) gives the envelope level crossing rate [4]

$$L(R) = \sqrt{2\pi(K+1)}f_d\rho e^{-K-(K+1)\rho^2}I_0\left(2\rho\sqrt{K(K+1)}\right) \tag{2.49}$$

where

$$\rho = \frac{R}{\sqrt{E_0^2}} = \frac{R}{R_{\mathrm{rms}}} \tag{2.50}$$

and $R_{\mathrm{rms}} \triangleq \sqrt{E_0^2}$ is the root mean square of the envelope level. For Rayleigh fading ($K = 0$) and isotropic scattering, the previous expression simplifies to

$$L(R) = \sqrt{2\pi}\, f_d\rho e^{-\rho^2} \tag{2.51}$$

For 2-D nonisotropic scattering channels that have both specular and scattered components, and have correlated in-phase and quadrature components of the complex faded envelope $h(t)$, the joint pdf $p(\beta, \dot{\beta})$ cannot be separated into the product of a pdf $p_{|h(t)|}(\beta)$ and a pdf $p_{|\dot{h}(t)|}(\dot{\beta})$. Instead, the joint pdf $p(\beta, \dot{\beta})$ for the complex faded envelope $h(t)$ with the correlated in-phase and quadrature components is calculated in [33] and can be written as

$$p(\beta,\dot{\beta}) = \frac{\beta e^{-(\beta^2+\delta^2)/(2b_0)}}{b_0\sqrt{\chi}(2\pi)^{3/2}} \int_0^{2\pi} e^{(\beta\delta/b_0)\cos\theta} e^{-\frac{1}{2\chi}[\dot{\beta}-\sqrt{2\chi}\xi\delta\sin\theta]^2} d\theta \tag{2.52}$$

where $\delta = \sqrt{K/(K+1)}$, $\chi = b_2 - b_1^2/b_0$, $\xi = (2\pi f_d \cos\alpha_0 - b_1/b_0)/\sqrt{2\chi}$, the parameters b_0, b_1, and b_2 are

$$b_0 \triangleq \mathrm{E}[h_i(t)^2] = \mathrm{E}[h_q(t)^2] \tag{2.53}$$

$$b_1 \triangleq \mathrm{E}[h_i(t)\dot{h}_q(t)] = \mathrm{E}[h_q(t)\dot{h}_i(t)] \tag{2.54}$$

$$b_2 \triangleq \mathrm{E}[\dot{h}_i(t)^2] = \mathrm{E}[\dot{h}_q(t)^2] \tag{2.55}$$

and $\dot{h}_i(t)$ and $\dot{h}_q(t)$ denote the first derivative of $h_i(t)$ and $h_q(t)$ with respect to time t. When 2-D nonisotropic scattering is characterized using the von Mises pdf, the parameters b_0, b_1, and b_2 can be evaluated and become

$$b_0 = \frac{E_0^2}{2}\frac{1}{K+1} \tag{2.56}$$

$$b_1 = b_0 \frac{2\pi f_d \cos(\mu) I_1(k)}{I_0(k)} \tag{2.57}$$

$$b_2 = b_0 \frac{2\pi^2 f_d^2 [I_0(k) + \cos(2\mu) I_2(k)]}{I_0(k)} \tag{2.58}$$

where $I_1(\cdot)$ and $I_2(\cdot)$ denote the first- and the second-order modified Bessel function of the first kind, respectively. By substituting (2.52) into (2.43) and solving the integral, the LCR for LoS conditions (Ricean fading, correlated in-phase and quadrature components) becomes [33]

$$L(R) = \frac{R\sqrt{2\chi}}{\pi^{3/2} b_0} e^{-(R^2+\delta^2)/(2b_0)} \int_0^{\pi/2} \cosh\left(\frac{R\delta}{b_0}\cos\theta\right)$$
$$\times \left[e^{-(\xi\delta\sin\theta)^2} + \sqrt{\pi}\,\xi\delta\sin\theta\ \mathrm{erf}(\xi\delta\sin\theta)\right] d\theta \tag{2.59}$$

where $\cosh(\cdot)$ is the hyperbolic cosine function and $\mathrm{erf}(\cdot)$ is the error function. The LCR for NLoS conditions can be obtained from (2.59) by setting $K = 0$; that is

$$L(R) = \sqrt{\frac{b_2}{b_0} - \frac{b_1^2}{b_0^2}} \frac{R e^{-R^2}}{\sqrt{\pi}} \tag{2.60}$$

2.3.5.2 Average Envelope Fade Duration

Another quantity of interest is the average duration that the envelope level remains below a specified level R. Although the pdf of the envelope fade duration is generally unknown, the average fade duration can still be calculated. Consider a very long time interval of length T and let t_i be the duration of the ith fade below the level R. The probability of the received envelope level being less than R is

$$P(\beta \le R) = \frac{1}{T}\sum_i t_i \tag{2.61}$$

The average envelope fade duration is equal to

$$T(R) = \frac{1}{TL(R)}\sum_i t_i = \frac{P(\beta \le R)}{L(R)} \tag{2.62}$$

where $L(R)$ denotes the envelope level crossing rate. If the envelope has the Rice distribution in (2.9), then

$$P(\beta \le R) = \int_0^R p(\beta)d\beta = 1 - Q\left(\sqrt{2K}, \sqrt{2(K+1)R^2}\right) \tag{2.63}$$

where $Q(a, b)$ is the Marcum Q function. When a LoS component is present, the average envelope fade duration becomes [4]

$$T(R) = \frac{1 - Q\left[\sqrt{2K}, \sqrt{2(K+1)R^2}\right]}{L(R)} \tag{2.64}$$

For NLoS conditions, the average envelope fade duration in (2.64) simplifies to

$$T(R) = \frac{1 - e^{-R^2}}{L(R)} \tag{2.65}$$

Note that both the level crossing rate and average fade duration depend on the velocity of the MS. Also note that very deep fades tend to occur infrequently and do not last very long. For example, at v = 60 miles/h and f_c = 900 MHz, the maximum Doppler frequency is f_d = 88 Hz [4]. Therefore, with isotropic scattering and Rayleigh fading (K = 0), there are $L(R)$ = 81 fade/s at R = 0 dB with an average fade duration of 7.8 ms. On the other hand, at R = –20 dB, there are only 2.2 fades/s with an average fade duration of 45 μs.

2.3.6 Modeling of F-to-M Frequency-Selective Multipath Fading Channels

Up to this point, we have considered channel models that are appropriate for narrowband transmission, where the inverse signal bandwidth is much greater than the time spread of the propagation path delays. Such a channel introduces very little or no distortion into the received signal and is said to exhibit flat fading. For digital communication systems, this means that the duration of a modulated symbol is much greater than the time spread of the propagation path delays. Under this condition, all frequencies in the transmitted signal will experience the same random attenuation and phase shift due to multipath fading.

On the other hand, if the range in the propagation path delays is large compared to the inverse signal bandwidth, then the frequency components in the transmitted signal will experience different phase shifts along the different paths. As differences between path delays become large, even closely separated frequencies in the transmitted signal can experience significantly different phase shifts. Under this condition, the channel introduces amplitude and phase distortion in the signal waveform. Such a channel is said to exhibit frequency-selective fading.

For frequency-selective fading, multipath-fading channels are modeled as *time-varying* in addition to being time-dispersive. The *input delay-spread function* function $h(t,\tau)$ is defined as the response of the channel at time t to a unit impulse at time $t-\tau$. This function is useful to analyze how rapidly the channel varies in time and quantify how dispersive the channel is.

If we take the Fourier transform of $h(t, \tau)$ with respect to time t, we can quantify how much the channel varies over time (delay). Analytically, the *delay Doppler-spread function* function $U(\nu, \tau)$ is defined as [35]

$$U(\nu,\tau) = \mathcal{F}_t[h(t,\tau)] = \int_{-\infty}^{\infty} h(t,\tau)e^{-j2\pi\nu t}dt \tag{2.66}$$

where ν is the Doppler frequency in hertz. This function provides a measure of the scattering amplitude of the channel in terms of the time delay τ and Doppler frequency ν.

On the other hand, if we take the Fourier transform of $h(t, \tau)$ with respect to delay τ, we can quantify how dispersive the channel is at a given time. Analytically, the *time-variant transfer function* $T(t, f)$ is defined as

$$T(t,f) = \mathcal{F}_\tau[h(t,\tau)] = \int_{-\infty}^{\infty} h(t,\tau)e^{-j2\pi f\tau}d\tau \tag{2.67}$$

This function characterizes the frequency selectivity of the channel at a given instant in time.

Finally, the *output Doppler-spread* function is obtained by taking the double Fourier transform of both time and delay simultaneously; that is,

$$G(\nu,f) = \mathcal{F}_{t,\tau}[h(t,\tau)] = \int_{-\infty}^{\infty}\int_{-\infty}^{\infty} h(t,\tau)e^{-j2\pi\nu t}e^{-j2\pi f\tau}dtd\tau \tag{2.68}$$

This function explicitly shows the effect of Doppler shift or spectral broadening on the output spectrum. Note that these four functions are related to each other through Fourier transform pairs.

2.3.7 Correlation Functions for F-to-M Frequency-Selective Multipath Fading Channels

Here we note that the channel impulse response $h(t,\tau) = h_i(t,\tau) + jh_q(t,\tau)$ can be modeled as a complex Gaussian random process, where the in-phase and quadrature components (i.e., $h_i(t,\tau)$ and $h_q(t,\tau)$) are correlated Gaussian random processes. Hence, all of the transmission functions defined in Section 2.3.6 are random processes. A thorough characterization of a channel requires knowledge of the joint pdf of all the transmission functions. However, this is a difficult task, and a more reasonable approach is to obtain statistical correlation functions for the individual transmission functions. If the underlying process is Gaussian, then a complete statistical description is provided by the means and autocorrelation functions. Here, we assume zero-mean Gaussian random processes so that only the autocorrelation functions are of interest. Since there are four transmission functions, there are four autocorrelation functions that can be defined as follows [4]:

$$R_h(t,s;\tau,\eta) = \frac{1}{2}\mathrm{E}[h(t,\tau)h^*(s,\eta)] \tag{2.69}$$

$$R_U(\tau,\eta;\nu,\mu) = \frac{1}{2}\mathrm{E}[U(\tau,\nu)U^*(\eta,\mu)] \tag{2.70}$$

$$R_T(f,m;t,s) = \frac{1}{2}\mathrm{E}[T(f,t)T^*(m,s)] \tag{2.71}$$

$$R_G(f,m;\nu,\mu) = \frac{1}{2}\mathrm{E}[G(f,\nu)G^*(m,\mu)] \tag{2.72}$$

where $(\cdot)^*$ denotes the complex conjugate operation. Note that these autocorrelation functions are related to each other through double Fourier transform pairs. The channel is said to be *wide sense stationary* if the fading statistics remain constant over short periods of time. This implies that the channel correlation functions depend on the time variables t and s only through the time difference $\Delta t = s - t$. Hence for WSS channels, the correlation functions become

$$R_h(t,t+\Delta t;\tau,\eta) = R_h(\Delta t;\tau,\eta) \tag{2.73}$$

$$R_U(\tau,\eta;\nu,\mu) = \psi_U(\tau,\eta;\nu)\delta(\nu-\mu) \tag{2.74}$$

$$R_T(f,m;t,t+\Delta t) = R_T(f,m;\Delta t) \tag{2.75}$$

$$R_G(f,m;\nu,\mu) = \psi_G(f,m;\nu)\delta(\nu-\mu) \tag{2.76}$$

where

$$\psi_U(\tau,\eta;\nu) = \int_{-\infty}^{\infty} R_h(\Delta t;\tau,\eta)e^{-j2\pi\nu\Delta t}d\Delta t \tag{2.77}$$

$$\psi_G(f,m;\nu) = \int_{-\infty}^{\infty} R_T(f,m;\Delta t)e^{-j2\pi\nu\Delta t}d\Delta t \tag{2.78}$$

are Fourier transform pairs. It can be shown that WSS channels give rise scattering with uncorrelated Doppler shifts. This behavior suggests that signal components with different Doppler shifts have uncorrelated attenuations and phase shifts.

The channel is said to exhibit *uncorrelated scattering* if the contributions from elemental scatterers corresponding to different delays are uncorrelated. This implies that the channel correlation functions depend on the frequency variables f and m only through the frequency difference $\Delta f = m - f$ [36]. Hence, for US channels, the correlation functions become

$$R_h(t,s;\tau,\eta) = \psi_h(t,s;\tau)\delta(\eta-\tau) \tag{2.79}$$

$$R_U(\tau,\eta;\nu,\mu) = \psi_U(\tau;\nu,\mu)\delta(\eta-\tau) \tag{2.80}$$

$$R_T(f, f+\Delta f; t, s) = R_T(\Delta f; t, s) \tag{2.81}$$

$$R_G(f, f+\Delta f; \nu, \mu) = R_G(\Delta f; \nu, \mu) \tag{2.82}$$

where

$$\psi_h(t; s, \tau) = \int_{-\infty}^{\infty} R_T(\Delta f; t, s) e^{j2\pi\Delta f\tau} d\Delta f \tag{2.83}$$

$$\psi_U(\tau; \nu, \mu) = \int_{-\infty}^{\infty} R_G(\Delta f; \nu, \mu) e^{j2\pi\Delta f\tau} d\Delta f \tag{2.84}$$

Wide sense stationary uncorrelated scattering (WSSUS) channels are a very special type of multipath-fading channels. These channels display uncorrelated scattering in both the time-delay and Doppler shift, and hence reduce to the following simple forms:

$$R_h(t, t+\Delta t; \tau, \eta) = \psi_h(\Delta t; \tau)\delta(\eta - \tau) \tag{2.85}$$

$$R_U(\tau, \eta; \nu, \mu) = \psi_U(\tau, \nu)\delta(\eta - \tau)\delta(\nu - \mu) \tag{2.86}$$

$$R_T(f, f+\Delta f; t, t+\Delta t) = R_T(\Delta f; \Delta t) \tag{2.87}$$

$$R_G(f, f+\Delta f; \nu, \mu) = \psi_G(\Delta f; \nu)\delta(\nu - \mu) \tag{2.88}$$

Fortunately, many radio channels can be accurately modeled as WSSUS channels.

2.3.8 Channel Characteristic Parameters—Delay Spread, Power Delay Profile, and Channel Coherence

In order to evaluate the characteristics of a propagation channel, a set of measured impulse responses needs to be analyzed. This section presents a number of parameters describing different aspects of the radio channel. In general, their computation requires the WSSUS assumption. As it is not practically feasible to collect a set of statistical channel realizations, we generally assume that the channel system functions are ergodic. Hence, it is for instance possible to approximate the statistical expectation using a temporal (or spatial) average over a set of successive measurements.

A channel's frequency selectivity is quantified by its span in the delay domain (i.e., its *delay spread*). As the channel's frequency response becomes more varied, an interacting signal will exhibit a broader envelope in the delay domain. In practice, delay spread is the result of different propagation delays of received signals from reflectors at large distances compared to the LoS path between the BS and MS.

To introduce the power delay profile, we observe the channel for a short period over which it is time-invariant (i.e., $h(t,\tau)$ does not depend on t and can be expressed as $h(\tau)$). The *power delay profile* (PDP) is defined as the envelope of $h(\tau)$,

$$P(\tau) = |h(\tau)|^2 \tag{2.89}$$

The *root means square* (RMS) delay spread is related to the statistics of the PDP. Specifically, the RMS delay spread σ_τ^2 is defined as the second central moment of the PDP; that is,

$$\sigma_\tau^2 = E_\tau[P(\tau)^2] - E_\tau[P(\tau)]^2 \tag{2.90}$$

where

$$E_\tau[P(\tau)^n] = \frac{\int \tau^n P(\tau)d\tau}{\int P(\tau)d\tau} \tag{2.91}$$

The RMS delay spread is a significant parameter for the analysis of the intersymbol interference.

Closely related to the RMS delay spread is the concept of *coherence bandwidth* W_c. Coherence bandwidth is inversely proportional to coherence time. The coherence bandwidth is a statistical measure of the bandwidth over which the channel exhibits approximately equal gain. In other words, the coherence bandwidth is the frequency bandwidth $W_c = f_1 - f_2$ within which the frequency autocorrelation function crosses a given threshold. For the 90 percent threshold, the coherence bandwidth can be approximated from the RMS delay spread as

$$W_c \approx \frac{1}{50\sigma_\tau} \tag{2.92}$$

In the literature, the coherence bandwidth is most often defined for the 50 percent correlation threshold, in which case

$$W_c \approx \frac{1}{5\sigma_\tau} \tag{2.93}$$

A channel's time variance is qualified by the *Doppler spread* (i.e., by the span of the channel's Doppler spectrum), because channels that vary faster exhibit a broader frequency range in the frequency domain. Similar to the RMS delay spread, the *RMS Doppler spread* σ_v^2 is the second central moment of the Doppler spectrum; that is,

$$\sigma_v^2 = \mathrm{E}_v[S(v)^2] - \mathrm{E}_v[S(v)]^2 \tag{2.94}$$

where

$$E_v[S(v)^n] = \frac{\int v^n S(v)dv}{\int S(v)dv} \tag{2.95}$$

Closely related to the Doppler spread is the *coherence time* T_c. The coherence time is a statistical measure of the time period over which the channel does not change appreciably. Put differently, the channel will affect two signals differently if the period between their arrival times exceeds T_c. The coherence time is inversely proportional to RMS Doppler spread. In the literature, the following ratio is used to estimate T_c:

$$T_c \approx \frac{1}{5\sigma_v} \tag{2.96}$$

In mobile channels where time–variation can be mainly attributed to movement at the transmitter or receiver, the coherence time is often estimated using the maximum Doppler shift f_d; that is,

$$T_c \approx \frac{0.423}{f_d} \tag{2.97}$$

2.3.9 MIMO F-to-M Cellular Radio Channels

In previous sections, we discussed cases where BS and MS are equipped with one antenna each. In the following, we consider the case where BS and MS are equipped with multiple antennas, as illustrated in Figure 2.8.

Wireless communication systems employing multiple antennas at both the BS and MS have been shown to offer significant gains over single-antenna systems. Often, especially in urban and indoor environments, there is no clear line of sight between the BS and MS. Instead, the signal is reflected along multiple paths before finally being received. Each of these bounces can introduce phase shifts, time delays, attenuations, and distortions that can destructively interfere with one another at the aperture of the receiving antenna. Antenna diversity is especially effective at mitigating these multipath situations. This is because multiple antennas provide the receiver with several observations of the same signal, as each antenna experiences a different interference environment. Thus, if one antenna is experiencing a deep fade, it is likely that another has a sufficient signal, providing the overall system with a more robust link.

There are several degrees of freedom in realizing antenna diversity: *polarization diversity* (i.e., uncorrelated fading between different antenna polarizations), *spatial diversity* (i.e., different fading for spaced-apart antennas because that changes the direction of the incoming waves), and *angular diversity* (i.e., directive antenna patterns can be rotated to receive from different directions). Note that more than one of these techniques can be deployed together.

When a system is equipped with multiple antennas at the BS and MS, a MIMO channel separates them. To simplify our discussion, the MIMO channel can be

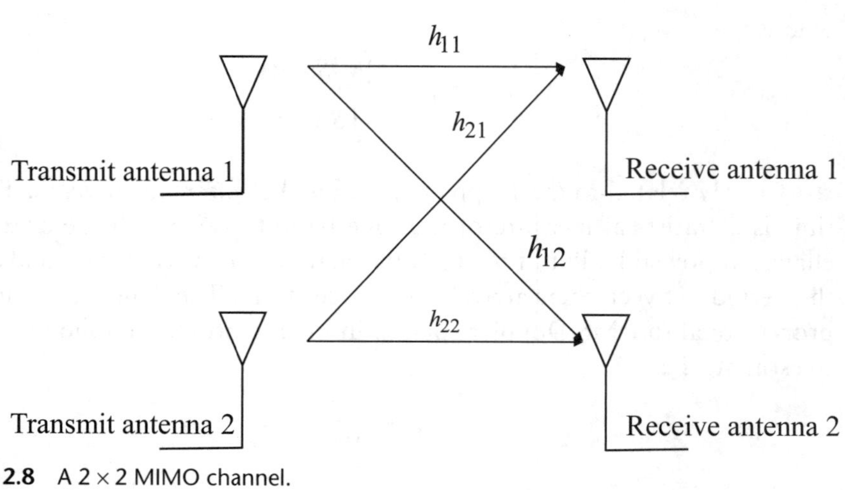

Figure 2.8 A 2 × 2 MIMO channel.

viewed as a collection of many, correlated, single-input single-output (SISO) channels. The narrowband (frequency-flat) MIMO channel can be described by an $L_r \times L_t$ matrix $\mathbf{H}(t) = [h_{ij}(t)]_{L_r \times L_t}$ of complex faded envelopes, where L_t and L_r enote the number of BS and MS antennas, respectively. Similarly, the wideband (frequency-selective) MIMO channel can be described by an an $L_r \times L_t$ matrix $\mathbf{H}(t,\tau) = [h_{ij}(t,\tau)]_{L_r \times L_t}$ of the input delay-spread functions.

2.4 Simulators for Fixed-to-Mobile Cellular Radio Channels

Channel models detailed in Section 2.3 are very useful mathematical tools for understanding propagation mechanisms and for statistical characterization of channels. However, these models assume large number of propagation rays, making these models impractical for simulators. The following sections describe practical simulators for frequency-flat and frequency-selective multipath fading with a small number of propagation rays that still match the statistical properties of the channel models detailed in Section 2.3.

2.4.1 Simulation Models for F-to-M Frequency-Flat Multipath Fading Channels

There are several different methods for simulating fading channels. The most accepted of these methods are filtered noise models [37–40] and sum-of-sinusoids (SoS) models [43–46]. The filtered noise models filter Gaussian noise through appropriately designed filters to generate the channel waveform with the desired channel power spectral density (the Doppler spectrum). The main limitation with this approach is that only rational forms of the Doppler spectrum can be produced, whereas the Doppler spectrum is typically nonrational [4]. To approximate the nonrational Doppler spectrum, a high-order filter is required. Unfortunately, a high-order filter has a long impulse response, which significantly increases run times for software simulation. Furthermore, the Doppler spectrum obtained using this method is not bandlimited because filters with sharp stop-bands are difficult to implement in practice. On the other hand, the SoS models generate the channel waveform by superimposing a finite number of properly selected sinusoids. In contrast to filtered noise models, SoS models have low complexity and produce channel waveforms that have high accuracy and a perfectly band-limited Doppler spectrum. In addition, SoS models can be easily extended to develop space-time correlated channel models for MIMO antenna systems. Due to these advantages, the next section focuses on SoS models.

2.4.2 Deterministic and Statistical SoS Simulation Models for F-to-M Frequency-Flat Multipath Fading Channels

Generally, SoS models can be classified as either deterministic or statistical. Deterministic SoS models have fixed random phases, amplitudes, and Doppler frequencies for all simulation trials. In contrast, statistical SoS models leave at least one of the parameter sets (amplitudes, phases, or Doppler frequencies) as random variables that vary with each simulation trial. The statistical properties of statistical SoS models also vary for each simulation trial, but converge to the desired properties when averaged over a number of simulation trials. A statistical model is said to be *ergodic* if it converges to the desired properties in a single simulation trial.

Clarke's reference model described in Section 2.3.2 assumes an infinite number of propagation rays, which makes this model impractical for use in simulators. Hence, it is desirable to design simulation models with a small number of propagation paths, while still matching the statistical properties of the Clarke's model; that is, [4]

$$R_{h_{i/q}h_{i/q}}(\tau) = \mathrm{E}\left[h_{i/q}(t+\tau)h_{i/q}(t)\right] = J_0(\omega_d\tau) \tag{2.98}$$

$$R_{h_ih_q}(\tau) = R_{h_qh_i}(\tau) = \mathrm{E}\left[h_{i/q}(t+\tau)h_{q/i}(t)\right] = 0 \tag{2.99}$$

$$R_{h_kh_l}(\tau) = \tfrac{1}{2}\mathrm{E}\left[h_k(t+\tau)h_l(t)^*\right] = \begin{cases} J_0(\omega_d\tau) & l = k \\ 0 & l \neq k \end{cases} \tag{2.100}$$

$$R_{|h_k|^2|h_k|^2}(\tau) = \mathrm{E}\left[\left|h_k(t+\tau)\right|^2\left|h_k(t)\right|^2\right] = 4 + 4J_0(\omega_d\tau) \tag{2.101}$$

where $E[\cdot]$ is the statistical expectation operator, $(\cdot)^*$ denotes the complex conjugate operator, and $J_0(\cdot)$ is the zeroth-order Bessel function of the first kind.

Many approaches have been suggested for deterministic SoS simulation of F-to-M Rayleigh fading channels. Several different approaches are described next.

2.4.2.1 Jakes's Model—Model I

Jakes derived his well-known deterministic simulation model for F-to-M Rayleigh fading channels [9] starting from (2.20) and assuming equal-strength multipath components ($C_n = 1$). Then, the complex low-pass impulse response has the form

$$h(t) = \sum_{n=1}^{N} e^{j(2\pi f_d \cos\alpha_n t + \phi_n)} \tag{2.102}$$

where N is the number of propagation rays and ϕ_n are random phases associated with each propagation ray. Jakes approximated a 2-D isotropic scattering environment by using a ring of discrete effective scatterers with the receiver at the center of the ring, as shown in Figure 2.9. The directions to the scatterers can be fixed, so that the moving MS "sees" each scatterer at the same bearing as it moves; or the effective scatterers are spatially fixed so that as the receiver moves, the angles to the sources change. The former technique is more computationally efficient, since the distances to the individual scatterers do not need to be calculated. In the latter technique, the distances to the scatterers can be made large so that it gives essentially the same situation as fixed bearing angle scatterers. Jakes's model uses the first technique (scatterers follow the MS) and chooses the N scatterers to be uniformly distributed in angle; that is,

$$\alpha_n = \frac{2\pi n}{N},\ n = 1,2,\ldots,N \tag{2.103}$$

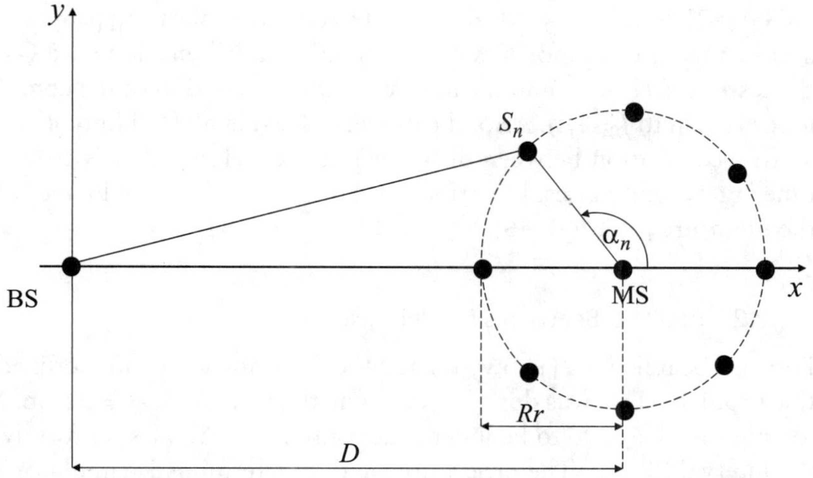

Figure 2.9 A receiver moving through directive scatterers is modeled by a ring of scatterers at a large radius.

Using the symmetry of the 2-D ring in Figure 2.9, Jakes further reduced the number of sinusoidal components necessary for simulation from N to $M = (N - 2)/4$. Then, the complex faded envelope can be written as $h(t) = h_i(t) + jh_q(t)$, where

$$h_i(t) = \sqrt{\frac{4}{N}}\cos(\beta_0)\cos(\omega_d t) + \sqrt{\frac{8}{N}}\sum_{n=1}^{M}\cos(\beta_n)\cos\left[\omega_d t\cos\left(\frac{2\pi n}{N}\right)\right] \tag{2.104}$$

$$h_q(t) = \sqrt{\frac{4}{N}}\sin(\beta_0)\cos(\omega_d t) + \sqrt{\frac{8}{N}}\sum_{n=1}^{M}\sin(\beta_n)\cos\left[\omega_d t\cos\left(\frac{2\pi n}{N}\right)\right] \tag{2.105}$$

and where ω_d is the maximum angular Doppler frequency. The parameter β_n is defined as $\beta n = \pi n/M$ for $n = 0, \ldots, M$.

It is often desirable to generate multiple uncorrelated faded envelopes (e.g., for simulation of MIMO systems), something that model I cannot do. Dent et al. [43] modified Model I by using orthogonal Walsh-Hadamard codewords to decorrelate the multiple faded envelopes. The kth complex faded envelope is defined as $h_k(t) = h_{ik}(t) + jh_{qk}(t)$, where

$$h_{ik}(t) = \sum_{n=1}^{M} A_k(n)\cos\left(\frac{\pi n}{M+1}\right)\cos\left\{\omega_d t\cos\left(\frac{2\pi n}{N}\right) + \frac{\pi n}{M+1} + \frac{2\pi n(k-1)}{M+1}\right\} \tag{2.106}$$

$$h_{qk}(t) = \sum_{n=1}^{M} A_k(n)\sin\left(\frac{\pi n}{M+1}\right)\cos\left\{\omega_d t\cos\left(\frac{2\pi n}{N}\right) + \frac{\pi n}{M+1} + \frac{2\pi n(k-1)}{M+1}\right\} \tag{2.107}$$

$M = (N - 2)/4$, $k = 1,\ldots,M$, and $A_k(n)$ is the nth element of the kth row of a Hadamard matrix $\mathbf{H}_M$ of dimension $M \times M$.

By reducing the number of sinusoidal components, Jakes simplified simulation of F-to-M Rayleigh fading channels. However, his model does not satisfy most of the statistical properties of Clarke's model [41], and it is not wide-sense stationary

(WSS) [42]. Specifically, Model I satisfies only the following properties: the in-phase and quadrature components of the complex faded envelope are Gaussian random processes for $N \to \infty$, and the auto-correlation function of the complex faded envelope is equal to $J_0(\omega_d \tau)$. Also, the model of Dent et al. [43] (modified Model I) yields a cross-correlation between different faded envelopes that is strictly zero only for time lag $\tau = 0$. As a result, various modifications of Model I have been proposed in the literature [42], [44–46].

2.4.2.2 Pop and Beaulieu's Model—Model II

Pop and Beaulieu [42] showed that Model I is not WSS and modified Model I to fix this problem. This was done by removing the constraint $\phi_n = 0$ from Model I and allowing the phases ϕ_n to be independent random variables uniformly distributed on the interval $[0, 2\pi)$. The procedure yields a deterministic simulator, since the same (random) phases are used for all simulation trials. The kth complex faded envelope is defined as $h_k(t) = h_{ik}(t) + jh_{qk}(t)$, where

$$h_{ik}(t) = \sqrt{\frac{4}{N}}\cos(\beta_0)\cos(\omega_d t + \phi_{0k}) + \sqrt{\frac{8}{N}}\sum_{n=1}^{M}\cos(\beta_n)\cos\left[\omega_d t\cos\left(\frac{2\pi n}{N}\right) + \phi_{nk}\right] \quad (2.108)$$

$$h_{qk}(t) = \sqrt{\frac{4}{N}}\sin(\beta_0)\cos(\omega_d t + \phi_{0k}) + \sqrt{\frac{8}{N}}\sum_{n=1}^{M}\sin(\beta_n)\cos\left[\omega_d t\cos\left(\frac{2\pi n}{N}\right) + \phi_{nk}\right] \quad (2.109)$$

$M = (N - 2)/4$, and $k = 0, \ldots, M - 1$. The parameter β_n is defined as $\beta_n = \pi n/M$ for $n = 0, \ldots, M$. Although it is WSS, Model II inherits its statistical properties from model I. Hence, this model does not satisfy most of the other statistical properties required by Clarke's model.

2.4.2.3 Li and Huang's Model—Model III

To improve model II, Li and Huang [44] proposed a deterministic model that generates multiple uncorrelated faded envelopes $h_k(t)$. Model III assumes P independent complex faded envelopes, each with $M = N/4$ sinusoidal terms in the I and Q components. The kth complex faded envelope is $h_k(t) = h_{ik}(t) + jh_{qk}(t)$, where

$$h_{ik}(t) = 2C\sum_{n=0}^{M-1}\cos\left(\omega_d t\cos\alpha_{nk} + \phi^i_{nk}\right) \quad (2.110)$$

$$h_{qk}(t) = 2C\sum_{n=0}^{M-1}\sin\left(\omega_d t\sin\alpha_{nk} + \phi^q_{nk}\right) \quad (2.111)$$

and where $k = 0, \ldots, P - 1$, ϕ^i_{nk} and ϕ^q_{nk} are independent random phases uniformly distributed on the interval $[0, 2\pi)$, α_{nk} is the nth angle of arrival in the kth complex faded envelope, and C is a constant gain. The angles of arrival are $\alpha_{nk} = (2\pi n)/N + (2\pi k)/(PN) + \alpha_{00}$ for $n = 0, \ldots, M - 1$, $k = 0, \ldots, P - 1$, where α_{00} is an initial angle of arrival, chosen to be $0 < \alpha_{00} < (2\pi)/(PN)$ and $\alpha_{00} \neq \pi/(PN)$.

Model III preserves the desirable statistical properties of Model II, while generating multiple uncorrelated faded envelopes. Compared to Model II, the in-phase and quadrature components of the complex faded envelope in Model III are less correlated. However, Model III still fails to satisfy equations (2.98) and (2.101) of Clarke's model.

2.4.2.4 MEDS Model—Model IV

To resolve the remaining disadvantages of Model II, Pätzold et al. [45] proposed a deterministic model called the Method of Exact Doppler Spreads (MEDS). The auto-correlation functions of the in-phase and quadrature components of the complex faded envelope are designed to satisfy (2.98). The kth complex faded envelope is defined as $h_k(t) = h_{ik}(t) + jh_{qk}(t)$, where

$$h_{(i/q)k}(t) = \sqrt{\frac{2}{N_{i/q}}} \sum_{n=1}^{Ni/q} \cos\left[\omega_d t \sin \alpha_n^{i/q} + \phi_{nk}^{i/q} \right] \tag{2.112}$$

for $k = 0, \ldots, P - 1$ and where $\phi_{nk}^{i/q}$ are independent random phases uniformly distributed on the interval $[0, 2\pi)$, P is the number of desired faded envelopes, and $N_{i/q}$ is the number of sinusoidal terms in the in-phase and quadrature components of $h_k(t)$, respectively. The nth angle of arrival is given by $\alpha_n^{i/q} = \pi(n - 0.5)/(2N_{i/q})$ for $n = 1, \ldots, N_{i/q}$.

Model IV preserves the desirable properties of Model II. In addition, the autocorrelation functions of the I and Q components in Model IV satisfy (2.98). Compared to Model II, the in-phase and quadrature components of the complex faded envelope in Model IV are less correlated if $N_q = N_i + 1$. However, Model IV produces multiple faded envelopes that are correlated.

2.4.2.5 Zajić and Stüber's Model—Model V

Models I – IV have difficulty creating uncorrelated in-phase and quadrature components of each complex faded envelope and generating multiple uncorrelated faded envelopes. Zajić and Stüber [46] proposed to solve these problems by using orthogonal functions for the in-phase and quadrature components of the complex faded envelope.

The following function is considered as the kth complex faded envelope

$$h_k(t) = \sum_{n=0}^{N-1} C_n e^{j(\omega_d t \cos \alpha_{nk} + \phi_{nk})} \tag{2.113}$$

where $C_n = (2e^{j\beta_n})/\sqrt{N}$, α_{nk}, ϕ_{nk}, and ω_d are the random path gain, the angle of arrival, the phase associated with the nth propagation path, and the maximum angular Doppler frequency, respectively. It is assumed that P independent complex faded envelopes are required ($k = 0, \ldots, P - 1$), each consisting of N sinusoidal components.

To reduce the number of sinusoidal components needed for simulation, this model uses a method similar to the one described in [44]. Specifically, by choosing $M = N/4$ to be an integer and by taking into account shifts of the angles α_{nk} and ϕ_{nk} in each quadrant of the circle, the sum in (2.113) can be split into four terms, viz.

$$h_k(t) = \frac{2}{\sqrt{N}} \sum_{n=1}^{M} e^{j\beta_n} e^{j[\omega_d t \cos\alpha_{nk} + \phi_{nk}]} + \frac{2}{\sqrt{N}} \sum_{n=1}^{M} e^{j\beta_n} e^{j[\omega_d t \cos(\alpha_{nk}+0.5\pi)+(\phi_{nk}+0.5\pi)]}$$
$$+ \frac{2}{\sqrt{N}} \sum_{n=1}^{M} e^{j\beta_n} e^{j[\omega_d t \cos(\alpha_{nk}+\pi)+(\phi_{nk}+\pi)]} + \frac{2}{\sqrt{N}} \sum_{n=1}^{M} e^{j\beta_n} e^{j[\omega_d t \cos(\alpha_{nk}+1.5\pi)+(\phi_{nk}+1.5\pi)]} \tag{2.114}$$

and then simplified to

$$h_k(t) = \frac{2}{\sqrt{N}} \sum_{n=1}^{M} 2\cos(\beta_n)\cos(\omega_d t \cos\alpha_{nk} + \phi_{nk})$$
$$+ j\frac{2}{\sqrt{N}} \sum_{n=1}^{M} 2\sin(\beta_n)\sin(\omega_d t \sin\alpha_{nk} + \phi_{nk}) \tag{2.115}$$

Then, the kth complex faded envelope is defined as $h_k(t) = h_{ik}(t) + jh_{qk}(t)$, where

$$h_{ik}(t) = \frac{2}{\sqrt{N}} \left[\sum_{n=0}^{M} a_n \cos(\omega_d t \cos\alpha_{nk} + \phi_{nk}) \right] \tag{2.116}$$

$$h_{qk}(t) = \frac{2}{\sqrt{N}} \left[\sum_{n=0}^{M} b_n \sin(\omega_d t \sin\alpha_{nk} + \phi_{nk}) \right] \tag{2.117}$$

The motivation for this model originates in Model II. As in Model II, the phases ϕ_{nk} are chosen to be independent random variables uniformly distributed on the interval $[0, 2\pi)$, and the path gains β_n are defined as $\beta_n = \pi n/M$ for $n = 0, ..., M$. By including $\beta_0 = 0$ in (2.115), the total number of propagation paths is increased slightly to $N = 4M + 2$. Parameters a_n and b_n are defined as follows:

$$a_n = \begin{cases} 2\cos(\beta_n), & n = 1, ..., M \\ \sqrt{2}\cos(\beta_n), & n = 0 \end{cases} \tag{2.118}$$

$$b_n = \begin{cases} 2\sin(\beta_n), & n = 1, ..., M \\ \sqrt{2}\sin(\beta_n), & n = 0 \end{cases} \tag{2.119}$$

The angles of arrival α_{nk} are defined as in Model III:

$$\alpha_{\text{nk}} = (2\pi n)/N + (2\pi k)/(PN) + \alpha_{00} \text{ for } n = 0, ..., M, k = 0, ..., P - 1.$$

This ensures an asymmetrical arrangement of arrival angles, which minimizes the cross-correlation between different faded envelopes. In addition, the initial angle of arrival α_{00} is optimized (through an exhaustive search) to minimize the cross-correlation between the in-phase and quadrature components of each complex faded envelope. The result is $\alpha_{00} = (0.2\pi)/(PN)$. In [46], if is shown that Model V satisfies (2.99) and (2.100) of Clarke's model. However, the auto-correlations of the quadrature components and the auto-correlation of the squared envelope do not satisfy (2.98) and (2.101), respectively.

2.4.2.6 Zheng and Xiao's Statistical Models

To improve on previously reported models, Zheng and Xiao proposed several new *statistical simulation models* [47–49]. By allowing all three parameter sets (amplitudes, phases, and Doppler frequencies) to be random variables, Zheng and Xiao's models obtain statistical properties similar to ones required by Clarke's model. However, the models are no longer ergodic. The statistical properties of these models vary for each simulation trial, but they converge to desired properties when averaged over 50 to 100 simulation trials. A detailed comparison of the statistical properties for Zheng and Xiao's models is presented in [50]. It is shown that the model presented in [48] has the statistical properties closest to those of the reference model and requires the fewest simulation trials (50). We will refer to this model as Model VI.

Model VI: The kth complex faded envelope is defined as $h_k(t) = h_{ik}(t) + jh_{qk}(t)$, where

$$h_{ik}(t) = \sqrt{\frac{2}{M}} \sum_{n=1}^{M} \cos\left[\omega_d t \cos\alpha_{nk} + \phi_{nk}^i\right] \tag{2.120}$$

$$h_{qk}(t) = \sqrt{\frac{2}{M}} \sum_{n=1}^{M} \cos\left[\omega_d t \sin\alpha_{nk} + \phi_{nk}^q\right] \tag{2.121}$$

for $k = 0, \ldots, P - 1$ and where ϕ_{nk}^i and ϕ_{nk}^q are independent random phases uniformly distributed on the interval $[-\pi, \pi)$. Model VI assumes P independent complex envelopes, each with $M = N/4$ sinusoidal terms in the in-phase and quadrature components. The nth angle of arrival in the kth complex envelope α_{nk} is $\alpha_{nk} = (2\pi n - \pi + \theta_k)/(4M)$ for $n = 1, \ldots, M$, where the θ_k are independent random variables uniformly distributed on the interval $[-\pi, \pi)$.

2.4.2.7 Zajić and Stüber's Statistical Model—Model VII

Model V (which is a deterministic model) can be modified to possess all statistical properties of Clarke's model by letting all three parameters $C_{nk} = (2e^{j\beta_{nk}})/\sqrt{N}$, α_{nk}, and ϕ_{nk} be random variables, similar to Model VI. The kth complex faded envelope is defined as $h_k(t) = h_{ik}(t) + jh_{qk}(t)$, where

$$h_{ik}(t) = \frac{2}{\sqrt{N}} \sum_{n=1}^{M} 2\cos(\beta_{nk})\cos(\omega_d t \cos\alpha_{nk} + \phi_{nk}) \tag{2.122}$$

$$h_{qk}(t) = \frac{2}{\sqrt{N}} \sum_{n=1}^{M} 2\sin(\beta_{nk})\sin(\omega_d t \sin\alpha_{nk} + \phi_{nk}) \tag{2.123}$$

It is assumed that P independent complex envelopes are desired ($k = 0, \ldots, P - 1$), each having $M = N/4$ sinusoidal terms in the in-phase and quadrature components. The parameters ϕ_{nk}, β_{nk} and θ are independent random variables uniformly distributed on the interval $[-\pi, \pi)$. The angles of arrival are chosen as follows: $\alpha_{nk} = (2\pi n)/N + (2\pi k)/(PN) + (\theta - \pi)/N$, for $n = 1, \ldots, M$, $k = 0, \ldots, P - 1$. The angles of arrival in the kth complex faded envelope are obtained by rotating the angles of arrival in the $(k - 1)$th complex envelope by $(2\pi)/(PN)$.

2.4.2.8 Performance Evaluation of Models I–VII

This section compares the performance and complexity of the models I – VII. In all simulations, a normalized sampling period $f_d T_s = 0.05$ (f_d is the maximum Doppler frequency and T_s is the sampling period) and $M = P = 8$ are used. However, for Model IV we use $N_i = 8$ and $N_q = 9$ to obtain uncorrelated quadrature components of the complex envelope. Note that in models I – V, the random phases associated with the nth propagation path are computed before the actual simulation starts, because a deterministic simulator needs only one simulation trial. During each simulation, all parameters are kept constant to provide simulation results that are always the same (i.e., deterministic).

To compare the complexity of Models I – VII, Table 2.1 summarizes the number of simulation trials required to obtain desired statistical properties, the number of operations needed to generate one sample of the complex faded envelope, and the relative simulation times needed to generate a sample of the complex faded envelope with desired statistical properties, in Matlab on a Pentium III laptop. Here, we count only the frequently executed operations and the number of random variables. Table 2.1 shows that the choice of cosine functions in models V and VII, which makes the in-phase and quadrature components of the complex faded envelope uncorrelated, slightly increasing the complexity of the models.

As pointed out earlier, Models I – IV have difficulty creating uncorrelated in-phase and quadrature components of each complex faded envelope and generating multiple uncorrelated faded envelopes. To illustrate this, Figures 2.10 and 2.11 compare the cross-correlation functions of the in-phase and quadrature components and the cross-correlation functions of two faded envelopes obtained by *deterministic models* (Models I – V), respectively. The results show that Model V yields the lowest cross-correlation between the in-phase and quadrature components of the complex faded envelope and the lowest cross-correlation between two different faded envelopes compared to Models I – IV.

Similarly, Figures 2.12 and 2.13 compare the cross-correlation functions of the in-phase and quadrature components and the cross-correlation functions of two faded envelopes obtained by Models VI and VII, respectively. For Model VI, we average over $N_{\text{stat}} = 100$ trials, while for Model VII, we average over $N_{\text{stat}} = 30$ and

Table 2.1 Complexity of Different SoS Simulation Models for F-to-M Frequency-Flat Multipath Fading Channels

Simulators	*Number of Simulation Trials*	*Relative Sim. Time to Generate a Sample of $h_k(t)$ with desired Stat Properties*	*Estimate Number of Computations Needed to Generate One Sample of $h_k(t)$*		
			Cosine	*Addition/ Multiplication*	*Number of Random Variables*
Model I	1	Tx	$4M$	$2M$ / $2M$	None
Model II	1	$1.03Tx$	$4M$	$3M$ / $2M$	$M + 1$
Model III	1	$1.52Tx$	$4M$	$4M$ / 0	$2M$
Model IV	1	$1.52Tx$	$4M$	$4M$ / 0	$2M$
Model V	1	$1.98Tx$	$6M$	$4M$ / $2M$	$M + 1$
Model VI	50	$4.85Tx$	$4M$	$4M$ / 0	$2M + 1$
Model VII	30	$3.15Tx$	$6M$	$4M$ / $2M$	$2M + 1$

N_{stat} = 50 trials. The results show that Model VII with N_{stat} = 30 has similar pefor-mance as models VI with N_{stat} = 100. An increase of the number of trials to N_{stat} = 50 yields significantly lower cross-correlations. Note that the scale on Figures 2.12 and 2.13 is different from that of Figures 2.10 and 2.11, and that statistical simula-tors tend to have significantly lower cross-correlations than deterministic models.

2.4.3 Simulation Models for F-to-M Frequency-Selective Multipath Fading Channels

In frequency-selective fading channels, frequency components in the transmitted signal experience different phase shifts along the different paths. As the differences in path delays become larger, even closely separated frequencies in the transmitted signal can experience significantly different phase shifts. This type of propagation can be characterized by a geometrical channel model illustrated in Figure 2.14. Considering only single reflections, all scatterers that are associated with a particu-lar path length are located on an ellipse with the transmitter and receiver located at the foci. Different delays correspond to different co-focal ellipses.

Frequency-selective channels have strong scatterers that are located on several ellipses that represent significantly different delays (compared to a symbol dura-tion). This means that, if the bandwidth of the signal is W, we can differentiate between signals arriving at least $\Delta\tau = 1/W$ s apart. As we move outward, energy reflected from successive ellipses takes longer to get to the receiver. In this way, at the receiver, we observe multiple copies of the transmitted signal, with each copy taking progressively longer to reach the receiver. This model does not consider second- or higher-order reflections, where energy is reflected between scatterers. Indeed, most ray-tracing models (such as Jake's model) are *single-bounced scatter-*

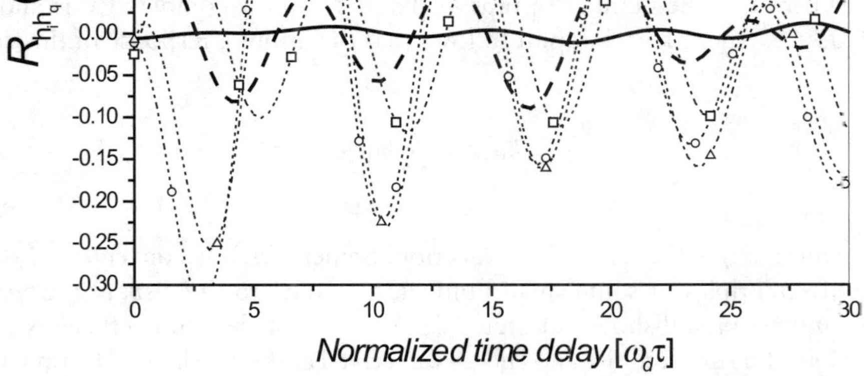

Figure 2.10 The normalized cross-correlation function of the in-phase and quadrature components of Models I – V.

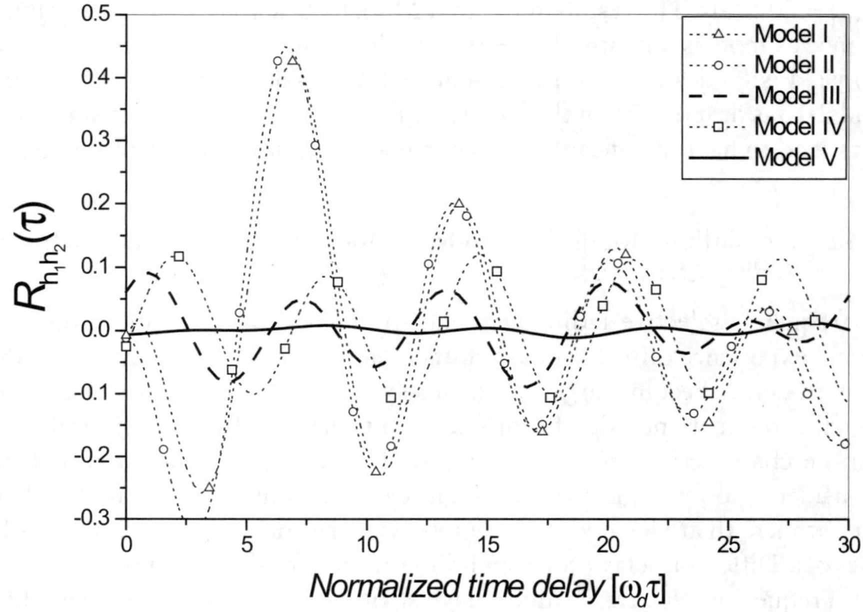

Figure 2.11 The normalized cross-correlation function of the first and the second complex envelope of Models I – V.

ing models—they assume that most of the scattered energy arriving at the receiver is due to single-bounced scattering.

This geometrical model can be described using a tapped-delay line with number of taps at different delays. Each tap is the result of a large number of multipath components, and, therefore, the taps will experience muptipath fading. If we use $\hat{s}\,(t)$ to denote the complex envelope of the transmitted signal, the complex envelope of the received signal can be written as

$$\hat{r}(t) = \sum_{l=1}^{L} h_l(t)\hat{s}(t - \tau_l) \tag{2.124}$$

where L is the number of taps $h_l(t)$ and τ_l are the complex gains and path delays associated with each tap. It follows that the impulse response of this channel is

$$h(t, \tau) = \sum_{l=1}^{L} h_l(t)\delta(t - \tau_l) \tag{2.125}$$

where $\delta(\cdot)$ is the Dirac delta function. Sometimes it is convenient if the tap delays are multiples of some small number τ, leading to the *τ-spaced tapped delay line channel model* shown in Figure 2.15. Many of the tap coefficients in the tapped delay line are zero (i.e., no energy is received at these delays). The time variant channel tap coefficients $\{h_l(t)\}$ can be generated using the channel models described in Section 2.4.2.

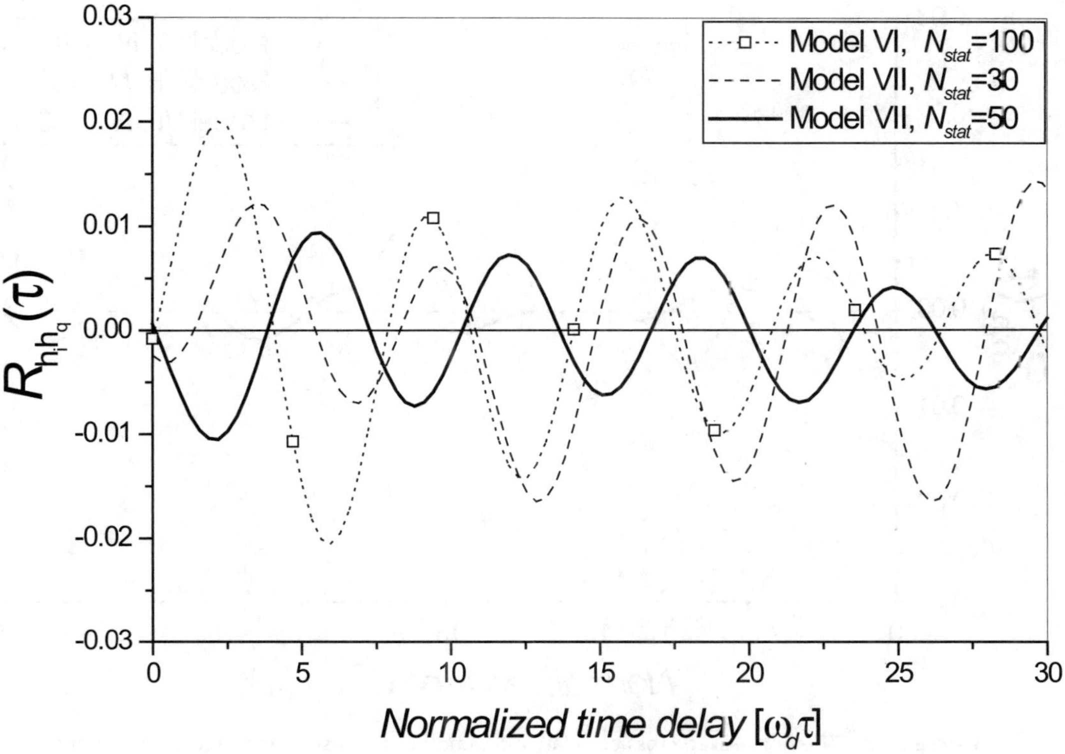

Figure 2.12 The normalized cross-correlation function of the in-phase and quadrature components of Models VI and VII.

In a typical digital communication system, data symbols are sent to the transmitter every T seconds, and T-spaced samples are taken at the output of the receiver, where T is the baud duration. Hence, it is desirable to have a channel model with T-spaced impulse response. The *T-spaced channel model* is similar to the τ-spaced channel model, except that the channel taps are T-spaced. This discrete system model is by far the most popular in the literature (see, for example, [4, 35]). This is largely due to the prevalence of digital hardware for measuring and/or computing the channel impulse response. Most channel sounding devices operate in the digital domain and thus sample the power delay profile at discrete intervals. Each delay bin is the sum of all energy received over the duration of that bin. This is analogous to discrete frequency spectra, in which each frequency bin is the sum of all energy over the bandwidth that bin occupies. Details on how the τ-spaced channel model can be converted into T-spaced channel model can be found in [4].

2.5 Small-Scale Channel Measurement Techniques

Because of the importance of the multipath structure in determining the small-scale fading effects, a number of channel sounding techniques have been developed. These techniques may be classified as *direct pulse–based channel impulse response measurements*, *spread spectrum–based channel impulse response measurements*, and *frequency domain–based channel impulse response measurements*.

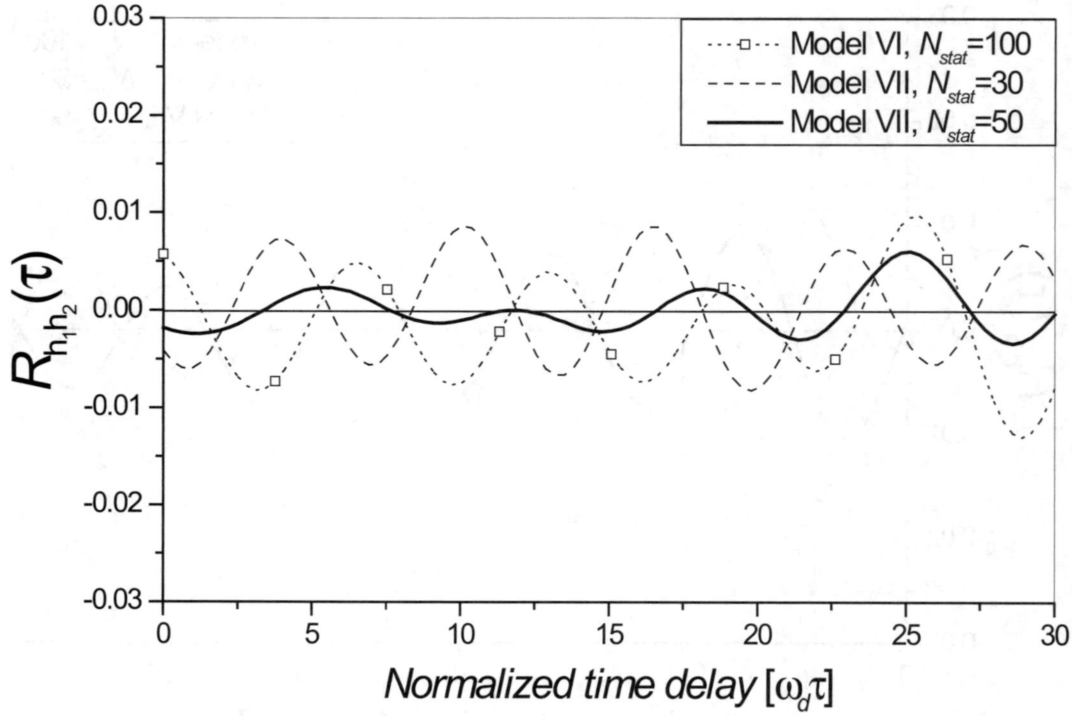

Figure 2.13 The normalized cross-correlation function of the first and the second fader of Models VI and VII.

2.5.1 Direct Pulse–Based Channel Impulse Response Measurements

A simple channel sounding approach is the direct pulse–based system, shown in Figure 2.16. This technique allows us to easily determine the power delay profile of any channel. This system transmits a repetitive pulse of width τ_p and uses a receiver with a wide bandpass filter ($B = 2/\tau_p$ Hz) to receive the signal. The signal is then amplified, detected with an envelope detector, and displayed and stored on a high speed oscilloscope. This gives an immediate measurement of the square of the channel impulse response convolved with the probing pulse. If the oscilloscope is set to averaging mode, this system can provide a local average power delay profile. Another attractive aspect of this system is the lack of complexity, since off-the-shelf equipment may be used.

The minimum resolvable delay between multipath components is equal to the probing pulse width τ_p. The main problem with this system is that it is subject to interference and noise, due to the wide pass-band filter required for multipath time resolution. Another disadvantage is that phases of the individual multipath components are not received, due to the use of an envelope detector. However, a coherent detector can be used with this technique to measure the multipath phases.

2.5.2 Spread Spectrum–Based Channel Impulse Response Measurements

A block diagram for a spread spectrum–based channel sounding system is shown in Figure 2.17. Compared to the direct pulse system, the advantage of a spread

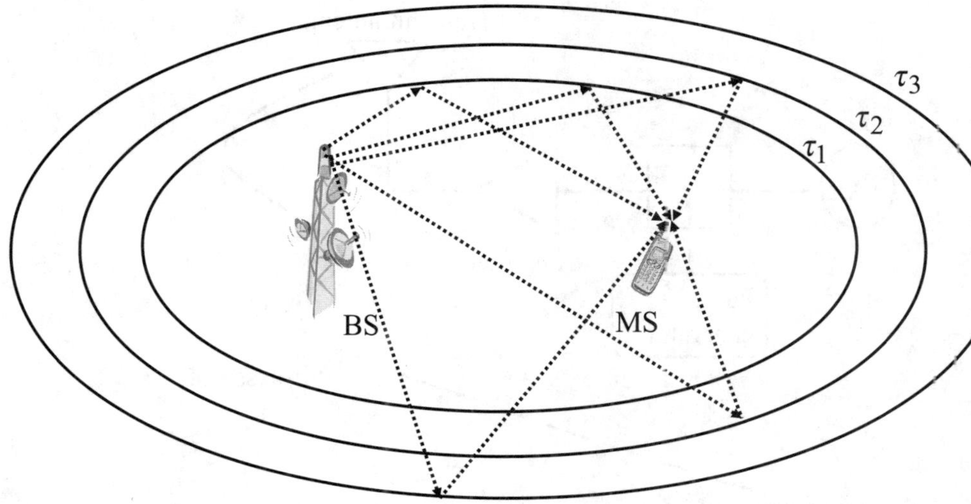

Figure 2.14 Geometrical model for F-to-M frequency-selective multipath fading channels.

spectrum–based system is that, while the probing signal may be wideband, it is possible to detect the transmitted signal using a narrowband receiver preceded by a wideband mixer, thus improving the dynamic range of the system.

In a spread spectrum–based channel sounder, a carrier signal is "spread" over a large bandwidth by mixing it with a binary pseudo-noise sequence having a chip duration T_c. The spread spectrum signal is then received, filtered, and despread using a pseudo-random sequence generator identical to that used at the transmitter.

Synchronization between the transmitter and the receiver is eliminated using a sliding correlator. Sensitivity is adjustable by changing the sliding factor and the post-correlator filter bandwidth. Also, required transmitter powers can be considerably lower than comparable direct pulse systems because the spread spectrum systems have better resilience to noise.

Compared to the direct pulse–based system, a disadvantage of the spread spectrum–based system is that measurements are not made in real time; instead, they are compiled as the pseudo-random codes slide past one another. Depending on system parameters and measurement objectives, the time required to make power delay profile measurements may be excessive. Another disadvantage of the system described here is that a noncoherent detector is used, so that phases of

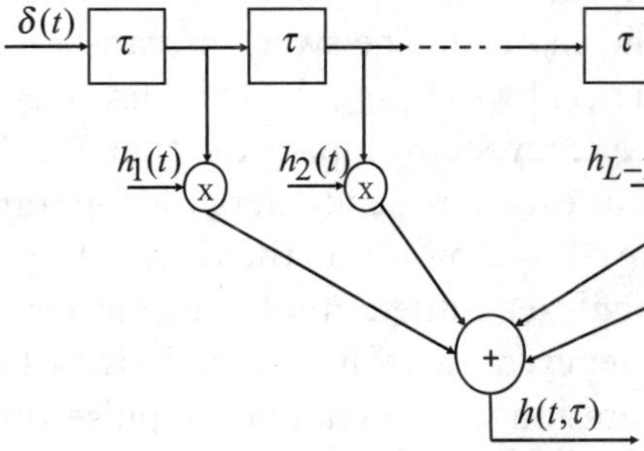

Figure 2.15 F-to-M frequency-selective multipath fading channel model with discrete multipath components.

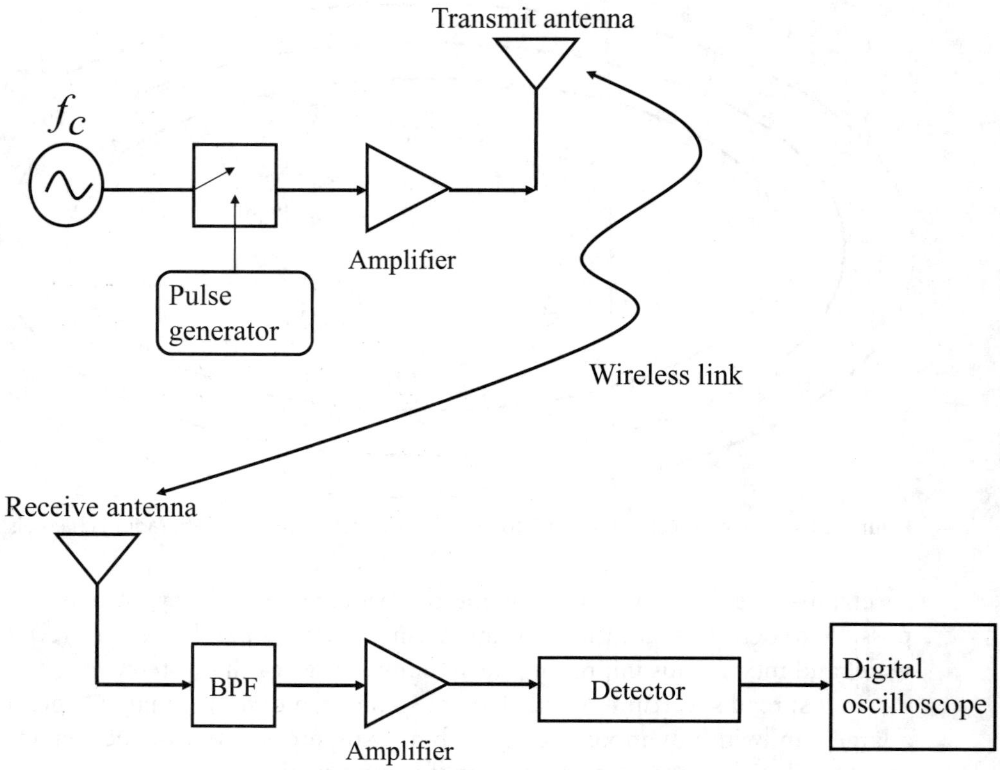

Figure 2.16 A direct pulse–based channel impulse response measurement system.

individual multipath components cannot be measured. Even if coherent detection is used, the sweep time of a spread spectrum signal induces delay such that the phases of individual multipath components with different time delays would be measured at substantially different times, during which the channel might change.

2.5.3 Frequency Domain–Based Channel Impulse Response Measurements

Because of the dual relationship between time-domain and frequency-domain techniques, it is possible to measure the channel impulse response in the frequency domain. Figure 2.18 shows a frequency domain–based channel sounder used for measuring channel impulse responses. A vector network analyzer controls a synthesized frequency sweeper, and an S-parameter test set is used to monitor the frequency response of the channel. The sweeper scans a particular frequency band (centered around the carrier) by stepping through discrete frequencies. The number and spacing of these frequency steps impact the time resolution of the impulse response measurement. Four each frequency step, the S-parameter test set transmits a known signal level at port 1 and monitors the received signal level at port 2. These signal levels allow the analyzer to determine the complex response of the channel over the measured frequency range. The transmissivity ($S_{21}(f)$) response is a frequency domain representation of the channel impulse response. This response is then converted to the time domain using inverse discrete Fourier transform processing, giving a band-limited version of the impulse response. In theory, this technique

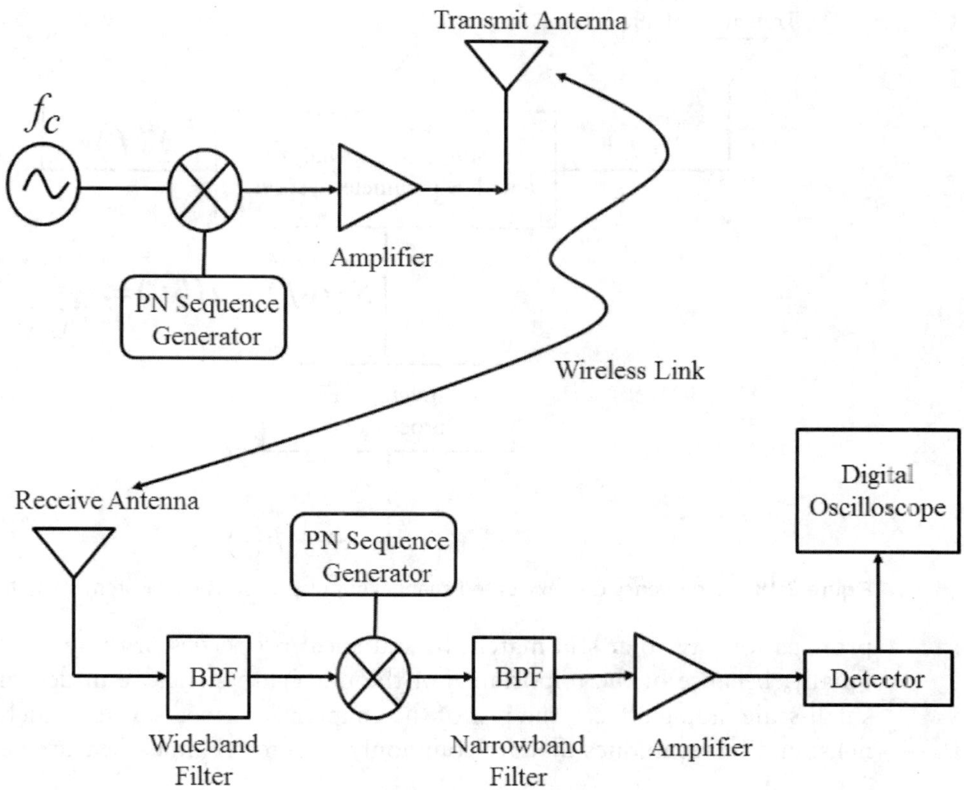

Figure 2.17 A spread spectrum–based channel impulse response measurement system.

works well and indirectly provides amplitude and phase information in the time domain. However, the system requires careful calibration and hardwired synchronization between the transmitter and receiver, making it useful only for very close measurements (e.g., indoor channel measurements). Another limitation with this system is the non-real-time nature of the measurement. For time varying channels, the channel frequency response can change rapidly, giving an erroneous impulse response measurement. To mitigate this effect, fast sweep times are necessary to keep the total swept frequency response measurement interval as short as possible. A faster sweep time can be accomplished by reducing the number of frequency steps, but this sacrifices time resolution and excess delay range in the time domain.

2.6 Summary

This chapter first briefly described the propagation principles that govern signal propagation in F-to-M cellular radio channels. The chapter proceeded to describe important concepts of propagation modeling such as large-scale and small-scale fading effects. Then, the statistical characterization of multipath propagation in F-to-M cellular radio channels is detailed. Several reference models and their statistics such as correlation function, Doppler spectrum, power delay profile, and level crossing rate are detailed. Furthermore, this chapter also described the simulators for frequency-flat and frequency-selective multipath fading with a small number of

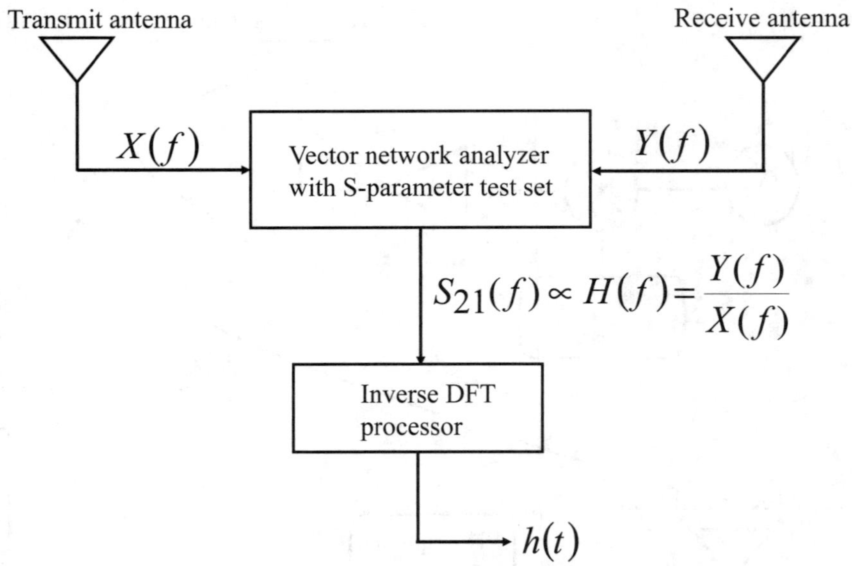

Figure 2.18 A frequency domain–based channel impulse response measurement system.

propagation rays that still match the statistical properties of the F-to-M channel. Finally, because of the importance of the multipath structure in determining the small-scale fading effects, the rest of the chapter described several wideband channel sounding techniques that are commonly used for channel measurements.

References

[1] Rappaport, T., *Wireless Communications: Principles and Practice*, Upper Saddle River, NJ: Pretince Hall PTR, 2001.

[2] Vaughan, R. and J. Andersen, *Channels, Propagation and Antennas for Mobile Communications*, The IEE, UK, 2003.

[3] Saunders, S. R., *Antennas and Propagation for Wireless Communication Systems*, New York: John Wiley and Sons, Inc., 1999.

[4] Stüber, G. L., *Principles of Mobile Communication*, 2nd ed., Norwell, MA: Kluwer, 2001.

[5] Pätzold, M., *Mobile Fading Channels*, John Wiley and Sons Inc., 2002.

[6] Gudmundson, M., "Correlation model for shadowing fading in mobile radio systems," *Electronic Letters*, Vol. 27, Nov. 1991, pp. 2145–2146.

[7] Pätzold M, U. Killat, and F. Laue, "A deterministic digital simulation model for Suzuki processes with application to a shadowed Rayleigh land mobile radio channel," *IEEE Transactions on Vehicular Technology*, Vol. 45, May 1996, pp. 318–331.

[8] Clarke, R. H., "A statistical theory of mobile-radio reception," *Bell Systems Technical Journal*, July 1968, pp. 957–1000.

[9] Jakes, W. C., *Microwave Mobile Communications*, 2nd ed. Piscataway, NJ: Wiley-IEEE Press, 1994.

[10] I. S. Gradshteyn and I. M. Ryzhik, *Table of Integrals, Series, and Products*, San Diego, CA: Academic Press, 1980.

[11] Sadowsky, J. S. and V. Kafedziski, "On the correlation and scattering functions of the WSSUS channel for mobile communication," *IEEE Transactions on Vehicular Technology*, Vol. 47, Feb. 1998, pp. 270–282.

[12] Pätzold M., Y. Li, and F. Laue, "A study of a land mobile satellite channel model with asymmetrical Doppler power spectrum and lognormally distributed line-of-sight component," *IEEE Transactions on Vehicular Technology*, Vol. 47, Feb. 1998, pp. 297–310.

[13] Pätzold M., U. Killat, Y. Li, and F. Laue, "Modeling, analysis, and simulation of nonfrequency-selective mobile radio channels with asymmetrical Doppler power spectral density shapes," *IEEE Transactions on Vehicular Technology*, Vol. 46, May 1997, pp. 494–507.

[14] Fontan F. P., M. A. V. Castro, J. Kunisch, J. Pamp, E. Zollinger, S. Buonomo, P. Baptista, and B. Arbesser, "A versatile framework for a narrow-and wide-band statistical propagation model for the LMS channel," *IEEE Transactions on Broadcasting*, Vol. 43, Dec. 1997, pp. 431–458.

[15] Braun W. R. and U. Dersch, "A physical mobile radio channel model," *IEEE Transactions on Vehicular Technology*, Vol. 40, May 1991, pp. 472–482.

[16] Kuchar A., E. A. Aparicio, J. P. Rossi, and E. Bonek, "Azimuth, elevation, and delay of signals at mobile station site," *Wireless Personal Communications: Emerging Technologies for Enhanced Communications*, Boston, MA: Kluwer, 1999, pp. 99–110.

[17] Rossi J. P., J. P. Barbot, and A. J. Levy, "Theory and measurement of the angle of arrival and time delay of UHF radio waves using a ring array," *IEEE Transactions on Antennas and Propagation*, Vol. 45, May 1997, pp. 876–884.

[18] Fuhl J., J. P. Rossi, and E. Bonek, "High-resolution 3-D direction-of-arrival determination for urban mobile radio," *IEEE Transactions on Antennas and Propagation*, Vol. 45, Apr. 1997, pp. 672–682.

[19] Fannin P. C. and A. Molina, "Analysis of mobile radio channel sounding measurements in inner city Dublin at 1.808 Ghz," *IEE Proceedings in Communications*, Vol. 143, Oct. 1996. pp. 311–316.

[20] Reudink D. O., "Large-scale variations of the average signal," *Microwave Mobile Communications* W. C. Jakes, Jr., Ed., New York: Wiley, 1974, pp. 79–131.

[21] Lee W. C. Y., "Finding the approximate angular probability density function of wave arrival by using a directional antenna," *IEEE Transactions on Antennas and Propagation*, Vol. 21, May 1973, pp. 328–334.

[22] Petrus P., J. H. Reed, and T. S. Rappaport, "Effects of directional antennas at the base station on the Doppler spectrum," *IEEE Commununications Letters*, Vol. 1, Mar. 1997, pp. 40–42.

[23] Anim-Appiah K., "Complex envelope correlations for non-isotropic scattering," *Electronic Letters*, Vol. 34, Apr. 1998, pp. 918–919.

[24] Austin M. D. and G. L. Stüber, "Velocity adaptive handoff algorithms for microcellular systems," *IEEE Transactions on Vehicular Technology*, Vol. 43, Aug. 1994, pp. 549–561.

[25] Pedersen K. I., P. E. Mogensen, and B. H. Fleury, "Power azimuth spectrum in outdoor environments," *IEEE Electronics Letters*, Vol. 33, Aug 1997 pp. 1583–1584.

[26] Abdi A., J. A. Barger, and M. Kaveh, "A parametric model for the distribution of the angle of arrival and the associated correlation function and power spectrum at the mobile station," *IEEE Transactions on Vehicular Technology*, Vol. 51, May 2002, pp. 425–434.

[27] Pedersen K. I., P. E. Mogensen, and B. H. Fleury, "A stochastic model of the temporal and azimuthal dispersion seen at the base station in outdoor propagation environments," *IEEE Transactions on Vehicular Technology*, Vol. 49, Mar. 2000, pp. 437–447.

[28] Von Mises R. "Über die "Ganzzahligkeit" der Atomgewicht und verwandte Fragen," *Physikal. Z.*, Vol. 19, 1918, pp. 490–500.

[29] Mardia K. V., *Statistics of Directional Data*, London, UK: Academic, 1972.

[30] Simon M. K., S. M. Hinedi, and W. C. Lindsey, *Digital Communication Techniques: Signal Design and Detection* Englewood Cliffs, NJ: Prentice-Hall, 1995.

[31] Viterbi, A. J., "Phase-locked loop dynamics in the presence of noise by Fokker–Planck techniques," *Proc. IEEE*, Vol. 51, Dec. 1963, pp. 1737–1753.

[32] Leib H. and S. Pasupathy, "The phase of a vector perturbed by Gaussian noise and differentially coherent receivers," *IEEE Transactions on Information Theory*, Vol. 34, Nov. 1988, pp. 1491–1501.

[33] Pätzold M., U. Killat, and F. Laue, "An extended Suzuki model for land mobile satellite channels and its statistical properties," *IEEE Transactions on Vehicular Technology*, Vol. 47, May 1998, pp. 617–630.

[34] Abdi A., H. Zhang, and C. Tepedelenlioglu, "A unified approach to the performance analysis of speed estimation techniques in mobile communication," *IEEE Transactions on Communications*, Vol. 56, Jan. 2008, pp. 126–135.

[35] Costa N. and S. Haykin, *Multiple-Input Multiple-Output Channel Models: Theory and Practice*, New York: John Wiley and Sons, Inc., 2010.

[36] Bello P., "Characterization of random time-variant linear channels," *IEEE Transactions on Communications*, Vol. 11, Dec. 1963, pp. 360–393.

[37] Young D. J. and N. C. Beaulieu, "A quantitative evaluation of generation methods for correlated Rayleigh random variates," *Proc. GLOBECOM '98*, Sydney, Australia, November 1998, pp. 3332–3337.

[38] Young D. J. and N. C. Beaulieu, "The generation of correlated Rayleigh random variates by inverse discrete Fourier transform," *IEEE Transactions on Communications*, Vol. 48, July 2000, pp. 1114–1127.

[39] Verdin D. and T. C. Tozer, "Generating a fading process for the simulation of land-mobile radio communications," *Electronics Letters*, Vol. 29, No. 23, November 1993, pp. 2011–2012.

[40] Fechtel S. A., "A novel approach to modeling and efficient simulation of frequency-selective fading radio channels," *IEEE Journal on Selected Areas in Communications*, Vol. 11, April 1993, pp. 422–431.

[41] Pätzold M. and F. Laue, "Statistical properties of Jakes' fading channel simulator," *Proc. IEEE VTC'98*, Ottawa, Canada, May 1998, pp. 712–718.

[42] Pop M. F. and N. C. Beaulieu, "Limitations of sum-of-sinusoids fading channel simulators," *IEEE Transactions on Communications*, Vol. 49, April 2001, pp. 699–708.

[43] Dent P. et al., "Jakes fading model revisited," *Electronics Letters*, Vol. 29, June 1993, pp. 1162–1163.

[44] Li Y. X. and X. Huang, "The simulation of independent Rayleigh faders," *IEEE Transactions on Communications*, Vol. 50, Sept. 2002, pp. 1503–1514.

[45] Pätzold M. et al., "On the statistical properties of deterministic simulation models for mobile fading channels," *IEEE Transactions on Vehicular Technology*, Vol. 47, Feb. 1998, pp. 254–269.

[46] Zajić A. G. and G. L. Stüber, "A efficient simulation of Rayleigh fading with enhanced de-correlation properties," *IEEE Transactions on Wireless Communications*, Vol. 5, July 2006, pp. 1866–1875.

[47] Zheng Y. R. and C. Xiao, "Simulation models with correct statistical properties for Rayleigh fading channels," *IEEE Transactions on Communications*, Vol. 51, June 2003, pp. 920–928.

[48] Zheng Y. R. and C. Xiao, "Improved models for the generation of multiple uncorrelated Rayleigh fading waveforms," *IEEE Communication Letters*, Vol. 6, June 2002, pp. 256–258.

[49] Zheng Y. R. and C. Xiao, "A statistical simulation model for mobile radio fading channels," *Proc. IEEE WCNC'03*, New Orleans, LA, March 2003, pp. 144–149.

[50] Patel C. S. et al., "Comparative analysis of statistical models for the simulation of Rayleigh faded cellular channels," *IEEE Transactions on Communicaitons*, Vol. 53, June 2005, pp. 1017–1026.

CHAPTER 3

Vehicle-to-Vehicle Radio Channels

Vehicle-to-vehicle (V-to-V) communication systems have recently drawn great attention, because they have the potential to improve convenience and safety of automobile transportation. For example, sensor-equipped cars that communicate via wireless links and thus create adhoc networks can be used to reduce traffic accidents and facilitate traffic flow. V-to-V communication systems also find applications in intelligent transportation systems, relay-based cellular networks, and future combat systems. The design of highly reliable V-to-V wireless communication systems requires a detailed understanding of multipath propagation among vehicles. V-to-V channels have characteristics that are very different from F-to-M cellular radio channels. In V-to-V channels, both the transmitter (T_x) and the receiver (R_x) are in motion, equipped with low elevation antennas, and surrounded by local scatterers, as shown in Figure 3.1. Furthermore, the channel characteristics of V-to-V channels are influenced by the properties of environment around the communicating vehicles and by typical traffic patterns. In general, the following environment categories can be distinguished: (a) highways: two to six lanes in each direction, few surrounding buildings, and high speeds; (b) urban streets: two to four lanes, tall buildings close to the curb, and high traffic density; (c) suburban streets: one to two lanes, low surrounding buildings, and low traffic density, and (d) rural streets: few or no surrounding buildings, but possibly hills or forests, and very light traffic. As shown in Figure 3.1, the T_x communicates with the R_x through either LoS link or through the NLoS link, because of natural and manmade objects that are situated between the T_x and R_x. This chapter details how V-to-V channels differ from F-to-M cellular radio channels and how to model and simulate V-to-V channels.

3.1 V-to-V Propagation Modeling

Similar to F-to-M cellular radio channels, propagation in V-to-V channels is governed by the four propagation mechanisms described in Section 2.1 (i.e., free-space propagation, reflection, diffraction, and scattering). The differences between F-to-M cellular and V-to-V channels arise due to the differences in propagation environments. Section 3.1 describes the modeling of large-scale and small-scale fading in V-to-V channels.

3.1.1 Large-Scale Fading—Pathloss and Shadowing

A popular way to model pathloss and shadowing is to use empirically fitted data. This approach is based on experimental measurements for a given environment.

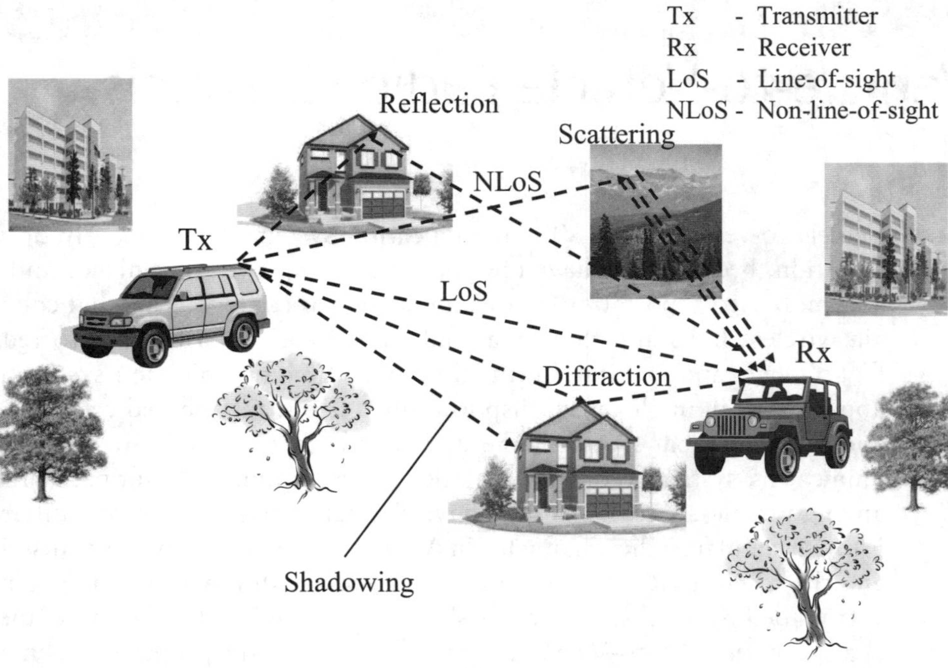

Figure 3.1 Typical vehicle-to-vehicle propagation environment.

Once the measurements are taken, statistical fitting techniques are used to obtain the pathloss and shadowing coefficients that match measured data.

There have been several measurement campaigns that resulted in pathloss and shadowing models for V-to-V channels [1–7]. In these measurements, it has been found that pathloss and shadowing coefficients depend on the type of the environment, so different models have been derived for highway [4–6], rural [4–6], urban [5, 6] and suburban [3–6] environments.

Here, we detail the most recent pathloss and shadowing models [6]. Propagation channel measurements were conducted in Lund, Sweden, at 5.2 GHz. Four different traffic environments were measured: *highway* measurements were performed on a 5.5 km-long two-lane highway strip; *rural* measurements were taken on a one-lane road, with the roadside environment contain mostly cornfields with a few houses and road signs; *urban* measurements were conducted along a one-lane city street, where four-story buildings were on both sides of the street; and *suburban* measurements were performed on a one-lane street that was 9m wide with buildings sparsely placed on both sides of the street. For each environment, measurements were made with cars driving in the same direction (convoy) or in the opposite directions. The measurements were mainly performed under LoS conditions between T_x and R_x, although occasionally obstruction of the LoS path occured.

The results in [5, 6] show that a single-slope pathloss model can be used to characterize the pathloss for highway, suburban, and urban environments. The pathloss and shadowing in dB can be written as [6]

$$PL(d) = PL(d_0) + 10n\log_{10}(d/d_0) + X_{\sigma_1} + \varsigma PL_c,\ d > d_0 \quad (3.1)$$

Table 3.1 The Estimated Parameters $PL(d_0)$, n, σ_1, and PL_c in Highway, Urban, and Suburban Environments

Scenario	$PL(d_0)$ [dB]	n	σ_1 [dB]	PL_c [dB]
Highway	63.3	1.77	3.1	3.3
Urban	62.0	1.68	1.7	1.5
Suburban	64.6	1.59	2.2	N/A

where n is the pathloss exponent, d is the propagation distance between the T_x and the R_x, $PL(d_0)$ is the pathloss at a reference distance $d_0 = 10$m, shadowing X_{σ_1} is modeled as a zero-mean normally distributed random variable with standard deviation σ_1, PL_c is a correlation term that accounts for the offset between forward pathloss (cars drive in opposite directions and away from each other) and reverse pathloss (cars drive in opposite directions and toward each other), and ζ is defined as

$$\varsigma = \begin{cases} 1, \text{for reverse pathloss} \\ -1, \text{ for forward pathloss} \\ 0, \text{ for convoy pathloss} \end{cases} \tag{3.2}$$

The estimated parameters $PL(d_0)$, n, σ_1, and PL_c are summarized in Table 3.1 for highway, urban, and suburban environments.

It is worth noting that all extracted pathloss exponents are lower than 2, which implies better-than-free-space propagation. This can be explained by observing that in addition to the LoS path energy, there is more energy available due to multipath propagation. Similar results are observed in [5, 7]. Note that the pathloss exponents in V-to-V channels differ from the typical pathloss exponents (2–4) in F-to-M cellular channels with LoS + multipath propagation. Possible explanation for this difference is that the communication distances in V-to-V channels are significantly shorter than those in F-to-M cellular channels (100m vs. 1000m) and the signal is less attenuated than when traveling longer distances.

For the rural environment, the pathloss and shadowing are represented by a two-ray model [6]; that is, the pathloss and shadowing are given by

$$\begin{aligned} PL(d) &= 20\log_{10}(4\pi f_c / c) - 7.3 + X_{\sigma_2} + \zeta PL_c \\ &-20\log_{10}\left(\frac{\exp(-jk_0 d)}{d} + \rho \frac{\exp\left(-jk_0\sqrt{d^2+4h^2}\right)}{\sqrt{d^2+4h^2}} \right) \end{aligned} \tag{3.3}$$

where d is the propagation distance between the T_x and the R_x, h = 2.53m is the antenna height (which is equal on both link ends), $\rho = 0.44e^{-j131}$ is the ground reflection coefficient, $k_0 = 2\pi f_c/c$ is the propagation number at the center frequency f_c, c is the speed of light, shadowing X_{σ_2} is modeled as a zero-mean normally distributed random variable with standard deviation σ_2 = 2.6 dB, PL_c = 2.3 dB is a correlation term that accounts for the offset between the forward and reverse pathloss, and ζ is defined as in (3.2). Since the two-ray behavior was not observed in the measurement data for small d, the validity of the model is limited to $d \geq 20$m.

In contrast to F-to-M cellular systems, V-to-V systems are composed of transmitting and receiving nodes with similar characteristics in terms of antenna height, mobility pattern, and so on. The shadowing fluctuations seen by the nodes at both ends of the radio link are therefore statistically identical [8]. This leads to a correlation function that is different from the the auto-correlated shadowing and the site-to-site correlated shadowing of the F-to-M cellular channels detailed in Section 2.2.2. In V-to-V channels, shadowing correlation function is referred to as joint correlation function (JCF). Extensive ray-tracing simulations have revealed that the JCF is well approximated by [8]

$$R(\Delta d_t, \Delta d_r) \propto e^{-\frac{\Delta d_t + \Delta d_r}{d_{\text{corr}}}} \tag{3.4}$$

where Δd_t and Δd_r are the displacement distances of the transmitter and receiver, respectively, and d_{corr} is the spatial correlation distance observed in the given environment. This spatial correlation distance is often smaller than in the F-to-M systems, where a typical mean value in urban environments is $d_{corr} = 40$m [1].

3.1.2 Small-Scale Fading—Multipath Propagation

Similarly to small-scale fading in F-to-M channels, small-scale fading in V-to-V channels is influenced by multipath propagation, speed of the T_x and R_x, speed of the surrounding objects, bandwidth of the transmitted signal, and so on. Hence, following the reasoning in Section 2.2.3, the V-to-V multipath channel can be modeled as a linear time-variant filter whose complex baseband impulse response is [9]

$$h(t,\tau) = \sum_{l=1}^{L} C_l e^{-j\phi_l(t)} \delta(\tau - \tau_l) \tag{3.5}$$

where $h(t, \tau)$ is the channel response at time t due to an impulse applied at time $t - \tau$, and $\delta(\cdot)$ is the Dirac delta function. The channel amplitude, phase, and time delay strongly depend on time and location and are modeled statistically.

It has been found [10,11] that the received envelope in V-to-V channels is Rayleigh faded under NLoS conditions and is Ricean faded under LoS conditions, but the statistical properties such as correlation function, Doppler spectrum, level crossing rate, and so on differ from F-to-M cellular channels. More details on Rayleigh and Ricean faded received signal envelopes can be found in Section 2.3.1.

3.1.3 Akki and Haber's Model for V-to-V Frequency-Flat Multipath Fading Channels

Akki and Haber were the first to propose a single-input single-output (SISO) V-to-V channel model [10]. This model is based on the following assumptions: (a) NLoS propagation conditions (i.e., there is no direct propagation path between the T_x and R_x); (b) all waves are double bounced (i.e., the waves from the T_x antenna elements impinge on the scatterers located around the T_x and scatter from the scatterers located around the R_x before arriving at the R_x antenna); (c) waves that are reflected off scatterers that are further away are negligible because they are significantly attenuated; (d) waves that experience more reflections are negligible because each reflection leads to significant power loss; and (e) the number of scatterers is very large (approaches infinity), so the power of each scattered wave is negligible compared to the total mean power.

According to this model, the complex faded envelope under frequency-flat fading can be written as [10]

$$h(t) = \sqrt{\Omega} \lim_{M,N\to\infty} \sum_{m=1}^{M}\sum_{n=1}^{N} a_{mn} \cdot e^{j(2\pi f_{mn}t+\phi_{mn})} \tag{3.6}$$

where Ω is the averaged total power ($\sqrt{\Omega}$ is assumed to be constant), and a_{mn} and ϕ_{mn} are the joint channel gain and phase shift caused by the interaction of the mth scatterer at the T_x and the nth scatterer at the R_x, respectively. Furthermore, f_{mn} is the Doppler shift caused by the movement of the T_x and R_x and experienced by the wave that is reflected of the mth and nth scatterers. The amplitudes, Doppler shift frequencies, and phases in this model can be calculated as follows:

- *Channel gain* a_{mn}. The channel gain a_{mn} can be split into two independent parts, a_m and a_n. The first part of the channel gain a_{mn} is introduced by the mth scatterer located near the T_x, while the second part of the channel gain a_{mn} is introduced by the nth scatterer located near the R_x. Furthermore, the gains a_m and a_n are normalized such that the ensemble averages of the a_ms and a_ns are given by

$$\sum_{m=1}^{M} E[a_m^2] = 1 \tag{3.7}$$

$$\sum_{n=1}^{N} E[a_n^2] = 1 \tag{3.8}$$

  Then the gain a_{mn} can be calculated as

$$a_{mn} = a_m \cdot a_n = \frac{1}{\sqrt{MN}} \tag{3.9}$$

  where the last part of (3.9) results from the normalizing conditions in (3.7) and (3.8).
- *Doppler shift* f_{mn}. The Doppler shift introduced by the mth scatterer located near the T_x is independent of the Doppler shift introduced by the nth scatterer located near the R_x; that is,

$$f_{mn} = f_m + f_n \tag{3.10}$$

  where

$$f_m = f_{\mathrm{Tmax}} \cdot \cos \alpha_{Tm} \tag{3.11}$$

$$f_n = f_{\mathrm{Rmax}} \cdot \cos \alpha_{Rn} \tag{3.12}$$

  α_{Tm} and α_{Rn} are the angles of departure (AoD) and the angles of arrival (AoA), respectively, and f_{Tmax} and f_{Rmax} are the maximum Doppler shifts experienced at the T_x and R_x respectively. These frequencies are given as $f_{\text{Tmax}} = f_c \cdot \upsilon_T / c$ and $f_{\text{Rmax}} = f_c \cdot \upsilon_R / c$, where f_c is the carrier frequency and υ_T and υ_R are the T_x and R_x velocities, respectively. Since the locations of the scatterers are not known a priori, the AoDs α_{Tm} and the AoAs α_{Rn} are discrete random variables. However, since the number of scatterers at each end is approaching infinity, these discrete random variables become continuous variables with probability density functions (pdfs) $p(\alpha_T)$ and $p(\alpha_R)$, respectively.

- *Phase* ϕ_{mn}. The phase shift introduced by the *m*th scatterer located near the T_x is independent of the phase shift introduced by the *n*th scatterer located near the R_x; that is,

$$\phi_{mn} = \phi_m + \phi_n \tag{3.13}$$

These phase shifts are ramdom, and, since the number of scatterers at either end is approaching infinity, these discrete ramdom variables become continuous with pdfs $p(\phi_T)$ and $p(\phi_R)$, respectively. Since all phase shifts are equally probable, it can be assumed that they are uniformly distributed in $[0,2\pi)$; that is, $p(\phi_T) = 1/(2\pi)$ and $p(\phi_R) = 1/(2\pi)$.

In the presence of a specular component (i.e., LoS or a strong reflected path), the complex baseband impulse response becomes a superposition of a strong specular component and scattered components in (3.6). To account for a specular component, the Akki and Haber's model can be modified as follows

$$\begin{aligned} h(t) = {} & \frac{\sqrt{K\Omega}}{\sqrt{K+1}} e^{j(2\pi f_{\text{T}\max} t \cos(\pi - \alpha_{LoS}) + 2\pi f_{\text{R}\max} t \cos \alpha_{LoS})} \\ & + \frac{\sqrt{\Omega}}{\sqrt{K+1}} \lim_{M,N\to\infty} \sum_{m=1}^{M} \sum_{n=1}^{N} a_{mn} \cdot e^{j(2\pi f_{mn} t + \phi_{mn})} \end{aligned} \tag{3.14}$$

where K is the ratio of the received specular to scattered power (i.e., Ricean factor) and α_{LoS} is the angle of arrival of the specular component.

3.1.4 Temporal Fading Characteristics of V-to-V Frequency-Flat Multipath Fading Channels

3.1.4.1 Temporal Autocorrelation Function and Doppler Power Spectrum for V-to-V Isotropic Scattering Channels

The temporal autocorrelation function (ACF) of the complex fading channel $h(t)$ is an important quantity since it allows the quantification of the required pilot density and interleaver depths for coherent systems or the performance degradation for both coherent and noncoherent systems.

The normalized ACF is defined as

$$R(\tau) = \frac{\mathrm{E}[h(t)h(t+\tau)^*]}{\sqrt{\mathrm{Var}[h(t)]\mathrm{Var}[h(t+\tau)^*]}} \tag{3.15}$$

where the variance Var[·] and the expectation E[·] are taken with respect to the set of variables τ, α_T, α_R, ϕ_T, and ϕ_R. Then, the normalized ACF can be written as

$$R(\tau) = \mathrm{E}_{\tau,\alpha_T,\alpha_R,\phi_T,\phi_R}\left[\lim_{M,N\to\infty}\sum_{m=1}^{M}\sum_{n=1}^{N}\frac{1}{\sqrt{MN}}e^{-j(2\pi(\bar{f}_m+f_n)(t+\tau)+\phi_m+\phi_n)}\right. \\ \left.\times \lim_{M,N\to\infty}\sum_{m'=1}^{M}\sum_{n'=1}^{N}\frac{1}{\sqrt{MN}}e^{j(2\pi(f_{m'}+f_{n'})t+\phi_{m'}+\phi_{n'})}\right] \tag{3.16}$$

It is reasonable to assume that the phases ϕ_m and $\phi_{m'}$ as well as ϕ_n and $\phi_{n'}$ are independent for $m \neq m'$ and $n \neq n'$, respectively, since their associated Doppler shifts are independent. Furthermore, since the phases are uniformly distributed over [0, 2π), the ACF $R(\tau)$ becomes

$$R(\tau) = E_{\alpha_{Tm}}\left[\lim_{M\to\infty}\sum_{m=1}^{M}\frac{1}{M}e^{-j2\pi f_{T\max}\cos\alpha_{Tm}\tau}\right] \\ \times E_{\alpha_{Rn}}\left[\lim_{N\to\infty}\sum_{n=1}^{N}\frac{1}{N}e^{-j2\pi f_{R\max}\cos\alpha_{Rn}\tau}\right] \tag{3.17}$$

For isotropic scattering, the model in Section 3.1.3 assumes that the planar waves depart and arrive from all directions with equal probability; that is, $p(\alpha_T) = 1/(2\pi)$ and $p(\alpha_R) = 1/(2\pi)$, for $\alpha_{T/R} \in [0, 2\pi)$. Then, the normalized ACF can be written as

$$\begin{aligned} R(\tau) &= \int_{-\pi}^{\pi} e^{-j2\pi f_{T\max}\cos\alpha_T\tau}p(\alpha_T)d\alpha_T \int_{-\pi}^{\pi} e^{-j2\pi f_{R\max}\cos\alpha_R\tau}p(\alpha_R)d\alpha_R \\ &= \int_{-\pi}^{\pi}\frac{1}{2\pi}e^{-j2\pi f_{T\max}\cos\alpha_T\tau}d\alpha_T \int_{-\pi}^{\pi}\frac{1}{2\pi}e^{-j2\pi f_{R\max}\cos\alpha_R\tau}d\alpha_R \\ &= J_0(2\pi f_{T\max}\tau)J_0(2\pi f_{R\max}\tau) \end{aligned} \tag{3.18}$$

where $J_0(\cdot)$ is the zeroth order Bessel function of the first kind. Note that ACF of V-to-V channel consists of a product of two Bessel functions in contrast to F-to-M cellular radio channel, where the ACF involves only a single Bessel function. Figure 3.2 compares the ACF for F-to-M cellular channels given in (2.29) of Section 2.3.3 with that of the V-to-V channel at f_c = 0.5 GHz carrier frequency. We assume that $v_t = 0$ and v_r = 20 m/s for F-to-M channel, while $v_t = v_r$ = 20 m/s for V-to-V channel. From Figure 3.2, we can observe that the V-to-V channel decorrelates faster than the F-to-M channel. This has several implications on the system design: (a) better temporal diversity is available in the channel, which leads to better system performance and reduced outages; and (b) noncoherent systems suffer from increased performance losses, while coherent systems may need to insert pilot symbols more often.

The normalized temporal autocorrelation function of the complex impulse response in the presence of both a strong specular component and a scatter component

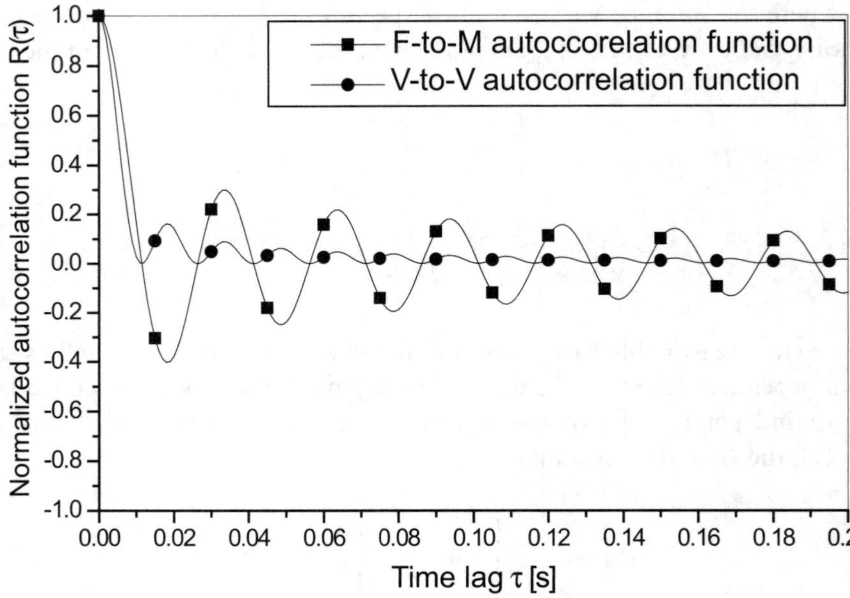

Figure 3.2 Normalized temporal autocorrelation functions of F-to-M cellular channel and V-to-V channel.

can be obtained by substituting (3.14) into (3.15) and evaluating the expectation, and the result is

$$R(\tau) = \frac{1}{K+1} J_0(2\pi f_{\mathrm{T\,max}}\tau) J_0(2\pi f_{\mathrm{R\,max}}\tau) + \frac{K}{K+1} e^{-j(2\pi f_{\mathrm{T\,max}}\tau\cos(\pi-\alpha_{LoS})+2\pi f_{\mathrm{R\,max}}\tau\cos\alpha_{LoS})} \tag{3.19}$$

The *Doppler power spectrum* of the complex channel $h(t)$ is the Fourier transform of its temporal autocorrelation function, $R(\tau)$; that is,

$$S(f) = \int_{-\infty}^{\infty} R(\tau) e^{-j2\pi f\tau} d\tau \tag{3.20}$$

Applying the Fourier transform to (3.17) yields [10]

$$S(f) = \frac{1}{\pi^2\sqrt{s} f_{\mathrm{T\,max}}} \kappa\left[\frac{1+s}{2\sqrt{s}}\sqrt{1-\left(\frac{f}{(1+s)f_{\mathrm{T\,max}}}\right)^2}\right] \tag{3.21}$$

where $\kappa[\cdot]$ is the complete elliptic integral of the first kind and s is the ratio of the maximum Doppler frequencies f_{Rmax} and f_{Tmax} (i.e., $s = f_{\mathrm{Rmax}}/f_{\mathrm{Tmax}}$). This function exhibits two peaks at $\pm(f_{\mathrm{Tmax}} - f_{\mathrm{Rmax}})$ and, for isotropic scattering, this function is symmetrical with respect to the motion of T_x or R_x. In the presence of both a strong specular component and a scatter component, the Doppler power spectrum can be obtained by substituting (3.19) into (3.20), and the result is

$$S(f) = \frac{1}{(K+1)\pi^2\sqrt{s}f_{\mathrm{T}\max}}\kappa\left[\frac{1+s}{2\sqrt{s}}\sqrt{1-\left(\frac{f}{(1+s)f_{\mathrm{T}\max}}\right)^2}\right] + \frac{K}{K+1}\delta(f-(f_{\mathrm{T}\max}\cos(\pi-\alpha_{LoS})+f_{\mathrm{R}\max}\cos\alpha_{LoS}) \tag{3.22}$$

where $\delta(\cdot)$ is the Dirac delta function.

3.1.4.2 Temporal Autocorrelation Function and Doppler Power Spectrum for V-to-V Nonisotropic Scattering Channels

One of the main assumptions in the Akki and Haber's channel model is that scattering is isotropic. However, it has been experimentally demonstrated [11] that scattering encountered in V-to-V environments is nonisotropic, resulting in nonuniform pdfs for AoDs and AoAs. Using similar reasoning as in Section 2.3.4, we use the von Mises pdf to characterize nonisotropic scattering.

The von Mises pdfs for the T_x and R_x scatterers can be written as [12]

$$p(\alpha_T) = \frac{\exp[k_T\cos(\alpha_T-\mu_T)]}{2\pi I_0(k_T)} \tag{3.23}$$

$$p(\alpha_R) = \frac{\exp[k_R\cos(\alpha_R-\mu_R)]}{2\pi I_0(k_R)} \tag{3.24}$$

where $\alpha_{T(R)} \in [-\pi, \pi)$ represents AoDs (AoAs), $I_0(\cdot)$ is the zeroth-order modified Bessel function of the first kind, $\mu_{T(R)} \in [-\pi, \pi)$ is the mean angle at which the scatterers are distributed, and $k_{T(\mathrm{R})}$ controls the spread of scatterers around the mean.

To evaluate the normalized temporal autocorrelation function, we start from (3.15) and follow the derivations in [14]. The nonisotropic scattering model assumes that planar waves propagate in the x-y plane and arrive from all directions with nonequal probability, where the von Mises pdf describes the angles of departure and the angles of arrival. Then, the normalized autocorrelation function can be calculated as follows:

$$\begin{aligned} R(\tau) &= \int_{-\pi}^{\pi} e^{-j2\pi f_{\mathrm{T}\max}\cos\alpha_t\tau}p(\alpha_t)d\alpha_t\int_{-\pi}^{\pi} e^{-j2\pi f_{\mathrm{R}\max}\cos\alpha_r\tau}p(\alpha_r)d\alpha_r \\ &= \int_{-\pi}^{\pi} e^{-j2\pi f_{\mathrm{T}\max}\cos\alpha_t\tau}e^{[k_T\cos(\alpha_T-\mu_T)]/(2\pi I_0(k_T))}d\alpha_t \\ &\times\int_{-\pi}^{\pi} e^{-j2\pi f_{\mathrm{R}\max}\cos\alpha_r\tau}e^{[k_R\cos(\alpha_R-\mu_R)]/(2\pi I_0(k_R))}d\alpha_r \end{aligned} \tag{3.25}$$

Using the equality $\int_{-\pi}^{\pi} e^{a\sin(c)+b\cos(c)}dc = 2\pi I_0\left(\sqrt{a^2+b^2}\right)$ (eq. 3.338-4 in [13]) to evaluate the integrals, the normalized autocorrelation function becomes [14]

$$R(\tau) = \frac{I_0\left(\sqrt{k_T^2-4\pi^2 f_{\mathrm{T}\max}^2\tau^2+j4\pi k_T f_{\mathrm{T}\max}\tau\cos\mu_T}\right)}{I_0(k_T)} \times\frac{I_0\left(\sqrt{k_R^2-4\pi^2 f_{\mathrm{R}\max}^2\tau^2+j4\pi k_R f_{\mathrm{R}\max}\tau\cos\mu_R}\right)}{I_0(k_R)} \tag{3.26}$$

In the presence of both a strong specular component and a scatter component, the temporal autocorrelation becomes

$$
\begin{aligned}
R(\tau) = {} & \frac{1}{K+1}\frac{I_0\left(\sqrt{k_T^2 - 4\pi^2 f_{\mathrm{T\,max}}^2\tau^2 + j4\pi k_T f_{\mathrm{T\,max}}\tau\cos\mu_T}\right)}{I_0(k_T)} \\
& \times\frac{I_0\left(\sqrt{k_R^2 - 4\pi^2 f_{\mathrm{R\,max}}^2\tau^2 + j4\pi k_R f_{\mathrm{R\,max}}\tau\cos\mu_R}\right)}{I_0(k_R)} \\
& + \frac{K}{K+1}e^{-j\left(2\pi f_{\mathrm{T\,max}}\tau\cos(\pi-\alpha_{LoS})+2\pi f_{\mathrm{R\,max}}\tau\cos\alpha_{LoS}\right)}
\end{aligned}
\tag{3.27}
$$

The Doppler power spectrum can be obtained as the Fourier transform of the temporal autocorrelation function R(τ) in (3.26). To calculate the Fourier transform, we first observe that Bessel functions in (3.26) can be written as

$$
I_0\left[j2\pi f_{\mathrm{T\,max}}\sqrt{\left(\tau-\frac{jk_T\cos\mu_T}{2\pi f_{\mathrm{T\,max}}}\right)^2+\left(\frac{-jk_T\sin\mu_T}{2\pi f_{\mathrm{T\,max}}}\right)^2}\right]
\tag{3.28}
$$

$$
I_0\left[j2\pi f_{\mathrm{R\,max}}\sqrt{\left(\tau-\frac{jk_R\cos\mu_R}{2\pi f_{\mathrm{Rmax}}}\right)^2+\left(\frac{-jk_R\sin\mu_R}{2\pi f_{\mathrm{Rmax}}}\right)^2}\right]
\tag{3.29}
$$

Then, using the equality $\int_0^\infty I_0\left(\alpha\sqrt{t^2+u^2}\right)\cos(\beta t)dt = \cosh\left(u\sqrt{\alpha^2-\beta^2}\right)/\sqrt{\alpha^2-\beta^2}$ (eq. 6.677-3 in [13]), the Doppler power spectrum becomes

$$
\begin{aligned}
S(f) = {} & \frac{e^{jk_T\cos\mu_T f/f_{\mathrm{T\,max}}}}{I_0(k_T)}\frac{\cosh\left[k_T\sin\mu_T\sqrt{1-(f/f_{\mathrm{T\,max}})^2}\right]}{\pi f_{\mathrm{T\,max}}\sqrt{1-(f/f_{\mathrm{T\,max}})^2}} \\
& \odot\frac{e^{jk_R\cos\mu_R f/f_{\mathrm{R\,max}}}}{I_0(k_R)}\frac{\cosh\left[k_R\sin\mu_R\sqrt{1-(f/f_{\mathrm{R\,max}})^2}\right]}{\pi f_{\mathrm{R\,max}}\sqrt{1-(f/f_{\mathrm{R\,max}})^2}}
\end{aligned}
\tag{3.30}
$$

where $\odot$ denotes convolution function and $|f| \le f_{\mathrm{Tmax}} + f_{\mathrm{Rmax}}$. To validate the derived Doppler power spectrum, we can compare this expression for $k_T = k_R = 0$ with the Doppler power spectrum for isotropic scattering given in (3.21). For different ratios $s = f_{\mathrm{Rmax}}/f_{\mathrm{Tmax}}$, Figure 3.3 shows good agreement between the Doppler spectra in (3.21) and (3.30).

In the presence of both a strong specular component and scattered components, the Doppler power spectrum can be obtained by substituting (3.27) into (3.20), and after integration the result is

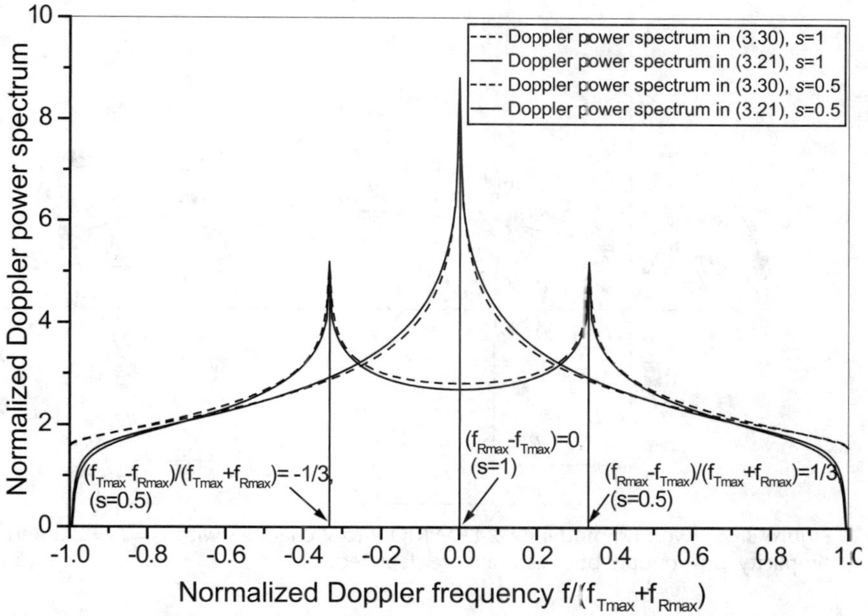

Figure 3.3 Comparison of the normalized Doppler power spectra in (3.21) and (3.30) for $k_T = k_R = 0$ and different ratios $s = f_{R\max}/f_{T\max}$.

$$
\begin{aligned}
S(f) = {} & \frac{1}{(K+1)} \frac{e^{jk_T \cos\mu_T f/f_{T\max}}}{I_0(k_T)} \frac{\cosh\left[k_T \sin\mu_T \sqrt{1-(f/f_{T\max})^2}\right]}{\pi f_{T\max}\sqrt{1-(f/f_{T\max})^2}} \\
& \odot \frac{e^{jk_R \cos\mu_R f/f_{R\max}}}{I_0(k_R)} \frac{\cosh\left[k_R \sin\mu_R \sqrt{1-(f/f_{R\max})^2}\right]}{\pi f_{R\max}\sqrt{1-(f/f_{R\max})^2}} \\
& + \frac{K}{K+1}\delta\left(f - \left(f_{T\max}\cos(\pi - \alpha_{LoS}) + f_{R\max}\cos(\alpha_{LoS})\right)\right)
\end{aligned}
\quad (3.31)
$$

where $\delta(\cdot)$ is the Dirac delta function.

3.1.5 Two-Dimensional Model for MIMO V-to-V Frequency-Flat Multipath Fading Channels

The design of high-speed wireless communication systems often requires the use of multiple antennas as well as multiple-input multiple-output (MIMO) diversity techniques. To model such a propagation environment, Pätzold et al. were the first to propose a 2-D model for MIMO V-to-V channels [15–17]. This model considers outdoor radio propagation where all rays are only double-bounced, similar to the SISO model detailed in Section 3.1.3. Zajić and Stüber [18] have noticed that if the double-bounced rays can be detected, then the single-bounced rays, which bear more energy than double-bounced rays, can also be detected. Hence, they extended this model to include the LoS, single-bounced rays at transmitter and single-bounced rays at receiver. The parametric nature of the model in [18] makes it more flexible and adaptable to a variety of

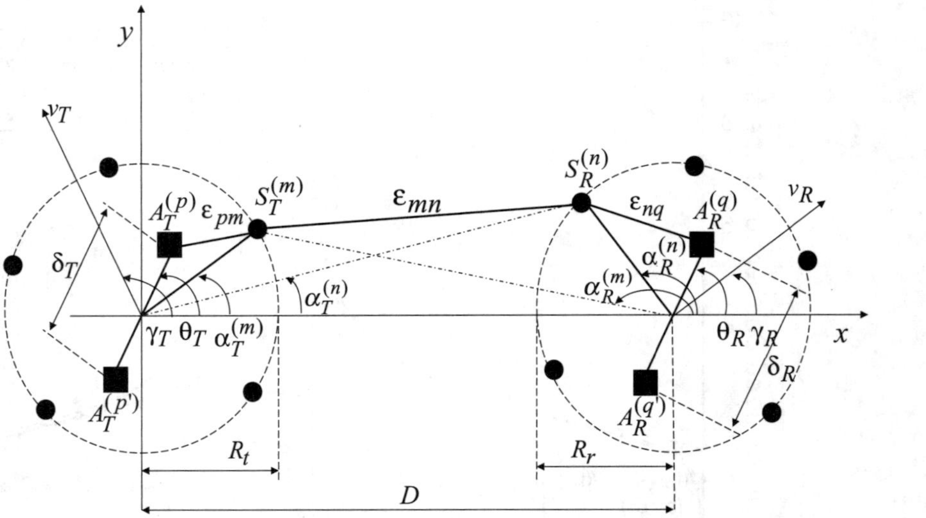

Figure 3.4 Two-ring model for 2-D MIMO V-to-V channels with $L_t = L_r = 2$ antenna elements. For simplicity, only double-bounced rays are depicted.

propagation environments. For example, this model describes the radio propagation on highways as a combination of single- and double-bounced rays, taking into account that single-bounced rays bear more energy than the double-bounced rays [11]. In particular, if there are no single-bounced and LoS rays, this model simplifies to the model proposed in [15–17]. Hence, we will detail the model in [18] here. This model follows the same assumptions as outlined in Section 3.1.3.

The geometry of the model in [18] originates in [19] and is shown in Figure 3.4. This elementary 2×2 antenna configuration is used later to construct uniform linear antenna arrays with arbitrary number of antennas. The spacing between two adjacent antenna elements at the T_x and R_x is denoted by δ_T and δ_R, respectively, while angles θ_T and θ_R describe the orientation of the T_x and R_x antenna array, respectively, relative to the x-axis.

The two-ring model defines two rings of fixed scatterers, one around the T_x and another around the R_x, as shown in Figure 3.4. Around the transmitter, M omnidirectional scatterers lie on a ring of radius R_t, and the mth transmit scatterer is denoted by $S_T^{(m)}$. Similarly, around the receiver, N omnidirectional scatterers lie on a ring of radius R_r and the nth receive scatterer is denoted by $S_R^{(n)}$. The T_x and R_x are moving with speeds v_T and v_R in directions described by angles γ_T and γ_R, respectively.

In Figure 3.4, the symbols $\alpha_T^{(m)}$ and $\alpha_T^{(n)}$ are the angles of departures (AoD) of the waves that bounce off the scatterers $S_T^{(m)}$ and $S_R^{(n)}$, respectively. Similarly, the symbols $\alpha_R^{(m)}$ and $\alpha_R^{(n)}$ are the angles of arrivals (AoA) of the waves scattered from $S_T^{(m)}$ and $S_R^{(n)}$, respectively. Finally, α_{Rq}^{LoS} denotes the AoA of the LoS path (not shown in Figure 3.4 to reduce clutter). This model assumes $\max\{\delta_T, \delta_R\} << \min\{R_t, R_r\}$, which allows us to approximate the angles of departure from different transmit antennas $\alpha_T^{(pm)}$ and $\alpha_T^{(pn)}$ with $\alpha_T^{(m)}$ and $\alpha_T^{(n)}$, respectively. Using similar reasoning, the angles of arrival from different receive antennas $\alpha_R^{(mq)}$ and $\alpha_R^{(nq)}$ are approximated with $\alpha_R^{(m)}$ and $\alpha_R^{(n)}$, respectively.

The distance between the T_x and R_x is D, and D is assumed to be smaller than $4R_tR_rL_r/(\lambda(L_t-1)(L_r-1))$ (the channel does not experience keyhole behavior [20]), where λ denotes the carrier wavelength. Distance ε_{pm} is the distance between the pth antenna $A_T^{(p)}$ and the mth scatterer placed around the T_x (i.e., $A_T^{(p)} - S_T^{(m)}$). Similarly, symbols ε_{mq}, ε_{pn}, ε_{nq}, and ε_{mn} denote distances $S_T^{(m)} - A_R^{(q)}$, $A_T^{(p)} - S_R^{(n)}$, $S_R^{(n)} - A_R^{(q)}$, and $S_T^{(m)} - S_R^{(n)}$, respectively. Finally, ε_{pq}^{LoS} denotes the distance $A_T^{(p)} - A_R^{(q)}$.

The complex faded envelope under frequency-flat fading can therefore be written as a superposition of (a) LoS, (b) single-bounced rays that have been reflected once at the transmitter (SBT) or at the receiver (SBR), and (c) double-bounced (DB) rays. For the channel between the pth transmit and qth receive antenna the channel impulse response is [18]

$$h_{pq}(t) = h_{pq}^{SBT}(t) + h_{pq}^{SBR}(t) + h_{pq}^{DB}(t) + h_{pq}^{LoS}(t) \tag{3.32}$$

where the respective components are

$$h_{pq}^{LoS}(t) = \sqrt{\Omega_{pq}}\, a_{pq}^{LoS} e^{\left(j2\pi f_{pq}^{LoS} t - j\frac{2\pi}{\lambda} d_{pq}^{LoS}\right)} \tag{3.33}$$

$$h_{pq}^{SBT}(t) = \sqrt{\Omega_{pq}} \lim_{M\to\infty} \sum_{m=1}^{M} a_{pq,m}^{SBT} e^{\left(j2\pi f_m^{SBT} t - j\frac{2\pi}{\lambda} d_{pq,m}^{SBT} + j\phi_{pq,m}^{SBT}\right)} \tag{3.34}$$

$$h_{pq}^{SBR}(t) = \sqrt{\Omega_{pq}} \lim_{N\to\infty} \sum_{n=1}^{N} a_{pq,n}^{SBR} e^{\left(j2\pi f_n^{SBR} t - j\frac{2\pi}{\lambda} d_{pq,n}^{SBR} + j\phi_{pq,n}^{SBR}\right)} \tag{3.35}$$

$$h_{pq}^{DB}(t) = \sqrt{\Omega_{pq}} \lim_{M,N\to\infty} \sum_{m=1}^{M}\sum_{n=1}^{N} a_{pq,mn}^{DB} e^{\left(j2\pi f_{mn}^{DB} t - j\frac{2\pi}{\lambda} d_{pq,mn}^{DB} + j\phi_{pq,mn}^{DB}\right)} \tag{3.36}$$

where Ω_{pq} is the power associated with the pqth planar wave. For $\max\{\delta_T, \delta_R\} \ll \min\{R_t, R_r\}$, we can assume that the total power is equally split among all antenna elements (i.e., $\Omega_{pq} = \Omega$ for all $p \in \{1,\ldots,L_t\}$ and $q \in \{1,\ldots,L_r\}$). The channel gains, Doppler shift frequencies, phases, and path lengths in this model can be calculated as follows:

- *Channel gains*. The channel gains in this model are normalized such that the ensemble averages of the $a_{pq,m}^{SBT}$s, $a_{pq,n}^{SBR}$s and $a_{pq,mn}^{DB}$s are given by

$$\sum_{m=1}^{M} E\left[\left(a_{pq,m}^{SBT}\right)^2\right] = 1 \tag{3.37}$$

$$\sum_{n=1}^{N} E\left[\left(a_{pq,n}^{SBR}\right)^2\right] = 1 \tag{3.38}$$

$$\sum_{m,n=1}^{M,N} E\left[\left(a_{pq,mn}^{DB}\right)^2\right] = 1 \tag{3.39}$$

In addition, K_{pq} describes how the total signal power between the pth transmit and the qth receive antenna is split between the LoS and NLoS components: the power allocated to the LoS component between the pth transmit and qth receive antenna is $K_{pq}/(1 + K_{pq})$, while the total power allocated to the NLoS components between the pth transmit and qth receive antenna is $1/(1 + K_{pq})$. For $\max\{\delta_T, \delta_R\} \ll \min\{R_t, R_r\}$, we can assume that the Ricean fading factor is the same between all antenna elements (i.e., $K_{pq} = K$ for all $p \in \{1,\ldots,L_t\}$ and $q \in \{1,\ldots,L_r\}$). Furthermore, it is assumed that the relative power allocated to the SBT, SBR, and DB components is respectively η_T, η_R, and η_{TR} such that $\eta_T + \eta_R + \eta_{TR} = 1$. These parameters have to be either set during simulations or estimated from measurements. Having all this in mind, the channel gains can be written as follows:

$$a_{pq}^{LoS} = \sqrt{\frac{K}{K+1}} \tag{3.40}$$

$$a_{pq,m}^{SBT} = \sqrt{\frac{1}{K+1}}\sqrt{\frac{\eta_T}{M}} \tag{3.41}$$

$$a_{pq,n}^{SBR} = \sqrt{\frac{1}{K+1}}\sqrt{\frac{\eta_R}{N}} \tag{3.42}$$

$$a_{pq,mn}^{DB} = \sqrt{\frac{1}{K+1}}\sqrt{\frac{\eta_{TR}}{MN}} \tag{3.43}$$

- *Doppler shifts.* Doppler shifts depend on the geometrical relation between directions of movement of T_x and R_x (γ_T and γ_R) and the directions of AoDs and AoAs. For this model, the Doppler shifts can be written as

$$f_{pq}^{LoS} = f_{\mathrm{Tmax}} \cos\left(\pi - \alpha_{Rq}^{LoS} - \gamma_T\right) + f_{\mathrm{Rmax}} \cos\left(\alpha_{Rq}^{LoS} - \gamma_R\right) \tag{3.44}$$

$$\begin{aligned} f_m^{SBT} &= f_{\mathrm{Tmax}} \cos\left(\alpha_T^{(m)} - \gamma_T\right) + f_{\mathrm{Rmax}} \cos\left(\alpha_R^{(m)} - \gamma_R\right) \\ &= f_{\mathrm{Tmax}} \cos\left(\alpha_T^{(m)} - \gamma_T\right) + f_{\mathrm{Rmax}} \left(\Delta_T \sin\gamma_R \sin\alpha_T^{(m)} - \cos\gamma_R\right) \end{aligned} \tag{3.45}$$

$$\begin{aligned} f_n^{SBR} &= f_{\mathrm{Tmax}} \cos\left(\alpha_T^{(n)} - \gamma_T\right) + f_{\mathrm{Rmax}} \cos\left(\alpha_R^{(n)} - \gamma_R\right) \\ &= f_{\mathrm{Tmax}} \left(\Delta_R \sin\gamma_T \sin\alpha_R^{(n)} + \cos\gamma_T\right) + f_{\mathrm{Rmax}} \cos\left(\alpha_R^{(n)} - \gamma_R\right) \end{aligned} \tag{3.46}$$

$$f_{mn}^{DB} = f_{\mathrm{Tmax}} \cos\left(\alpha_T^{(m)} - \gamma_T\right) + f_{\mathrm{Rmax}} \cos\left(\alpha_R^{(n)} - \gamma_R\right) \tag{3.47}$$

where frequencies $f_{\mathrm{Tmax}} = v_T/\lambda$ and $f_{\mathrm{Rmax}} = v_R/\lambda$ are the maximum Doppler frequencies associated with the T_x and R_x, respectively, and λ is the carrier wavelength. For $\max\{\delta_T, \delta_R\} \ll \min\{R_t, R_r\}$, the LoS AoAs are approximately equal to π (i.e., $\alpha_{Rq}^{LoS} = \pi$). Since the location of the scatterers is not known a priori, the AoDs ($\alpha_T^{(m)}$ and $\alpha_T^{(n)}$) and the AoAs ($\alpha_R^{(m)}$ and $\alpha_R^{(n)}$) are modeled as random variables. However, note that when the number of scatterers at each end is approaching infinity, these discrete random variables can be characterized as continuous random variables with given pdfs. Note that double-bounced rays have the AoD, $\alpha_T^{(m)}$, independent from the AoA, $\alpha_R^{(n)}$ [20], whereas single-bounced rays have the AoA, $\alpha_R^{(m)}$, dependent on the AoD, $\alpha_T^{(m)}$, and the AoD, $\alpha_T^{(n)}$, dependent on the AoA $\alpha_R^{(n)}$ [18]. The geometry that propagating waves create traveling from the transmitter to the receiver dictates the relationship between the AoDs and AoAs. Since single-bounced rays form a triangle between the transmit and the receive antennas, only one angle can be independently selected—either AoDs or AoAs. On the other hand, double-bounced rays form a quadrilateral between the transmit and receive antennas, and there can be two independently selected angles—AoDs and AoAs.

- *Phases*. First, we note that the phase shift introduced by the mth scatterer located near the T_x is independent of the phase shift introduced by the nth scatterer located near the R_x. Then, without loss of generality, we can assume that the phases $\phi_{pq,m}^{SBT}$, $\phi_{pq,n}^{SBR}$, $\phi_{pq,mn}^{DB}$ and are independent random variables. Here, it is assumed that they are uniformly distributed on the interval $[0, 2\pi)$ and independent from any other random variable.
- *Path lengths*. From Figure 3.4, the path lengths of the respective waves can be written as

$$d_{pq}^{LoS} = \varepsilon_{pq} \tag{3.48}$$

$$d_{pq,m}^{SBT} = \varepsilon_{pm} + \varepsilon_{mq} \tag{3.49}$$

$$d_{pq,n}^{SBR} = \varepsilon_{pn} + \varepsilon_{nq} \tag{3.50}$$

$$d_{pq,mn}^{DB} = \varepsilon_{pm} + \varepsilon_{mn} + \varepsilon_{nq} \tag{3.51}$$

For outdoor propagation,which is typical for V-to-V communications, we can assume the radii R_t and R_r are much smaller than the distance D between the T_x and R_x; that is, $\max\{R_t, R_r\} \ll D$ (local scattering condition). Under this assumption, the distances ε_{pq}, ε_{pm}, ε_{mq}, ε_{pn}, ε_{nq}, and ε_{mn} can be approximated as follows [18]:

$$\begin{aligned}\varepsilon_{pq} \approx D &- (0.5L_t + 0.5 - p)\delta_T \cos\theta_T \\ &- (0.5L_r + 0.5 - q)\delta_R \cos(\pi - \theta_R)\end{aligned} \tag{3.52}$$

$$\varepsilon_{pm} \approx R_t - (0.5L_t + 0.5 - p)\delta_T \cos\left(\theta_T - \alpha_T^{(m)}\right) \tag{3.53}$$

$$\varepsilon_{mq} \approx D - (0.5L_r + 0.5 - q)\delta_R \left(\Delta_T \sin\theta_R \sin\alpha_T^{(m)} - \cos\theta_R\right) \tag{3.54}$$

$$\varepsilon_{pn} \approx D - (0.5L_t + 0.5 - p)\delta_T \left(\Delta_R \sin\theta_T \sin\alpha_R^{(n)} + \cos\theta_T\right) \tag{3.55}$$

$$\varepsilon_{nq} \approx R_r - (0.5L_r + 0.5 - q)\delta_R \cos\left(\alpha_R^{(n)} - \theta_R\right) \tag{3.56}$$

$$\varepsilon_{mn} \approx D \tag{3.57}$$

where parameters p and q take values from the sets $p \in \{1,\ldots,L_t\}$ and $q \in \{1, \ldots, L_r\}$, respectively, $\Delta_T = R_t/D$, and $\Delta_R = R_r/D$.

Here, we note that the complex faded envelope in (3.32) also can be written as $h_{pq}(t) = h_{pq}^{(I)}(t) + jh_{pq}^{(Q)}(t)$ where

$$h_{pq}^{(I)}(t) = \Re\{h_{pq}^{SBT}(t)\} + \Re\{h_{pq}^{SBR}(t)\} + \Re\{h_{pq}^{DB}(t)\} + \Re\{h_{pq}^{LoS}(t)\} \tag{3.58}$$

$$h_{pq}^{(Q)}(t) = \Im\{h_{pq}^{SBT}(t)\} + \Im\{h_{pq}^{SBR}(t)\} + \Im\{h_{pq}^{DB}(t)\} + \Im\{h_{pq}^{LoS}(t)\} \tag{3.59}$$

are the in-phase (I) and quadrature (Q) components of the complex faded envelope, and $\Re\{\cdot\}$ and $\Im\{\cdot\}$ denote the real and imaginary operation, respectively.

3.1.6 Spatial-Temporal Fading Characteristics of V-to-V Frequency-Flat Multipath Fading Channels

Using the model described in Section 3.1.5, we can now derive the key temporal and spatial characteristics of V-to-V frequency-flat multipath fading channels.

3.1.6.1 Space-Time Correlation Function

Assuming a 2-D nonisotropic scattering environment characterized by von Mises pdfs (defined in 3.23 and 3.24), we derive the space-time correlation function (ST-CF) of the complex faded envelope in (3.32).

The normalized space-time correlation function between two complex faded envelopes $h_{pq}(t)$ and $h_{\tilde{p}\tilde{q}}(t)$ is defined as

$$R_{pq,\tilde{p}\tilde{q}}(\delta_T, \delta_R, \tau) = \frac{\mathrm{E}\left[h_{pq}(t) h_{\tilde{p}\tilde{q}}(t+\tau)^*\right]}{\sqrt{\mathrm{Var}\left[h_{pq}(t)\right]\mathrm{Var}\left[h_{\tilde{p}\tilde{q}}(t+\tau)^*\right]}} \tag{3.60}$$

where $(\cdot)^*$ denotes complex conjugate operation, E[·] is the statistical expectation operator, and Var[·] is the statistical variance operator. Since $h_{pq}^{SBT}(t)$, $h_{pq}^{SBR}(t)$, and $h_{pq}^{DB}(t)$ are independent complex Gaussian random processes with zero means, (3.60) can be simplified to

$$\begin{aligned} R_{pq,\tilde{p}\tilde{q}}(\delta_T, \delta_R, \tau) = {} & R_{pq,\tilde{p}\tilde{q}}^{LoS}(\delta_T, \delta_R, \tau) + R_{pq,\tilde{p}\tilde{q}}^{SBT}(\delta_T, \delta_R, \tau) \\ & + R_{pq,\tilde{p}\tilde{q}}^{SBR}(\delta_T, \delta_R, \tau) + R_{pq,\tilde{p}\tilde{q}}^{DB}(\delta_T, \delta_R, \tau) \end{aligned} \tag{3.61}$$

where $R_{pq,\tilde{p}\tilde{q}}^{LoS}(\delta_T, \delta_R, \tau)$, $R_{pq,\tilde{p}\tilde{q}}^{SBT}(\delta_T, \delta_R, \tau)$, $R_{pq,\tilde{p}\tilde{q}}^{SBR}(\delta_T, \delta_R, \tau)$, and $R_{pq,\tilde{p}\tilde{q}}^{DB}(\delta_T, \delta_R, \tau)$ denote the normalized ST-CFs of the LoS, single-bounced transmit, single-bounced receive, and double-bounced components, respectively.

Following the reasoning similar to one in Section 3.1.4.2, the space-time correlation functions of the LoS, SBT, SBR, and DB components can be obtained as follows [18]

$$R_{pq,\tilde{p}\tilde{q}}^{LoS}(\delta_T,\delta_R,\tau)=\frac{\mathrm{E}\left[h_{pq}^{LoS}(t)h_{\tilde{p}\tilde{q}}^{LoS}(t+\tau)^*\right]}{\sqrt{\mathrm{Var}\left[h_{pq}^{LoS}(t)\right]\mathrm{Var}\left[h_{\tilde{p}\tilde{q}}^{LoS}(t+\tau)^*\right]}}$$
$$=\frac{K}{K+1}e^{j\frac{2\pi}{\lambda}(\tilde{p}-p)\delta_T\cos\theta_T-j\frac{2\pi}{\lambda}(\tilde{q}-q)\delta_R\cos\theta_R+j2\pi\tau(f_{T\max}\cos\gamma_T-f_{R\max}\cos\gamma_R)} \tag{3.62}$$

$$R_{pq,\tilde{p}\tilde{q}}^{SBT}(\delta_T,\delta_R,\tau)=\frac{E\left[h_{pq}^{SBT}(t)h_{\tilde{p}\tilde{q}}^{SBT}(t+\tau)^*\right]}{\sqrt{\mathrm{Var}\left[h_{pq}^{SBT}(t)\right]\mathrm{Var}\left[h_{\tilde{p}\tilde{q}}^{SBT}(t+\tau)^*\right]}}$$
$$=\lim_{M\to\infty}\frac{1}{K+1}\frac{\eta_T}{M}\sum_{m=1}^{M}$$
$$E_{\alpha_T^{(m)}}\left[e^{j\frac{2\pi}{\lambda}\left[(\tilde{p}-p)\delta_T\cos\left(\theta_T-\alpha_T^{(m)}\right)+(\tilde{q}-q)\delta_R\left(\Delta_T\sin\theta_R\sin\alpha_T^{(m)}-\cos\theta_R\right)\right]}\right.$$
$$\left.\times e^{-j2\pi\tau\left[f_{T\max}\cos\left(\alpha_T^{(m)}-\gamma_T\right)+f_{R\max}\left(\Delta T\sin\gamma R\sin\alpha_T^{(m)}-\cos\gamma R\right)\right]}\right]$$
$$=\frac{\eta_T}{K+1}e^{-j\frac{2\pi}{\lambda}(\tilde{q}-q)\delta_R\cos\theta_R+j2\pi\tau f_{R\max}\cos\gamma_R}\frac{I_0\left(\sqrt{x_{SBT}^2+y_{SBT}^2}\right)}{I_0(k_T)} \tag{3.63}$$

$$R_{pq,\tilde{p}\tilde{q}}^{SBR}(\delta_T,\delta_R,\tau)=\frac{\mathrm{E}\left[h_{pq}^{SBR}(t)h_{\tilde{p}\tilde{q}}^{SBR}(t+\tau)^*\right]}{\sqrt{\mathrm{Var}[h_{pq}^{SBR}(t)]\mathrm{Var}[h_{\tilde{p}\tilde{q}}^{SBR}(t+\tau)^*]}}$$
$$=\frac{1}{K+1}\lim_{N\to\infty}\frac{\eta_R}{N}\sum_{n=1}^{N}$$
$$E_{\alpha_R^{(n)}}\left[e^{j\frac{2\pi}{\lambda}[(\tilde{q}-q)\delta_R\cos(\alpha_R^{(n)}-\theta_R)+(\tilde{p}-p)\delta_T(\Delta_R\sin\theta_T\sin\alpha_R^{(n)}+\cos\theta_T)]}\right.$$
$$\left.\times e^{-j2\pi\tau[f_{T\max}(\Delta_R\sin\gamma_T\sin\alpha_R^{(n)}+\cos\gamma_T)+f_{R\max}\cos(\alpha_R^{(n)}-\gamma_R)]}\right]$$
$$=\frac{\eta_R}{K+1}e^{j\frac{2\pi}{\lambda}(\tilde{p}-p)\delta_T\cos\theta_T-j2\pi\tau f_{T\max}\cos\gamma_T}\frac{I_0\left(\sqrt{x_{SBR}^2+y_{SBR}^2}\right)}{I_0(k_R)} \tag{3.64}$$

$$R_{pq,\tilde{p}\tilde{q}}^{DB}(\delta_T,\delta_R,\tau)=\frac{\mathrm{E}\left[h_{pq}^{DB}(t)h_{\tilde{p}\tilde{q}}^{DB}(t+\tau)^*\right]}{\sqrt{\mathrm{Var}\left[h_{pq}^{DB}(t)\right]\mathrm{Var}\left[h_{\tilde{p}\tilde{q}}^{DB}(t+\tau)^*\right]}}$$
$$=\frac{1}{K+1}\lim_{M,N\to\infty}\frac{\eta_{TR}}{MN}\sum_{m=1}^{M}\sum_{n=1}^{N} \tag{3.65}$$

$$\mathrm{E}_{\alpha_T^{(m)},\alpha_R^{(n)}}\left[e^{-j2\pi\tau\left(f_{T\max}\cos\left(\alpha_T^{(m)}-\gamma_T\right)+f_{R\max}\cos\left(\alpha_R^{(n)}-\gamma_R\right)\right)}\right.$$
$$\left.\times e^{j\frac{2\pi}{\lambda}\left[(q-\tilde{q})\delta_R\cos\left(\alpha_R^{(n)}-\theta_R\right)+(\tilde{p}-p)\delta_T\cos\left(\theta_T-\alpha_T^{(m)}\right)\right]}\right]$$
$$=\frac{\eta_{TR}}{K+1}\frac{I_0\left(\sqrt{x_{DB}^2+y_{DB}^2}\right)I_0\left(\sqrt{z_{DB}^2+w_{DB}^2}\right)}{I_0(k_T)I_0(k_R)} \tag{3.66}$$

where $I_0(\cdot)$ is the zeroth-order modified Bessel function of the first kind and the parameters x_{SBT}, y_{SBT}, x_{SBR}, y_{SBR}, x_{DB}, y_{DB}, z_{DB}, and w_{DB} are

$$x_{SBT}=j\frac{2\pi}{\lambda}(\tilde{p}-p)\delta_T\cos\theta_T-j2\pi\tau f_{T\max}\cos\gamma_T+k_T\cos\mu_T \tag{3.67}$$

$$y_{SBT}=j\frac{2\pi}{\lambda}\left((\tilde{p}-p)\delta_T\sin\theta_T+(\tilde{q}-q)\delta_R\Delta_T\sin\theta_R\right)$$
$$-j2\pi\tau\left(f_{T\max}\sin\gamma_T+f_{R\max}\Delta_T\sin\gamma_R\right)+k_T\sin\mu_T \tag{3.68}$$

$$x_{SBR}=j\frac{2\pi}{\lambda}(\tilde{q}-q)\delta_R\cos\theta_R-j2\pi\tau f_{R\max}\cos\gamma_R+k_R\cos\mu_R \tag{3.69}$$

$$y_{SBR}=j\frac{2\pi}{\lambda}\left((\tilde{p}-p)\delta_T\Delta_R\sin\theta_T+(\tilde{q}-q)\delta_R\sin\theta_R\right)$$
$$-j2\pi\tau\left(f_{T\max}\Delta_R\sin\gamma_T+f_{R\max}\sin\gamma_R\right)+k_R\sin\mu_R \tag{3.70}$$

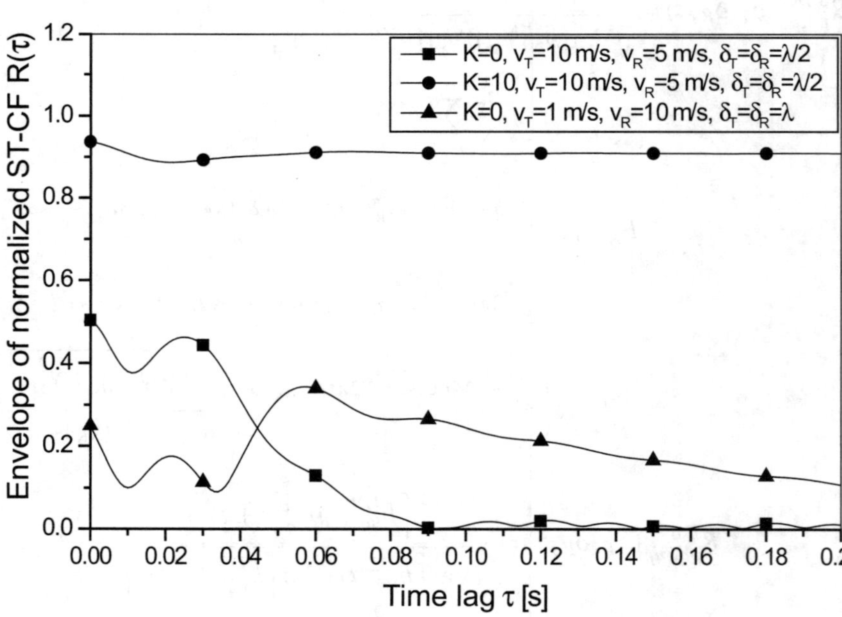

Figure 3.5 Envelope of the correlation function versus time lag for several scenarios: (a) K = 0, v_T = 10 m/s, υ_R = 5 m/s, $\delta_T = \delta_R = \lambda/2$, (b) K = 10, v_T = 10 m/s, v_R = 5 m/s, $\delta_T = \delta_R = \lambda/2$, and (c) K = 0, v_T = 1 m/s, υ_R = 10 m/s, $\delta_T = \delta_R = \lambda$.

$$x_{DB} = j\frac{2\pi}{\lambda}(\tilde{p}-p)\delta_T \cos\theta_T - j2\pi\tau f_{T\max}\cos\gamma_T + k_T\cos\mu_T \tag{3.71}$$

$$y_{DB} = j\frac{2\pi}{\lambda}(\tilde{p}-p)\delta_T \sin\theta_T - j2\pi\tau f_{T\max}\sin\gamma_T + k_T\sin\mu_T \tag{3.72}$$

$$z_{DB} = j\frac{2\pi}{\lambda}(\tilde{q}-q)\delta_R \cos\theta_R - j2\pi\tau f_{R\max}\cos\gamma_R + k_R\cos\mu_R \tag{3.73}$$

$$w_{DB} = j\frac{2\pi}{\lambda}(\tilde{q}-q)\delta_R \sin\theta_R - j2\pi\tau f_{R\max}\sin\gamma_R + k_R\sin\mu_R \tag{3.74}$$

The MIMO V-to-V space-time correlation function in (3.61) can be simplified for many special cases. The simplest special case is Clarke's temporal correlation function $J_0(2\pi f_{R\max}\tau)$ [21], obtained for $K = 0$ (NLoS condition), $k_R = 0$ (2-D isotropic scattering around R_x), $f_{T\max} = 0$, $k_T = 0$, $\eta_T = 0$, $\eta_{TR} = 0$ (fixed T_x, no scattering around T_x), and $\delta_T = \delta_R = 0$ (single-antenna T_x and R_x), where $J_0(\cdot)$ is the first kind zeroth-order Bessel function. Expressions for other space-time correlation functions based on the "one-ring" model [22, 23] can be similarly obtained. The temporal correlation function for V-to-V channels, assuming 2-D isotropic scattering, $J_0(2\pi f_{T\max}\tau)\, J_0(2\pi f_{R\max}\tau)$ [10] is obtained for $K = 0$, $\eta_T = \eta_R = 0$, $k_T = k_R = 0$, and $\delta_T = \delta_R = 0$. Similarly, the spatial correlation function for V-to-V channel $J_0(2\pi(\tilde{p}-p)\,\delta_T/\lambda)J_0(2\pi(\tilde{q}-q)\delta_R/\lambda)$ [24] is obtained for $K = 0$, $\eta_T = 0$, $\eta_R = 0$, $k_T = 0$, $k_R = 0$, and $\tau = 0$. Finally, the space-time correlation function for V-to-V channels, when isotropic scattering is assumed and there is only the double-bounced component, is obtained for $K = 0$, $\eta_T = \eta_R = 0$, $k_T = k_R = 0$, $p = 1$, $\tilde{p} = 2$, $q = 1$, and $\tilde{q} = 2$, and can be written as [15, 16]

$$R(\delta_T,\delta_R,\tau) = J_0\left(2\pi\sqrt{(\delta_T/\lambda)^2 + (f_{T\max}\tau)^2 - 2\frac{\delta_T}{\lambda} f_{T\max}\tau\cos(\gamma_T-\theta_T)}\right) \times J_0\left(2\pi\sqrt{(\delta_R/\lambda)^2 + (f_{R\max}\tau)^2 - 2\frac{\delta_R}{\lambda} f_{R\max}\tau\cos(\gamma_R-\theta_R)}\right) \tag{3.75}$$

The space-time correlation function in (3.61) exhibits some interesting properties, which are illustrated in Figure 3.5. As an example, we plot three ST-CFs that are obtained with the following parameters: $\lambda = 0.3$m, $\theta_T = \pi/3$, $\theta_R = \pi/4$, $\gamma_T = \gamma_R = \pi/2$, $\eta_T = \eta_R = 0.45$, $\eta_{TR} = 0.1$, $k_T = 3$, $k_R = 5$, $\mu_T = \mu_R = 0$, $\Delta_T = \Delta_R = 0.001$, $p = 1$, $\tilde{p} = 2$, $q = 1$, and $\tilde{q} = 2$. The first curve plots the ST-CF for the propagation environment with high mobility and NLoS (i.e., $K = 0$, $v_T = 10$ m/s, $v_R = 5$ m/s, $\delta_T = \delta_R = \lambda/2$). The second curve plots the ST-CF for the propagation environment with high mobility and LoS (i.e., $K = 10$, $v_T = 10$ m/s, $v_R = 5$ m/s, $\delta_T = \delta_R = \lambda/2$). Finally, the third curve plots the ST-CF for the propagation environment with low mobility, NLoS, and larger antenna spacings i.e., $K = 0$, $v_T = 1$ m/s, $v_R = 10$ m/s, $\delta_T = \delta_R = \lambda$. It can be observed that the largest space-time correlation value is not necessarily at zero delay, as usually observed in F-to-M cellular channels. Furthermore, it can be observed that the Rice faded channel is more correlated than Rayleigh faded channel.

The space-time correlation functions of the I and Q components of the complex faded envelope can be obtained by substituting (3.58) and (3.59) into (3.61). Using similar derivations as discussed earlier, it can be shown that for a large number of scatterers ($M, N \to \infty$) and nonisotropic scattering, the space-time auto- and cross-correlation functions of the I and Q components of the complex faded envelope are

$$
\begin{aligned}
R_{pq,\tilde{p}\tilde{q}}^{(I,I)}(\delta_T,\delta_R,\tau) &= R_{pq,\tilde{p}\tilde{q}}^{(Q,Q)}(\delta_T,\delta_R,\tau) \\
&= \Re\left\{R_{pq,\tilde{p}\tilde{q}}(\delta_T,\delta_R,\tau)\right\} \\
&= \frac{K}{K+1}\cos\left(\frac{2\pi}{\lambda}(\tilde{p}-p)\delta_T\cos\theta_T - \frac{2\pi}{\lambda}(\tilde{q}-q)\delta_R\cos\theta_R\right. \\
&\qquad\left. + 2\pi\tau f_{T\max}\cos\gamma_T - 2\pi\tau f_{R\max}\cos\gamma_R\right) \\
&+ \frac{\eta_T}{K+1}\cos\left(2\pi\tau f_{R\max}\cos\gamma_R - \frac{2\pi}{\lambda}(\tilde{q}-q)\delta_R\cos\theta_R\right) \\
&\times \Re\left\{\frac{I_0\left(\sqrt{x_{SBT}^2+y_{SBT}^2}\right)}{I_0(k_T)}\right\} \\
&- \frac{\eta_T}{K+1}\sin\left(2\pi\tau f_{R\max}\cos\gamma_R - \frac{2\pi}{\lambda}(\tilde{q}-q)\delta_R\cos\theta_R\right) \\
&\times \mathfrak{F}\left\{\frac{I_0\left(\sqrt{x_{SBT}^2+y_{SBT}^2}\right)}{I_0(k_T)}\right\} \\
&+ \frac{\eta_R}{K+1}\cos\left(\frac{2\pi}{\lambda}(\tilde{p}-p)\delta_T\cos\theta_T - 2\pi\tau f_{T\max}\cos\gamma_T\right) \\
&\times \Re\left\{\frac{I_0\left(\sqrt{x_{SBR}^2+y_{SBR}^2}\right)}{I_0(k_R)}\right\} \\
&- \frac{\eta_R}{K+1}\sin\left(\frac{2\pi}{\lambda}(\tilde{p}-p)\delta_T\cos\theta_T - 2\pi\tau f_{T\max}\cos\gamma_T\right) \\
&\times \mathfrak{F}\left\{\frac{I_0\left(\sqrt{x_{SBR}^2+y_{SBR}^2}\right)}{I_0(k_R)}\right\} \\
&+ \frac{\eta_{TR}}{K+1}\Re\left\{\frac{I_0\left(\sqrt{x_{DB}^2+y_{DB}^2}\right)}{I_0(k_T)}\right\}\Re\left\{\frac{I_0\left(\sqrt{z_{DB}^2+w_{DB}^2}\right)}{I_0(k_R)}\right\} \\
&- \frac{\eta_{TR}}{K+1}\mathfrak{F}\left\{\frac{I_0\left(\sqrt{x_{DB}^2+y_{DB}^2}\right)}{I_0(k_T)}\right\}\mathfrak{F}\left\{\frac{I_0\left(\sqrt{z_{DB}^2+w_{DB}^2}\right)}{I_0(k_R)}\right\}
\end{aligned}
\tag{3.76}
$$

$$
\begin{aligned}
R_{pq,\tilde{p}\tilde{q}}^{(I,Q)}(\delta_T,\delta_R,\tau) &= \mathfrak{F}\left\{R_{pq,\tilde{p}\tilde{q}}(\delta_T,\delta_R,\tau)\right\} \\
&= \frac{K}{K+1}\sin\left(\frac{2\pi}{\lambda}(\tilde{p}-p)\delta_T\cos\theta_T - \frac{2\pi}{\lambda}(\tilde{q}-q)\delta_R\cos\theta_R\right. \\
&\qquad \left. + 2\pi\tau f_{T\max}\cos\gamma_T - 2\pi\tau f_{R\max}\cos\gamma_R\right) \\
&+ \frac{\eta_T}{K+1}\sin\left(2\pi\tau f_{R\max}\cos\gamma_R - \frac{2\pi}{\lambda}(\tilde{q}-q)\delta_R\cos\theta_R\right) \\
&\times \mathfrak{R}\left\{\frac{I_0\left(\sqrt{x_{SBT}^2+y_{SBT}^2}\right)}{I_0(k_T)}\right\} \\
&+ \frac{\eta_T}{K+1}\cos\left(2\pi\tau f_{R\max}\cos\gamma_R - \frac{2\pi}{\lambda}(\tilde{q}-q)\delta_R\cos\theta_R\right) \\
&\times \mathfrak{F}\left\{\frac{I_0\left(\sqrt{x_{SBT}^2+y_{SBT}^2}\right)}{I_0(k_T)}\right\} \\
&+ \frac{\eta_R}{K+1}\sin\left(\frac{2\pi}{\lambda}(\tilde{p}-p)\delta_T\cos\theta_T - 2\pi\tau f_{T\max}\cos\gamma_T\right) \\
&\times \mathfrak{R}\left\{\frac{I_0\left(\sqrt{x_{SBR}^2+y_{SBR}^2}\right)}{I_0(k_R)}\right\} \\
&+ \frac{\eta_R}{K+1}\cos\left(\frac{2\pi}{\lambda}(\tilde{p}-p)\delta_T\cos\theta_T - 2\pi\tau f_{T\max}\cos\gamma_T\right) \\
&\times \mathfrak{F}\left\{\frac{I_0\left(\sqrt{x_{SBR}^2+y_{SBR}^2}\right)}{I_0(k_R)}\right\} \\
&+ \frac{\eta_{TR}}{K+1}\mathfrak{R}\left\{\frac{I_0\left(\sqrt{x_{DB}^2+y_{DB}^2}\right)}{I_0(k_T)}\right\}\mathfrak{F}\left\{\frac{I_0\left(\sqrt{z_{DB}^2+w_{DB}^2}\right)}{I_0(k_R)}\right\} \\
&+ \frac{\eta_{TR}}{K+1}\mathfrak{F}\left\{\frac{I_0\left(\sqrt{x_{DB}^2+y_{DB}^2}\right)}{I_0(k_T)}\right\}\mathfrak{R}\left\{\frac{I_0\left(\sqrt{z_{DB}^2+w_{DB}^2}\right)}{I_0(k_R)}\right\}
\end{aligned} \tag{3.77}
$$

3.1.6.2 Doppler Power Spectrum

The Doppler power spectrum can be obtained by taking the Fourier transform of the ST-CF in (3.61) with respect to time lag τ, which essentially leads to a space-Doppler power spectral density (sD-psd). Applying (3.20) to (3.61) yields the following results [18]:

$$
\begin{aligned}
S_{pq,\tilde{p}\tilde{q}}(\delta_T,\delta_R,f) &= S_{pq,\tilde{p}\tilde{q}}^{LoS}(\delta_T,\delta_R,f) + S_{pq,\tilde{p}\tilde{q}}^{SBT}(\delta_T,\delta_R,f) \\
&\quad + S_{pq,\tilde{p}\tilde{q}}^{SBR}(\delta_T,\delta_R,f) + S_{pq,\tilde{p}\tilde{q}}^{DB}(\delta_T,\delta_R,f)
\end{aligned} \tag{3.78}
$$

where $S^{LoS}_{pq,\tilde{p}\tilde{q}}(\delta_T,\delta_R,f), S^{SBT}_{pq,\tilde{p}\tilde{q}}(\delta_T,\delta_R,f), S^{SBR}_{pq,\tilde{p}\tilde{q}}(\delta_T,\delta_R,f)$, and $S^{DB}_{pq,\tilde{p}\tilde{q}}(\delta_T,\delta_R,f)$ denote the normalized sD-psds of the LoS, single-bounced transmit, single-bounced receive, and double-bounced components, respectively.

By calculating the Fourier transformation of the space-time correlation function in (3.62), we obtain the space-Doppler power spectral density of the LoS component:

$$S^{LoS}_{pq,\tilde{p}\tilde{q}}(\delta_T,\delta_R,f)=\frac{K}{K+1}e^{j\frac{2\pi}{\lambda}[(\tilde{p}-p)\delta_T\cos\theta_T-(\tilde{q}-q)\delta_R\cos\theta_R]}$$
$$\times\,\delta(f-f_{T\max}\cos\gamma_T+f_{R\max}\cos\gamma_R) \tag{3.79}$$

where $(\cdot)$ denotes the Dirac delta function.

By observing that functions $I_0\left(\sqrt{x^2_{SBT}+y^2_{SBT}}\right)$ and $I_0\left(\sqrt{x^2_{SBR}+y^2_{SBR}}\right)$ can be written as

$$I_0\left[j2\pi f_{T\max}\sqrt{(\tau-A_{SBT})^2+\left(\frac{p_{xSBT}q_{ySBT}}{f_{T\max}}-\frac{p_{ySBT}q_{xSBT}}{f_{T\max}}+\frac{w_{ySBT}}{2\pi f_{T\max}}\right)^2}\right]$$

$$I_0\left[j2\pi f_{R\max}\sqrt{(\tau-A_{SBR})^2+\left(\frac{p_{xSBR}q_{ySBR}}{f_{R\max}}-\frac{p_{ySBR}q_{xSBR}}{f_{R\max}}+\frac{w_{ySBR}}{2\pi f_{R\max}}\right)^2}\right]$$

and by using the equality $\int_0^\infty I_0\left(\alpha\sqrt{t^2+u^2}\right)\cos(\beta t)dt=\cosh\left(u\sqrt{\alpha^2+\beta^2}\right)/\sqrt{\alpha^2-\beta^2}$ (eq. 6.677-3 in [13]), we obtain sD-psds of the SBT and SBR components as follows:

$$S^{SBT}_{pq,\tilde{p}\tilde{q}}(\delta_T,\delta_R,f)=\frac{\eta_T}{(K+1)I_0(k_T)}\frac{\exp\{-j2\pi p_{x_{SBR}}-j2\pi(f-f_{R\max}\cos\gamma_R)A_{SBT}\}}{\pi f_{T\max}\sqrt{1-[(f-f_{R\max}\cos\gamma_R)/f_{T\max}]^2}}$$
$$\times\cosh\left[(k_T\sin(\mu_T-\gamma_T)+j2\pi p_{x_{SBT}}q_{y_{SBT}}-j2\pi p_{y_{SBT}}q_{x_{SBT}})\right.$$
$$\left.\times\sqrt{1-[(f-f_{R\max}\cos\gamma_R)/f_{T\max}]^2}\right] \tag{3.80}$$

$$S^{SBR}_{pq,\tilde{p}\tilde{q}}(\delta_T,\delta_R,f)=\frac{\eta_R}{(K+1)I_0(k_R)}\frac{\exp\{j2\pi p_{x_{SBT}}-j2\pi(f+f_{T\max}\cos\gamma_T)A_{SBR}\}}{\pi f_{R\max}\sqrt{1+[(f+f_{T\max}\cos\gamma_T)/f_{R\max}]^2}}$$
$$\times\cosh\left[(k_R\sin(\mu_R-\gamma_R)+j2\pi p_{x_{SBR}}q_{y_{SBR}}-j2\pi p_{y_{SBR}}q_{x_{SBR}})\right.$$
$$\left.\times\sqrt{1-[(f+f_{T\max}\cos\gamma_T)/f_{R\max}]^2}\right] \tag{3.81}$$

where $\cosh(\cdot)$ is the hyperbolic cosine, $|f-f_{R\max}\cos\gamma_R|\le f_{T\max}$, $|f+f_{T\max}\cos\gamma_T|\le f_{R\max}$, and the auxiliary variables are

$$p_{xSBT} = (\tilde{p} - p)(\delta_T / \lambda)\cos\theta_T \tag{3.82}$$

$$q_{xSBT} = \cos\gamma_T \tag{3.83}$$

$$A_{SBT} = (2\pi p_{xSBT} q_{xSBT} + 2\pi p_{ySBT} q_{ySBT} + w_{xSBT}) / 2\pi f_{T\max} \tag{3.84}$$

$$p_{ySBT} = [(\tilde{p} - p)\delta_T \sin\theta_T + (\tilde{q} - q)\delta_R \Delta_T \sin\theta_R] / \lambda \tag{3.85}$$

$$q_{ySBT} \approx \sin\gamma_T \tag{3.86}$$

$$A_{SBR} = (2\pi p_{xSBR} q_{xSBR} + 2\pi p_{ySBR} q_{ySBR} + w_{xSBR}) / 2\pi f_{R\max} \tag{3.87}$$

$$w_{xSBT} = -jk_T \cos(\gamma_T - \mu_T) \tag{3.88}$$

$$w_{ySBT} = -jk_T \sin(\gamma_T - \mu_T) \tag{3.89}$$

$$p_{xSBR} = (\tilde{q} - q)(\delta_R / \lambda)\cos\theta_R \tag{3.90}$$

$$q_{xSBR} = \cos\gamma_R \tag{3.91}$$

$$w_{xSBR} = -jk_R \cos(\gamma_R - \mu_R) \tag{3.92}$$

$$p_{ySBR} = [(\tilde{p} - p)\delta_T \Delta_R \sin\theta_T + (\tilde{q} - q)\delta_R \sin\theta_R] / \lambda \tag{3.93}$$

$$q_{ySBR} \approx \sin\gamma_R \tag{3.94}$$

$$w_{ySBR} = -jk_R \sin(\gamma_R - \mu_R) \tag{3.95}$$

Similarly, the space-Doppler power spectral density of the double-bounced component can be written as [18]

$$\begin{aligned} S^{DB}_{pq,\tilde{p}\tilde{q}}(\delta_T, \delta_R, f) = & \frac{\eta_{TR}}{(K+1) I_0(k_T) \pi f_{T\max} \sqrt{1 - (f / f_{T\max})^2}} \\ & \times e^{-j(2\pi p_{xDB} q_{xDB} + 2\pi p_{yDB} q_{yDB} + jk_T \cos(\gamma_T - \mu_T)) f / f_{T\max}} \\ & \times \cosh\left[(k_T \sin(\mu_T - \gamma_T) + j2\pi p_{xDB} q_{yDB} - j2\pi p_{yDB} q_{xDB}) \right. \\ & \left. \times \sqrt{1 - (f/f_{T\max})^2} \right] \\ & \odot \frac{e^{-j(2\pi p_{zDB} q_{zDB} + 2\pi p_{wDB} q_{wDB} + jk_R \cos(\gamma_R - \mu_R)) f / f_{R\max}}}{I_0(k_R) \pi f_{R\max} \sqrt{1 - (f/f_{R\max})^2}} \\ & \times \cosh\left[(k_R \sin(\mu_R - \gamma_R) + j2\pi p_{zDB} q_{wDB} - j2\pi p_{wDB} q_{zDB}) \right. \\ & \left. \times \sqrt{1 - (f/f_{R\max})^2} \right] \end{aligned} \tag{3.96}$$

where $\odot$ is the convolution operation, $|f| \leq f_{T\max} + f_{R\max}$ and the auxiliary variables are

$$p_{xDB} = (\tilde{p} - p)(\delta_T / \lambda)\cos\theta_T \tag{3.97}$$

$$q_{xDB} = \cos\gamma_T \tag{3.98}$$

$$p_{yDB} = (\tilde{p} - p)(\delta_T / \lambda)\sin\theta_T \tag{3.99}$$

$$q_{yDB} = \sin\gamma_T \tag{3.100}$$

$$p_{zDB} = (\tilde{q} - q)(\delta_R / \lambda)\cos\theta_R \tag{3.101}$$

$$q_{zDB} = \cos\gamma_R \tag{3.102}$$

$$p_{wDB} = (\tilde{q} - q)(\delta_R / \lambda)\sin\theta_R \tag{3.103}$$

$$q_{wDB} = \sin\gamma_R \tag{3.104}$$

In Figures 3.6 and 3.7, we plot the sD-psd, $S_{pq,\tilde{p}\tilde{q}}(\delta_T, \delta_R, f)$, for several MIMO systems with parameters $f_{R\max} = f_{T\max} = 200$ Hz, $\delta_T = \delta_R \in \{0.5\lambda, 1\lambda, 5\lambda, 10\lambda\}$, $\theta_T = \pi/3$, $\theta_R = \pi/4$, $\gamma_T = \gamma_R = \pi/2$, and $L_t = L_r = 2$. In these figures, we analyze the radio propagation in outdoor urban environments, assuming 2-D isotropic scattering ($k_T = k_R = 0$) and LoS ($K = 10$) conditions between the transmitter and receiver. In Figure 3.6, we assume that the single-bounced rays bear more energy than the double-bounced rays (i.e., $\eta_T = \eta_R = 0.4$ and $\eta_{TR} = 0.2$), which is characteristic for propagation on highways. We can observe that this spectrum is similar to the U-shaped spectrum of F-to-M cellular channels. In Figure 3.7, we consider propagation on surface streets where the double-bounced rays bear more energy than the single-bounced rays ($\eta_T = \eta_R = 0.1$ and $\eta_{TR} = 0.8$). In this case, the sD-psd differs from the U-shaped spectrum of cellular channels. Note that this agrees with measurements, because similar effects are observed with measured Doppler spectra [25, 26].

3.1.6.3 Level Crossing Rate and Average Fade Duration for V-to-V Channels

The level crossing rate (LCR), $L(R)$, is introduced in Section 2.3.5.1 and defined as the rate at which the signal envelope crosses level R in the positive direction. The LCR can be written as [9]

$$L(R) = \int_0^\infty \dot{\beta} p_{\beta,\dot{\beta}}(R, \dot{\beta}) d\dot{\beta} \tag{3.105}$$

where $\beta = |h(t)|$ is the envelope level, $\dot{\beta} = |\dot{h}(t)|$ is the envelope slope, and $p_{\beta,\dot{\beta}}(R, \dot{\beta})$ is the joint probability density function (pdf) of the envelope level and the envelope slope.

Here, we will deploy the model described in Section 3.1.5 to calculate the LCR. First, we observe that the joint pdf $p_{\beta,\dot{\beta}}(\beta, \dot{\beta})$ cannot be represented as a product of

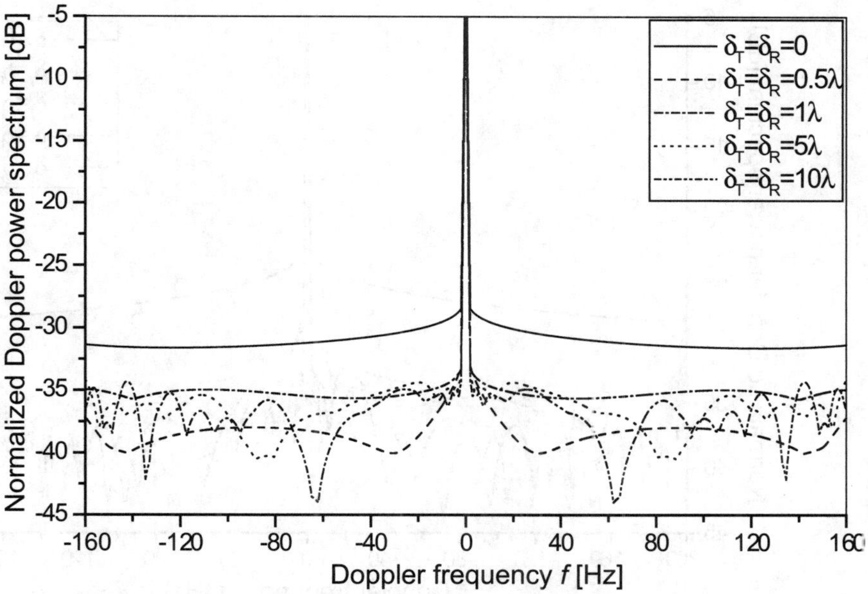

Figure 3.6 The normalized space-Doppler power spectra characteristic for the outdoor M-to-M micro-cells. The curves are obtained with parameters $s = f_{Rmax}/f_{Tmax} = 1$, $\theta_T = \pi/3$, $\theta_R = \pi/4$, $\gamma_T = \gamma_R = \pi/2$, $\eta_T = \eta_R = 0.4$, $\eta_{TR} = 0.2$, $k_T = k_R = 0$, $K = 10$, and $L_t = L_r = 2$.

a Rice pdf $p_{|h(t)|}(\beta)$ and a Gaussian pdf $p_{|\dot{h}(t)|}(\dot{\beta})$, because the I and Q components of the complex faded envelope $h(t)$ are correlated (i.e., $R_{\dot{h}i,hq}(\tau)$ given in (3.77) is not zero). Then, the LCR for LoS conditions (Ricean fading, correlated in-phase and quadrature components) can be written as [27]

$$L(R) = \frac{R\sqrt{2\chi}}{\pi^{3/2} b_0} e^{-(R^2+\delta^2)/(2b_0)}$$

$$\int_0^{\pi/2} \cosh\left(\frac{R\delta}{b_0}\cos\theta\right)\left[e^{-(\xi\delta\sin\theta)^2} + \sqrt{\pi}\,\xi\delta\sin\theta\,\mathrm{erf}(\xi\delta\sin\theta)\right]d\theta \qquad (3.106)$$

where $\cosh(\cdot)$ is the hyperbolic cosine function and $\mathrm{erf}(\cdot)$ is the error function. The rest of the parameters are $\delta = \sqrt{K/(K+1)}$, $\chi = b_2 - b_1^2/b_0$, $\xi = (2\pi f_{pq}^{LoS} - b_1/b_0)/\sqrt{2\chi}$, and

$$b_0 \triangleq E\left[h_i(t)^2\right] = E\left[h_q(t)^2\right] \qquad (3.107)$$

$$b_1 \triangleq E\left[h_i(t)\dot{h}_q(t)\right] = E\left[h_q(t)\dot{h}_i(t)\right] \qquad (3.108)$$

$$b_2 \triangleq E\left[\dot{h}_i(t)^2\right] = E\left[\dot{h}_q(t)^2\right] \qquad (3.109)$$

where $\dot{h}_i(t)$ and $\dot{h}_q(t)$ are first derivatives of $h_i(t)$ and $h_q(t)$ with respect to time t. When 2-D nonisotropic scattering is characterized using the von Mises pdfs defined in (3.23) and (3.24), closed-form expressions for b_0, b_1, and b_2 can be obtained by substituting (3.58) into (3.107). Parameter b_0 then becomes

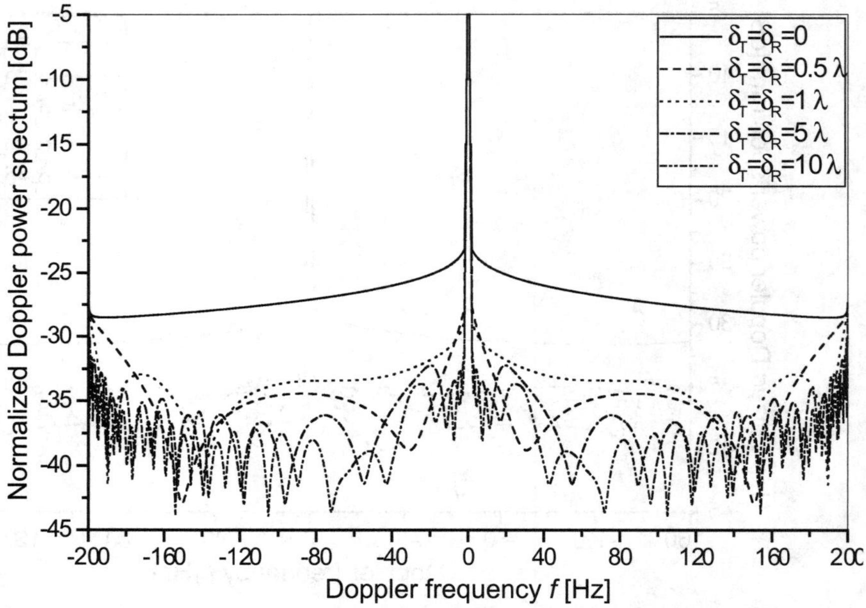

Figure 3.7 The normalized space-Doppler power spectra characteristic for the outdoor M-to-M macro-cells. The curves are obtained with parameters $s = f_{R\max}/f_{T\max} = 1$, $\theta_T = \pi/3$, $\theta_R = \pi/4$, $\gamma_T = \gamma_R = \pi/2$, $\eta_T = \eta_R = 0.1$, $\eta_{TR} = 0.8$, $k_T = k_R = 0$, K=10, and $L_t = L_r = 2$.

$$
\begin{aligned}
2b_0 &= 2b_0^{SBT} + 2b_0^{SBR} + 2b_0^{DB} \\
&= \frac{1}{K+1}
\end{aligned}
\tag{3.110}
$$

where b_0^{SBT}, b_0^{SBR}, and b_0^{DB} are, respectively,

$$
\begin{aligned}
2b_0^{SBT} &= \frac{\eta_T}{K+1}\int_{-\pi}^{\pi} \frac{\exp[k_T\cos(\alpha_T - \mu_T)]}{2\pi I_0(k_T)} d\alpha_T \\
&= \frac{\eta_T}{K+1}
\end{aligned}
\tag{3.111}
$$

$$
\begin{aligned}
2b_0^{SBR} &= \frac{\eta_R}{K+1}\int_{-\pi}^{\pi} \frac{\exp[k_R\cos(\alpha_R - \mu_R)]}{2\pi I_0(k_R)} d\alpha_R \\
&= \frac{\eta_R}{K+1}
\end{aligned}
\tag{3.112}
$$

$$
\begin{aligned}
2b_0^{DB} &= \frac{\eta_{TR}}{K+1}\int_{-\pi}^{\pi}\int_{-\pi}^{\pi} \frac{\exp[k_T\cos(\alpha_T - \mu_T)]}{2\pi I_0(k_T)} \frac{\exp[k_R\cos(\alpha_R - \mu_R)]}{2\pi I_0(k_R)} d\alpha_T d\alpha_R \\
&= \frac{\eta_{TR}}{K+1}
\end{aligned}
\tag{3.113}
$$

Similarly, parameters b_1 and b_2 can be obtained as follows:

$$
b_n = b_n^{SBT} + b_n^{SBR} + b_n^{DB}
\tag{3.114}
$$

where $n \in \{1, 2\}$ and b_n^{SBT}, b_n^{SBR}, and b_n^{DB} are, respectively,

$$b_n^{SBT} = (2\pi)^n\, b_0^{SBT} \int_{-\pi}^{\pi} \frac{\exp[k_T \cos(\alpha_T - \mu_T)]}{2\pi I_0(k_T)} \times \left[f_{T\max} \cos(\alpha_T - \gamma_T) + f_{R\max}(\cos\gamma_R - \Delta_T \sin\gamma_R \sin\alpha_T)\right]^n d\alpha_T \tag{3.115}$$

$$b_n^{SBR} = (2\pi)^n\, b_0^{SBR} \int_{-\pi}^{\pi} \frac{\exp[k_R \cos(\alpha_R - \mu_R)]}{2\pi I_0(k_R)} \times \left[f_{R\max} \cos(\alpha_R - \gamma_R) + f_{T\max}(\cos\gamma_T + \Delta_R \sin\gamma_T \sin\alpha_R)\right]^n d\alpha_R \tag{3.116}$$

$$b_n^{DB} = (2\pi)^n\, b_0^{DB} \int_{-\pi}^{\pi}\int_{-\pi}^{\pi} \frac{\exp[k_T \cos(\alpha_T - \mu_T)]}{2\pi I_0(k_T)} \frac{\exp[k_R \cos(\alpha_R - \mu_R)]}{2\pi I_0(k_R)} \times \left[f_{R\max} \cos(\alpha_R - \gamma_R) + f_{T\max} \cos(\alpha_T - \gamma_T)\right]^n d\alpha_R d\alpha_T \tag{3.117}$$

Using the equality $\int_{-\pi}^{\pi} e^{-jm\theta + jz\sin\theta} d\theta = 2\pi J_m(z)$ (eq. 8.411 in [13]), where $J_m(\cdot)$ is the mth-order Bessel function of the first kind, parameters b_1^{SBT}, b_1^{SBR}, b_1^{DB}, b_2^{SBT}, b_2^{SBR}, and b_2^{DB} become:

$$b_1^{SBT} = 2\pi b_0^{SBT} \left\{ f_{T\max} \cos(\mu_T - \gamma_T) \frac{I_1(k_T)}{I_0(k_T)} + f_{R\max}\left[\cos\gamma_R + \Delta_T \sin\gamma_R \sin\mu_T \frac{I_1(k_T)}{I_0(k_T)}\right]\right\} \tag{3.118}$$

$$b_1^{SBR} = 2\pi b_0^{SBR} \left\{ f_{R\max} \cos(\mu_R - \gamma_R) \frac{I_1(k_R)}{I_0(k_R)} + f_{T\max}\left[\cos\gamma_T - \Delta_R \sin\gamma_T \sin\mu_R \frac{I_1(k_R)}{I_0(k_R)}\right]\right\} \tag{3.119}$$

$$b_1^{DB} = 2\pi b_0^{DB} \left\{ f_{T\max} \cos(\mu_T - \gamma_T) \frac{I_1(k_T)}{I_0(k_T)} + f_{R\max} \cos(\mu_R - \gamma_R) \frac{I_1(k_R)}{I_0(k_R)}\right\} \tag{3.120}$$

$$\begin{aligned} b_2^{SBT} = (2\pi)^2 b_0^{SBT} \Bigg\{ & f_{T\max}^2 \frac{I_0(k_T) + \cos(2(\mu_T - \gamma_T)) I_2(k_T)}{2I_0(k_T)} \\ & + f_{T\max} f_{R\max} \Delta_T \sin\gamma_R \left(\frac{I_2(k_T)}{I_0(k_T)} \sin(2\mu_T - \gamma_T) + \sin\gamma_T \right) \\ & + 2 f_{T\max} f_{R\max} \cos\gamma_R \cos(\mu_T - \gamma_T) \frac{I_1(k_T)}{I_0(k_T)} + f_{R\max}^2 \cos\gamma_R^2 + f_{R\max}^2 \Delta_T^2 \sin\gamma_R^2 \\ & \times \frac{I_0(k_T) - \cos(2\mu_T) I_2(k_T)}{2I_0(k_T)} + f_{R\max}^2 \Delta_T \sin(2\gamma_R) \sin\mu_T \frac{I_1(k_T)}{I_0(k_T)} \Bigg\} \end{aligned} \tag{3.121}$$

$$b_2^{SBR} = (2\pi)^2 b_0^{SBR} \left\{ f_{\mathrm{R\,max}}^2 \frac{I_0(k_R) + \cos(2(\mu_R - \gamma_R))I_2(k_R)}{2I_0(k_R)} \right.$$

$$+ f_{\mathrm{R\,max}} f_{\mathrm{T\,max}} \Delta_R \sin\gamma_T \left(\frac{I_2(k_R)}{I_0(k_R)} \sin(2\mu_R - \gamma_R) + \sin\gamma_R \right)$$

$$- 2 f_{\mathrm{R\,max}} f_{\mathrm{T\,max}} \cos\gamma_T \cos(\mu_R - \gamma_R) \frac{I_1(k_R)}{I_0(k_R)} + f_{\mathrm{T\,max}}^2 \cos\gamma_T^2 + f_{\mathrm{T\,max}}^2 \Delta_R^2 \sin\gamma_T^2$$

$$\left. \times \frac{I_0(k_R) - \cos(2\mu_R) I_2(k_R)}{2I_0(k_R)} - f_{\mathrm{T\,max}}^2 \Delta_R \sin(2\gamma_T) \sin\mu_R \frac{I_1(k_R)}{I_0(k_R)} \right\} \tag{3.122}$$

$$b_2^{DB} = (2\pi)^2 b_0^{DB} \left\{ f_{\mathrm{Tmax}}^2 \frac{I_0(k_T) + \cos(2(\mu_T - \gamma_T))I_2(k_T)}{2I_0(k_T)} \right.$$

$$+ 2 f_{\mathrm{Tmax}} f_{\mathrm{Rmax}} \cos(\mu_T - \gamma_T) \frac{I_1(k_T)}{I_0(k_T)} \cos(\mu_R - \gamma_R) \frac{I_1(k_R)}{I_0(k_R)}$$

$$\left. + f_{\mathrm{Rmax}}^2 \frac{I_0(k_R) + \cos(2(\mu_R - \gamma_R))I_2(k_R)}{2I_0(k_R)} \right\} \tag{3.123}$$

The derived LCR in (3.106) can be simplified for many special cases. For example, the LCR for NLoS conditions can be obtained from (3.106) by setting $K = 0$; that is,

$$L(R) = \sqrt{\frac{b_2}{b_0} - \frac{b_1^2}{b_0^2}} \frac{R\, e^{-R^2}}{\sqrt{\pi}} \tag{3.124}$$

Furthermore, for an NLoS isotropic scattering environment where only double-bounced rays are present, the LCR further simplifies to [28]

$$L(R) = \sqrt{2\pi (f_{\mathrm{T\,max}}^2 + f_{\mathrm{R\,max}}^2)} \cdot R \cdot e^{R^2} \tag{3.125}$$

by setting $K = 0$, $k_T = 0$, $k_R = 0$.

The average fade duration (AFD) is the average time that the envelope spends below level R. When a LoS component is present, the AFD can be written as [9]

$$T(R) = \frac{1 - Q\left(\sqrt{2K}, \sqrt{2(K+1)R^2}\right)}{L(R)} \tag{3.126}$$

where $Q(a,b)$ is the Marcum Q function. For NLoS conditions, the AFD in (3.126) simplifies to

$$T(R) = \frac{1 - e^{-R^2}}{L(R)} \tag{3.127}$$

Furthermore, for an NLoS isotropic scattering environment where only double-bounced rays are present, the AFD further simplifies to

$$T(R) = \frac{1}{\sqrt{2\pi(f_{\text{Tmax}}^2 + f_{\text{Rmax}}^2)}} e^{R^2-1} \tag{3.128}$$

3.1.7 Estimation of Mobile Velocities in V-to-V Channels

Estimation of a mobile user's speed plays an important role in a variety of wireless applications. For example, it is used for energy-efficient routing in wireless networks [29], for mobility and resource management in wireless multimedia networks [30], and for handoff, adaptive transmission, dynamic channel assignment, localization [9, 31–33]. In the last two decades, a F-to-M cellular mobile user's speed estimation has been extensively studied in the literature (e.g., [9, 31–41] and references therein. Speed estimation techniques can be classified into four major classes: *crossing-based* [9, 34], *covariance-based* [31, 3], *power spectrum-based* [39], and *maximum likelihood-based* [33, 40] methods. Among these, crossing-based and covariance-based methods have low complexity and have been extensively used to estimate the speed of the mobile user (relative to the fixed base-station).

For V-to-V channels, the first crossing-based estimator that jointly estimates velocities of both the T_x and R_x in V-to-V systems has recently been proposed in [41]. In adhoc networking, where no node is a master node, knowing the actual (not just relative to one node) velocities of other nodes in the network is very important. Without knowing the velocity and direction of other nodes, it is hard to predict which route is the optimal choice. Similarly, if the goal is energy-efficient routing, it is useful to have a list of all velocities in the network, rather than having only relative velocities with respect to a given node. In addition, in safety applications, knowing the velocity of the car that is approaching us at the intersection may help us decide to speed up or to slow down to avoid a collision. Knowing only the relative velocity between the vehicles cannot help in this situation (i.e., we would not know whether we will reach the intersection simultaneously).

The velocities estimator in [41] is designed for narrowband wireless V-to-V communications. To derive closed-form expressions for estimation of T_x and R_x velocities, a noise-free isotropic scattering environment with no LoS component is assumed and the SISO version of the model in Section 3.1.5 is utilized. In the absence of noise and LoS component, the received complex faded envelope $h(t)$ is equal to the multipath component of the channel impulse response $h^d(t) = h^{SET}(t) + h^{SBR}(t) + h^{DB}(t)$, i.e., $h(t) = h^d(t)$. Furthermore, since the T_x and R_x velocities v_T and v_R are proportional to the maximum Doppler angular frequencies $\omega_{\text{Tmax}} = 2\pi f_{\text{Tmax}}$ and $\omega_{\text{Rmax}} = 2\pi f_{\text{Rmax}}$, respectively, in the rest of this section we will focus on the estimation of ω_{Tmax} and ω_{Rmax} to simplify the notation.

First, we define $L_{h_i^d}(0,T)$ as the count of how many times the in-phase component $h_i^d(t) = \Re\{h^d(t)\}$ crosses the zero-level threshold with a positive slope, counted over the time interval $(0,T]$. Similarly, we define $L_{h_q^d}(0,T)$ as the count of positive-slope zero-level crossings for the quadrature component $h_q^d(t) = \Im\{h^d(t)\}$ counted over the time interval $(0,T]$. Here, $\Re\{\cdot\}$ denotes the real part operation and $\Im\{\cdot\}$ denotes the imaginary part operation. Also, we define $M_{h_i^d}(T)$ as the number of maxima in $h_i^d(t)$ during the time interval $(0,T]$. Similarly, $M_{h_q^d}(T)$ is the number of maxima in $h_q^d(t)$ during the time interval $(0,T]$.

To estimate the T_x and R_x angular frequencies ω_{Tmax} and ω_{Rmax}, we combine the number of zero-crossings with the number of maxima in the in-phase and the quadrature component of $h^d(t)$. This crossing-based estimation approach is chosen because it does not depend on the type of propagation environment (i.e., it does not depend on the parameters η_T, η_R, and η_{TR}).

For stationary zero-mean Gaussian processes $h_i^d(t)$ and $h_q^d(t)$, the number of zero crossings can be written as [34]

$$E[L_{h_i^d}(0,T)] = E[L_{h_q^d}(0,T)] = \frac{T}{2\pi}\sqrt{\frac{b_2}{b_0}} \tag{3.129}$$

and the number of maxima can be written as [34]

$$E[M_{h_i^d}(T)] = E[M_{h_q^d}(T)] = \frac{T}{2\pi}\sqrt{\frac{b_4}{b_2}} \tag{3.130}$$

The parameters b_0, b_2, and b_4 are defined as [42]

$$b_0 \triangleq \mathrm{E}\left[h_i^d(t)^2\right] = \mathrm{E}\left[h_q^d(t)^2\right] \tag{3.131}$$

$$b_2 \triangleq \mathrm{E}\left[\dot{h}_i^d(t)^2\right] = \mathrm{E}\left[\dot{h}_q^d(t)^2\right] \tag{3.132}$$

$$b_4 \triangleq \mathrm{E}\left[\ddot{h}_i^d(t)^2\right] = \mathrm{E}\left[\ddot{h}_q^d(t)^2\right] \tag{3.133}$$

where $\mathrm{E}[\cdot]$ denotes the statistical expectation operator, $\dot{h}_i^d(t)$ and $\dot{h}_q^d(t)$ denote the first derivative of $h_i^d(t)$ and $h_q^d(t)$ with respect to time t, and $\ddot{h}_i^d(t)$ and $\ddot{h}_q^d(t)$ denote the second derivative of $h_i^d(t)$ and $h_q^d(t)$ with respect to time t. Using the model in Section 3.1.5, the parameters b_0, b_2, and b_4 can be written as [41]

$$b_0 = \frac{1}{2(K+1)} \tag{3.134}$$

$$b_2 = \frac{b_0}{2}\left[\omega_{\mathrm{Tmax}}^2 + \omega_{\mathrm{Rmax}}^2\right] \tag{3.135}$$

$$b_4 = b_0\left[\frac{3}{8}\omega_{\mathrm{Tmax}}^4 + \frac{3}{8}\omega_{\mathrm{Rmax}}^4 + \frac{3}{2}\omega_{\mathrm{Rmax}}^2\omega_{\mathrm{Tmax}}^2\right] \tag{3.136}$$

Substituting (3.134), (3.135), and (3.136) into (3.129) and (3.130), the number of zero crossings and the number of maxima in $h_i^d(t)$ or $h_q^d(t)$ become, respectively,

$$\begin{aligned} E[L_{h_i^d}(0,T)] &= E[L_{h_q^d}(0,T)] \\ &= \frac{T}{2\sqrt{2}\pi}\sqrt{\omega_{\mathrm{Tmax}}^2 + \omega_{\mathrm{Rmax}}^2} \end{aligned} \tag{3.137}$$

$$E[M_{h_i^d}(T)] = E[M_{h_q^d}(T)] = \frac{T}{2\pi}\sqrt{\frac{\frac{3}{8}\omega_{T\max}^4 + \frac{3}{8}\omega_{R\max}^4 + \frac{3}{2}\omega_{T\max}^2\omega_{R\max}^2}{\frac{1}{2}\omega_{T\max}^2 + \frac{1}{2}\omega_{R\max}^2}} \tag{3.138}$$

Solving this set of equations with respect to $\omega_{T\max}$ and $\omega_{R\max}$, the following expressions for potential estimators can be obtained:

$$\hat{\omega}_{T_i} = \frac{2\pi}{T}\sqrt{L_{h_i^d}(0,T)^2 + \sqrt{3L_{h_i^d}(0,T)^4 - 4L_{h_i^d}(0,T)^2 M_{h_i^d}(T)^2/3}} \tag{3.139}$$

$$\hat{\omega}_{R_i} = \frac{2\pi}{T}\sqrt{L_{h_i^d}(0,T)^2 - \sqrt{3L_{h_i^d}(0,T)^4 - 4L_{h_i^d}(0,T)^2 M_{h_i^d}(T)^2/3}} \tag{3.140}$$

$$\hat{\omega}_{T_q} = \frac{2\pi}{T}\sqrt{L_{h_q^d}(0,T)^2 + \sqrt{3L_{h_q^d}(0,T)^4 - 4L_{h_q^d}(0,T)^2 M_{h_q^d}(T)^2/3}} \tag{3.141}$$

$$\hat{\omega}_{R_q} = \frac{2\pi}{T}\sqrt{L_{h_q^d}(0,T)^2 - \sqrt{3L_{h_q^d}(0,T)^4 - 4L_{h_q^d}(0,T)^2 M_{h_q^d}(T)^2/3}} \tag{3.142}$$

From the analysis of candidate estimators in (3.139)–(3.142), it is not clear if the estimator in (3.139) outperforms the estimator in (3.141) or vice versa. Furthermore, the estimated velocities have to have nonnegative, noncomplex values. To ensure that the estimated velocities have nonnegative, noncomplex values and to capture the best performance between the estimators in (3.139) and (3.141), it has been empirically found that the best results are obtained by combining the estimators in (3.139) and (3.141) as described in the following algorithm: if

$\Re\left\{\sqrt{3L_{h_i^d}(0,T)^4/T^4 - 4L_{h_i^d}(0,T)^2 M_{h_i^d}(T)^2/(3T^4)}\right\} < 2$

or $\Re\left\{\sqrt{3L_{h_q^d}(0,T)^4/T^4 - 4L_{h_q^d}(0,T)^2 M_{h_q^d}(T)^2/(3T^4)}\right\} < 2$, the angular velocity $\hat{\omega}_T$ is estimated as [41]

$$\hat{\omega}_T = \Re\left\{\sqrt{\min(\hat{\omega}_{Ti}^2, \hat{\omega}_{Tq}^2)}\right\} \tag{3.143}$$

where $\hat{\omega}_{Ti}$ and $\hat{\omega}_{Tq}$ are defined as in (3.139) and (3.141), respectively, and $\min(\cdot)$ denotes the minimum function. On the other hand, if

$\Re\left\{\sqrt{3L_{h_i^d}(0,T)^4/T^4 - 4L_{h_i^d}(0,T)^2 M_{h_i^d}(T)^2/(3T^4)}\right\} \geq 2$ or

$\Re\left\{\sqrt{3L_{h_q^d}(0,T)^4/T^4 - 4L_{h_q^d}(0,T)^2 M_{h_q^d}(T)^2/(3T^4)}\right\} \geq 2$, the angular Doppler frequency $\hat{\omega}_T$ is estimated as

$$\hat{\omega}_T = \Re\left\{\sqrt{\max(\hat{\omega}_{Ti}^2, \hat{\omega}_{Tq}^2)}\right\} \tag{3.144}$$

where max (·) denotes the maximum function. Using the obtained results , the angular Doppler frequency $\hat{\omega}_R$ is estimated as

$$\hat{\omega}_R = \Re\left\{\sqrt{\frac{4}{T^2}[\max(\mathrm{E}[L_{h_i^d}(0,T)],\mathrm{E}[L_{h_q^d}(0,T)])]^2 - \mathrm{E}[\hat{\omega}_T]^2}\right\} \tag{3.145}$$

To illustrate the performance of the velocities estimator, Figure 3.8 shows the effect of white Gaussian noise on the velocities estimator. Here, an isotropic scattering environment with no LoS component is assumed. The estimation error for the velocities estimator is evaluated using the root mean square error (RMSE) given by

$$\mathrm{E}[(\hat{f}_T - f_{T\max})^2]^{1/2} = \left(\mathrm{Var}[\hat{f}_T] + \mathrm{E}[\hat{f}_T] - f_{T\max})^2\right)^{1/2} \tag{3.146}$$

where Var[·] denotes the variance operation, $(\mathrm{E}[\hat{f}_T] - f_{T\max})^2$ represents the bias, and $\hat{f}_T = \hat{\omega}_T/(2\pi)$ and $f_{T\max} = \omega_{T\max}/(2\pi)$ denote the estimated and actual Doppler frequencies at the T_x, respectively. Figure 3.8 plots the RMSE of estimated Doppler frequency $\hat{f}_T$ versus actual Doppler frequency, when signal-to-noise (SNR) ratio is set to SNR = 10 dB. The results show that the velocities estimator performs relatively well in the presence of noise for all estimated velocities.

3.1.8 Estimation of Direction of Movement in V-to-V Channels

As pointed out in Section 3.1.7, knowing velocities of both the T_x and R_x plays an important role in adhoc networking, energy efficient routing, and safety applications. Furthermore, these applications also require knowledge of the relative direction of movement of the T_x and R_x.

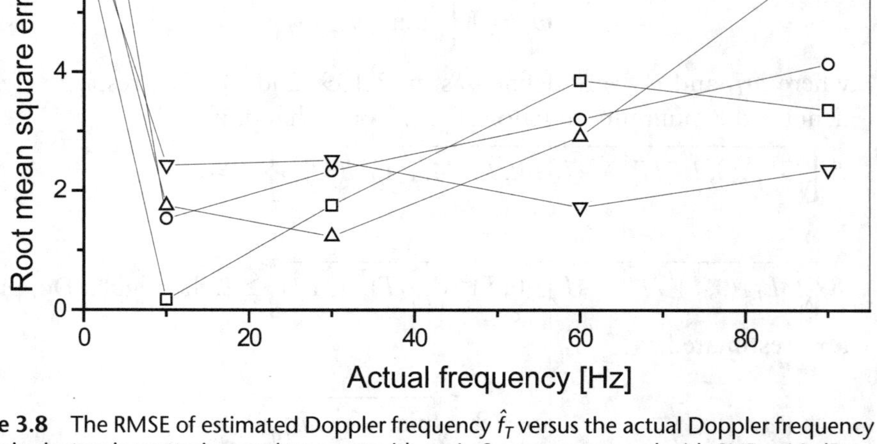

Figure 3.8 The RMSE of estimated Doppler frequency $\hat{f}_T$ versus the actual Doppler frequency $f_{T\max}$, in a noisy isotropic scattering environment with no LoS component and with SNR = 10 dB.

This section describes the first direction-of-movement estimator, which was recently proposed in [41]. To derive the closed-form expression for the estimation of the relative direction of movement, we make similar assumptions to those we used for the velocities estimator. A noise-free isotropic scattering environment with no LoS component is assumed and the SISO version of model in Section 3.1.5 is utilized. The cross-correlation between the in-phase component of $h(t)$ (i.e., $h_i(t) = \Re\{h(t)\}$) and the quadrature component of $h(t)$ (i.e., $h_q(t) = \Im\{h(t)\}$) is used to estimate the relative direction of movement of T_x and R_x. The cross-correlation between the in-phase and and quadrature component of $h(t)$ is given in (3.77). For a small time-lag $\tau \to 0$, the cross-correlation function in (3.77) can be approximated as [41]

$$R^{IQ}(\tau) \approx \frac{0.5}{K+1}\Bigg[K\omega_T\tau\cos\gamma_T - K\omega_R\tau\cos\gamma_R + \eta_T\tau\omega_{R\max}\cos\gamma_R - \eta_R\tau\omega_{T\max}\cos\gamma_T + (\eta_R + \eta_{TR})\Im\left\{\frac{I_0\left(\sqrt{k_R^2 - \tau^2\omega_{R\max}^2 - j2k_R\tau\omega_{R\max}\cos(\mu_R - \gamma_R)}\right)}{I_0(k_R)}\right\}\Bigg] \tag{3.147}$$

Without loss of generality, we assume that the coordinate system is aligned with the receiver (i.e., $\gamma_R = 0$) and that the angle γ_T represents the relative direction of movement of T_x and R_x. Then, (3.147) can be simplified to

$$R^{IQ}(\tau) = \frac{0.5}{K+1} \times \left[K\omega_T\tau\cos\gamma_T - K\omega_R\tau + \eta_T\tau\omega_{R\max} - \eta_R\tau\omega_{T\max}\cos\gamma_T + (\eta_R + \eta_{TR})\Im\{B\}\right] \tag{3.148}$$

where $B = I_0\left(\sqrt{k_R^2 - \tau^2\omega_{R\max}^2 - j2k_R\tau\omega_{R\max}\cos(\mu_R - \gamma_R)}\right)\Big/ I_0(k_R)$. Finally, solving (3.148) with respect to γ_T, the relative direction of movement of T_x and R_x is estimated as

$$\hat{\gamma}_T = \arccos\left(\frac{-2(K+1)R^{IQ}(\tau) + \eta_T\hat{\omega}_R\tau - K\hat{\omega}_R\tau + (\eta_T + \eta_{TR})\Im\{\hat{B}\}}{\eta_R\hat{\omega}_T\tau - K\hat{\omega}_T\tau}\right) \tag{3.149}$$

where $0 \le \hat{\gamma}_T \le 180°$ and the angular Doppler frequencies $\hat{\omega}_T$ and $\hat{\omega}_R$ are estimated using (3.143), (3.144), and (3.145), respectively.

To illustrate the performance of this estimator, Figure 3.9 shows the effect of white Gaussian noise on the direction estimator. Here, an isotropic scattering environment with no LoS component is assumed. The estimation error for the direction estimator is evaluated using the the root mean square error (RMSE) given by

$$\mathrm{E}\left[(\hat{\gamma}_T - \gamma_T)^2\right]^{1/2} = \left(\mathrm{Var}[\hat{\gamma}_T] + \left(\mathrm{E}[\hat{\gamma}_T] - \gamma_T\right)^2\right)^{1/2} \tag{3.150}$$

where $\left(\mathrm{E}[\hat{\gamma}_T] - \gamma_T\right)^2$ represents the bias. Figure 3.9 plots the RMSE of the estimated direction $\hat{\gamma}_T$ versus the actual direction γ_T, when signal-to-noise (SNR) ratio is SNR = {5,10,15,20} dB and the estimated results for Doppler frequencies

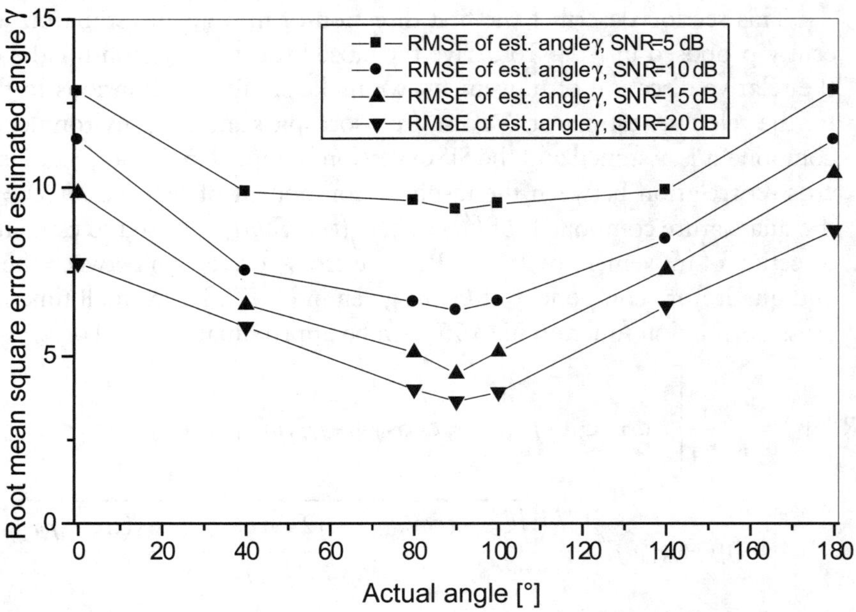

Figure 3.9 The RMSE of the estimated direction $\hat{\gamma}_T$ versus the actual direction γ_T, in a noisy isotropic scattering environment with no LoS component and with SNR = {5,10, 15, 20} dB.

$f_{T\max} = f_{R\max} = 100$ Hz are selected. As expected, the performance of the direction estimator improves as the SNR increases. However, even for low SNR, (e.g., SNR= 5 dB), the RMSE does not exceed 15°. Although a 15° error may not be acceptable for localization purposes, it can be tolerable in adhoc or energy-efficient routing algorithms, where only general orientation is needed (e.g., we typically need to know if vehicles are moving in the same, orthogonal, or opposite directions).

3.1.9 Modeling of V-to-V Frequency-Selective Multipath Fading Channels

The V-to-V channel models described Sections 3.1.3 and 3.1.5 assume that the fields incident on the T_x or R_x antennas are composed of a number of waves traveling only in the *horizontal* plane. This assumption is valid only for certain environments (e.g., rural areas) and seems inappropriate for urban environments where the T_x and R_x antenna arrays are often located in close proximity to and lower than the surrounding buildings. Scattered waves may propagate by diffraction from the edges of buildings down to the street and, thus, not necessarily travel horizontally.

Zajić et al. were the first to propose a 3-D model for MIMO V-to-V multipath fading channels [43]. They introduced a *two-cylinder* geometrical propagation model for *narrowband* MIMO V-to-V multipath fading channels. To completely characterize MIMO V-to-V multipath fading channels, they also developed the 3-D model for *wideband* MIMO V-to-V channels [44] that will be detailed in this section.

The concentric-cylinders model [44] with LoS, single-bounced at transmitter (SBT), single-bounced at the receiver (SBR), and double-bounced (DB) rays for a MIMO V-to-V channel with $L_t = L_r = 2$ antenna elements is shown in Figure 3.10.

Vertical concentric cylinders are chosen to model the scattering surfaces and diffracting edges because most of the structures in outdoor metropolitan environments (e.g., buildings, highway dividers) have vertical surfaces.

The concentric-cylinders model defines four vertical cylinders, two around the T_x and another two around the R_x, as shown in Figure 3.10. Around the transmitter, M fixed omnidirectional scatterers occupy the volume between cylinders of radii R_{t1} and R_{t2}. The model assumes that the M scatterers lie on L cylindric surfaces of radii $R_{t1} \le R_t^{(l)} \le R_{t2}$, where $1 \le l \le L$. The lth cylindric surface contains $M^{(l)}$ fixed omnidirectional scatterers, and the (m,l)th transmit scatterer is denoted by $S_T^{(m,l)}$, where $1 \le m \le M^{(l)}$. Similarly, around the receiver, N fixed omnidirectional scatterers occupy a volume between cylinders of radii R_{r1} and R_{r2}, where they lie on F cylindric surfaces of radii $R_{r1} \le R_r^{(k)} \le R_{r2}$, where $1 \le k \le F$. The kth cylindric surface contains $N^{(k)}$ fixed omnidirectional scatterers, and the (n,k)th receive scatterer is denoted by $S_R^{(n,k)}$, where $1 \le n \le N^{(k)}$.

The distance between the centers of the T_x and R_x cylinders is D. It is also assumed that the radii R_{t2} and R_{r2} are sufficiently smaller than the distance D (local scattering condition), while the distance D is smaller than $4R_{t1}R_{r1}L_r/(\lambda(L_t - 1)(L_r - 1))$, where λ denotes the carrier wavelength, so that the channel does not experience keyhole behavior [20]. Distance $\varepsilon_{p,m,l}$ is the distance between the pth antenna $A_T^{(p)}$ and the (m,l)th scatterer placed around the T_x (i.e., $A_T^{(p)} - S_T^{(m,l)}$). Similarly,

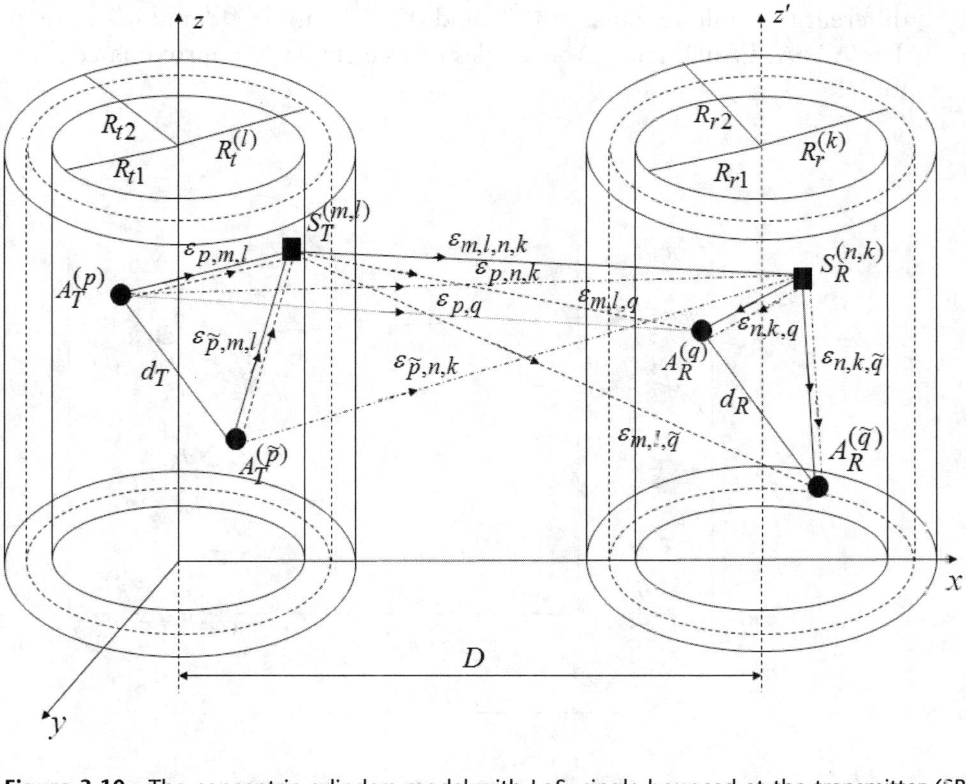

Figure 3.10 The concentric-cylinders model with LoS, single-bounced at the transmitter (SBT), single-bounced at the receiver (SBR), and double-bounced (DB) rays for a wideband MIMO V-to-V channel with $L_t = L_r = 2$ antenna elements.

symbols $\varepsilon_{m,l,q}$, $\varepsilon_{p,n,k}$, $\varepsilon_{n,k,q}$, $\varepsilon_{m,l,n,k}$, and $\varepsilon_{p,q}$ denote distances $S_T^{(m,l)} - A_R^{(q)}$, $A_T^{(p)} - S_R^{(n,k)}$, $S_R^{(n,k)} - A_R^{(q)}$, $S_T^{(m,l)} - S_R^{(n,k)}$, and $A_T^{(p)} - A_R^{(q)}$, respectively, as shown in Figure 3.10.

For ease of reference, Figure 3.11 shows one of the L cylindric surfaces around the T_x and one of F cylindric surfaces around the R_x, and, to reduce clutter, only shows the geometry of double-bounced rays. The spacing between antenna elements at the T_x and R_x is denoted by d_T and d_R, respectively. It is assumed that d_T and d_R are much smaller than the radii R_{t1} and R_{r1} (i.e., the T_x and R_x antenna arrays are much smaller than the T_x-to-scatterer and scatterer-to-R_x distances). The T_x and R_x antenna heights are denoted h_T and h_R, respectively; angles θ_T and θ_R in Figure 3.11 describe the orientation of the T_x and R_x antenna array in the x-y plane, respectively; and angles ψ_T and ψ_R describe the elevation of the T_x's antenna array and the R_x's antenna array relative to the x-y plane, respectively. The T_x and R_x are moving with speeds v_T and v_R in directions described by angles γ_T and γ_R in the x-y plane (relative to the x-axis), respectively.

Symbols $\alpha_T^{(m,l)}$ and $\alpha_T^{(n,k)}$ represent the azimuth angles of departures (AAoD) of the waves that impinge on the scatterers $S_T^{(m,l)}$ and $S_R^{(n,k)}$, respectively, whereas $\alpha_R^{(m,l)}$ and $\alpha_R^{(n,k)}$ are the azimuth angles of arrivals (AAoA) of the waves scattered from $S_T^{(m,l)}$ and $S_R^{(n,k)}$, respectively. Similarly, the symbols $\beta_T^{(m,l)}$ and $\beta_R^{(n,k)}$ denote the elevation angle of departure (EAoD) and the elevation angle of arrival (EAoA), respectively. Finally, symbols α_{Tp}^{LoS} (not shown in Figure 3.11 for figure clarity) and α_{Rq}^{LoS} denote the AAoDs and AAoAs of the LoS paths. This model assumes $\{d_T, d_R\} \ll \{R_{t1}, R_{r1}\}$, which allows us to approximate the angles of departure from different transmit antennas $\alpha_T^{(p,m,l)}$ and $\alpha_T^{(p,n,k)}$ with $\alpha_T^{(m,l)}$ and $\alpha_T^{(n,k)}$, respectively. The AAoA, EAoD, and EAoA angles can be similarly approximated. For ease of

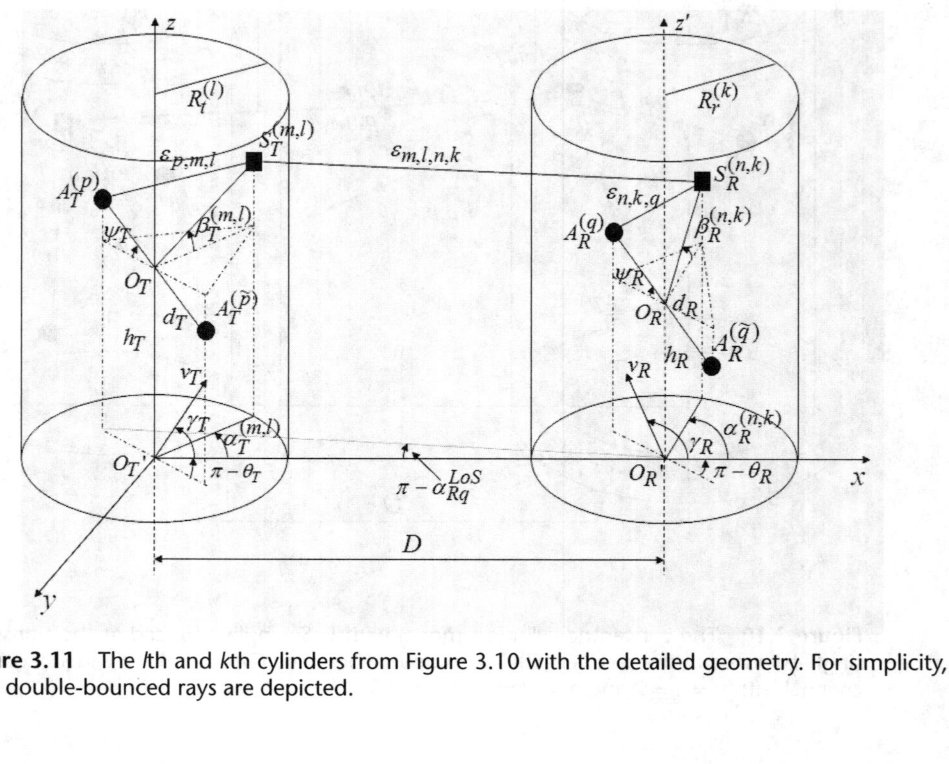

Figure 3.11 The lth and kth cylinders from Figure 3.10 with the detailed geometry. For simplicity, only double-bounced rays are depicted.

reference, the parameters introduced in this model are summarized in Table 3.2. We can observe from the 3-D geometrical model in Figure 3.10 that some waves from the T_x antenna elements may traverse directly to the R_x antenna elements (LoS rays), while others are single-bounced at the T_x (i.e., waves from the T_x antenna elements scatter from the scatterers located around the T_x before arriving at the R_x antenna elements), single-bounced at the R_x (i.e., waves from the T_x antenna elements scatter from the scatterers located around the R_x before arriving at the R_x antenna elements), and/or double-bounced (i.e., waves from the T_x antenna elements impinge on the scatterers located around the T_x and scatter from the scatterers located around the R_x before arriving at the R_x antenna elements). Hence, the input delay-spread function of the link $A_T^{(p)} - A_R^{(q)}$ can be written as a superposition of the LoS, SBT, SBR, and DB rays:

$$h_{pq}(t,\tau) = h_{pq}^{LoS}(t,\tau) + h_{pq}^{SBT}(t,\tau) + h_{pq}^{SBR}(t,\tau) + h_{pq}^{DB}(t,\tau) \tag{3.151}$$

To simplify further analysis, the time-variant transfer function is used instead of the input delay-spread function. The time-variant transfer function is the Fourier transform of the input delay-spread function (see Section 2.3.6 for more details) and can be written as

Table 3.2 Definition of the Parameters Used in the Concentric-Cylinders Geometrical Model

D	Distance between the center of the Tx and Rx cylinders
$R_t^{(l)}, R_r^{(k)}$	Radius of the lth Tx and kth Rx cylinder, respectively.
d_T, d_R	Spacing between two adjacent antenna elements at the T_x and R_x, respectively.
θ_T, θ_R	Orientation of the T_x and R_x antenna array in the x-y plane (relative to the x-axis), respectively.
ψ_T, ψ_R	Elevation of the T_x's and R_x's antenna array relative to the x-y plane, respectively.
v_T, v_R	Velocities of the T_x and R_x, respectively.
γ_T, γ_R	Moving directions of the T_x and R_x, in the x-y plane (relative to the x-axis), respectively.
h_T, h_R	Distances $d(O_T, O_T')$ and $d(O_R, O_R')$, respectively.
$R_{t1}, R_{t2}, R_{r1}, R_{r2}$	Minimum and maximum radii of the cylinders around the T_x and R_x respectively.
$\alpha_T^{(m,l)}, \alpha_T^{(n,k)}$	Azimuth angles of departure (AAoD) of the waves that impinge on the scatters $S_T^{(m,l)}$ and $S_R^{(n,k)}$, respectively.
$\alpha_R^{(m,l)}, \alpha_R^{(n,k)}$	Azimuth angles of arrival (AAoA) of the waves scattered from $S_T^{(m,l)}$ and $S_R^{(n,k)}$ respectively.
$\beta_T^{(m,l)}, \beta_T^{(n,k)}$	Elevation angles of departure (EAoD) of the waves that impinge on the scattered from $S_T^{(m,l)}$ and $S_R^{(n,k)}$, respectively.
$\beta_R^{(m,l)}, \beta_R^{(n,k)}$	Elevation angles of arrival (EAoA) of the waves scattered from $S_T^{(m,l)}$ and $S_R^{(n,k)}$, respectively.
$\alpha_{Rq}^{LoS}, \alpha_{Tp}^{LoS}$	AAoA and the AAoD of the LoS paths, respectively.
$\varepsilon_{p,m,l}, \varepsilon_{m,l,q}$, $\varepsilon_{p,n,k}, \varepsilon_{n,k,q}$, $\varepsilon_{m,l,n,k}$ and ε_{pq}	Distances $d\left(A_T^{(p)}, S_T^{(m,l)}\right)$, $d\left(S_T^{(m,l)}, A_R^{(q)}\right)$, $d\left(A_T^{(p)}, S_R^{(n,k)}\right)$, $d\left(S_R^{(n,k)}, A_R^{(q)}\right)$, $d\left(S_T^{(m,l)}, S_R^{(n,k)}\right)$, and $d\left(A_T^{(p)}, A_R^{(q)}\right)$, respectively.

$$T_{pq}(t,f)=\mathcal{F}_{\tau}\left\{h_{pq}(t,\tau)\right\}=T_{pq}^{LoS}(t,f)+T_{pq}^{SBT}(t,f)+T_{pq}^{SBR}(t,f)+T_{pq}^{DB}(t,f) \tag{3.152}$$

where the four components (LoS, SBT, SBR, DB) are

$$T_{pq}^{LoS}(t,f)=a_{pq}^{LoS}e^{\left(j2\pi f_{pq}^{LoS}t-j2\pi f\tau_{pq}^{LoS}-j\frac{2\pi}{\lambda}d_{pq}^{LoS}\right)} \tag{3.153}$$

$$T_{pq}^{SBT}(t,f)=\lim_{M\to\infty}\sum_{l=1}^{L}\sum_{m=1}^{M^{(l)}}a_{pq,ml}^{SBT}e^{\left(j2\pi f_{ml}^{SBT}t-j2\pi f\tau_{ml}^{SBT}-j\frac{2\pi}{\lambda}d_{pq,ml}^{SBT}+j\phi_{pq,ml}^{SBT}\right)} \tag{3.154}$$

$$T_{pq}^{SBR}(t,f)=\lim_{N\to\infty}\sum_{k=1}^{F}\sum_{n=1}^{N^{(k)}}a_{pq,nk}^{SBR}e^{\left(j2\pi f_{nk}^{SBR}t-j2\pi f\tau_{nk}^{SBR}-j\frac{2\pi}{\lambda}d_{pq,nk}^{SBR}+j\phi_{pq,nk}^{SBR}\right)} \tag{3.155}$$

$$T_{pq}^{DB}(t,f)=\lim_{M,N\to\infty}\sum_{l,m=1}^{L,M^{(l)}}\sum_{k,n=1}^{F,N^{(k)}}a_{pq,mlnk}^{DB}e^{\left(j2\pi f_{mlnk}^{DB}t-j2\pi f\tau_{mlnk}^{DB}-j\frac{2\pi}{\lambda}d_{pq,mlnk}^{DB}+j\phi_{pq,mlnk}^{DB}\right)} \tag{3.156}$$

and the amplitudes, Doppler shift frequencies, phrases, path lengths, and time delays in this model can be calculated as follows:

- *Amplitudes*. The total signal power between the pth transmit and the qth receive antenna is split between the LoS and NLoS components: the power allocated to the LoS component between the pth transmit and qth receive antenna is $K_{pq}/(1 + K_{pq})$, while the total power allocated to the NLoS components between the pth transmit and qth receive antenna is $1/(1 + K_{pq})$. For $\{d_T, d_R\} \ll \{R_{t1}, R_{r1}\}$, we can assume that the Ricean fading factor is the same between all antenna elements (i.e., $K_{pq} = K$ for all $p \in \{1, \dots, L_t\}$ and $q \in \{1, \dots, L_r\}$). Furthermore, it is assumed that the relative power allocated to the SBT, SBR, and DB components is, respectively, η_T, η_R, and η_{TR} such that $\eta_T + \eta_R + \eta_{TR} = 1$. We also assume that a power P_{pq} is channelled from transmit antenna p to receive antenna q. These parameters have to be either set during simulations or estimated from measurements. Having all this in mind, the amplitudes can be written as follows:

$$a_{pq}^{LoS}=\Omega_{pq}\sqrt{\frac{K}{K+1}} \tag{3.157}$$

$$\begin{aligned}a_{pq,ml}^{SBT}&=\sqrt{\frac{1}{K+1}}\sqrt{\frac{\eta_T}{M}}\frac{\sqrt{P_{pq}}\lambda}{4\pi}\left[\left|d\left(O_T,S^{(m,l)}\right)\right|+\left|d\left(S^{(m,l)},O_R\right)\right|\right]^{-\gamma/2}\\&\approx\sqrt{\frac{1}{K+1}}\sqrt{\frac{\eta_T}{M}}\Omega_{pq}\left(1-\frac{\gamma}{2}\frac{R_t^{(l)}}{D}\right)\end{aligned} \tag{3.158}$$

$$a_{pq,nk}^{SBR} = \sqrt{\frac{1}{K+1}}\sqrt{\frac{\eta_R}{N}}\frac{\sqrt{P_{pq}}\lambda}{4\pi}\left[\left|d\left(O_T, s^{(n,k)}\right)\right| + \left|d\left(s^{(n,k)}, O_R\right)\right|\right]^{-\gamma/2}$$
$$\approx \sqrt{\frac{1}{K+1}}\sqrt{\frac{\eta_R}{N}}\Omega_{pq}\left(1 - \frac{\gamma}{2}\frac{R_r^{(k)}}{D}\right) \tag{3.159}$$

$$a_{pq,mlnk}^{DB} = \sqrt{\frac{1}{K+1}}\sqrt{\frac{\eta_{TR}}{MN}}\frac{\sqrt{P_{pq}}\lambda}{4\pi}$$
$$\times\left[\left|d\left(O_T', s^{(m,l)}\right)\right| + \left|d\left(s^{(m,l)}, s^{(n,k)}\right)\right| + \left|d\left(s^{(n,k)}, O_R'\right)\right|\right]^{\gamma/2}$$
$$\approx \sqrt{\frac{1}{K+1}}\sqrt{\frac{\eta_{TR}}{MN}}\Omega_{pq}\left(1 - \frac{\gamma}{2}\frac{R_t^{(l)} + R_r^{(k)}}{2D}\right) \tag{3.160}$$

where γ is the pathloss exponent, $d(\cdot,\cdot)$ denotes distance between two points, and $\Omega_{pq} = D^{-\gamma/2}\sqrt{P_{pq}}\,\lambda/4\pi$.

- *Doppler shifts*. Doppler shifts depend on the geometrical relation between directions of movement of T_x and R_x (i.e., γ_T and γ_R and the directions of AAoDs, EAoDs, AAoAs, and EAoAs). For this model, the Doppler shifts can be written as

$$f_{pq}^{LoS} = f_{T\max}\cos\left(\pi - \alpha_{Rq}^{LoS} - \gamma_T\right) + f_{R\max}\cos\left(\alpha_{Rq}^{LoS} - \gamma_R\right) \tag{3.161}$$

$$f_{ml}^{SBT} = f_{T\max}\cos\left(\alpha_T^{(m,l)} - \gamma_T\right)\cos\beta_T^{(m,l)} + f_{R\max}\cos\left(\alpha_R^{(m,l)} - \gamma_R\right)\cos\beta_R^{(m,l)}$$
$$= f_{T\max}\cos\left(\alpha_T^{(m,l)} - \gamma_T\right)\cos\beta_T^{(m,l)}$$
$$+ f_{R\max}\left(\cos\gamma_R - \Delta_T^{(l)}\sin\gamma_R\sin\alpha_T^{(m,l)}\right)\cos\left(\Delta_T^{(l)}\beta_T^{(m,l)} + \Delta_H/D\right) \tag{3.162}$$

$$f_{nk}^{SBR} = f_{T\max}\cos\left(\alpha_T^{(n,k)} - \gamma_T\right)\cos\beta_T^{(n,k)} + f_{R\max}\cos\left(\alpha_R^{(n,k)} - \gamma_R\right)\cos\beta_R^{(n,k)}$$
$$= f_{T\max}\left(\Delta_R^{(k)}\sin\gamma_T\sin\alpha_R^{(n,k)} + \cos\gamma_T\right)\cos\left(\Delta_R^{(k)}\beta_R^{(n,k)} - \Delta_H/D\right)$$
$$+ f_{R\max}\cos\left(\alpha_R^{(n,k)} - \gamma_R\right)\cos\beta_R^{(n,k)} \tag{3.163}$$

$$f_{mlnk}^{DB} = f_{T\max}\cos\left(\alpha_T^{(m,l)} - \gamma_T\right)\cos\beta_T^{(m,l)} + f_{R\max}\cos\left(\alpha_R^{(n,k)} - \gamma_R\right)\cos\beta_R^{(n,k)} \tag{3.164}$$

where frequencies $f_{T\max} = v_T/\lambda$ and $f_{R\max} = v_R/\lambda$ are the maximum Doppler frequencies associated with the T_x and R_x, respectively, $\Delta_T^{(l)} = R_t^{(l)}/D$, $\Delta_R^{(k)} = R_r^{(k)}/D$, $\Delta_H = h_T - h_R$, and λ is the carrier wavelength. For $\{d_T, d_R\} \ll \{R_{t1}, R_{r1}\}$, the LoS AAoDs are $\alpha_{Tp}^{LoS} = \pi - \alpha_{Rq}^{LoS}$, and AAoAs are approximately equal to π (i.e., $\alpha_{Rq}^{LoS} = \pi$). Since the location of the scatterers is not known a priori, the AAoDs ($\alpha_T^{(m,l)}$ and $\alpha_T^{(n,k)}$), the EAoDs ($\beta_T^{(m,l)}$ and $\beta_T^{(n,k)}$), the AAoAs ($\alpha_R^{(m,l)}$ and $\alpha_R^{(n,k)}$), and the EAoAs ($\beta_R^{(m,l)}$ and $\beta_R^{(n,k)}$)

are modeled as random variables. However, note that when the number of scatterers at each end is approaching infinity, these discrete random variables can be characterized as continuous random variables with given pdfs. As in Section 3.1.5, note that double-bounced rays have the angles of departure ($\alpha_T^{(m,l)}$ and $\beta_T^{(m,l)}$) independent from the angles of arrival ($\alpha_R^{(n,k)}$ and $\beta_R^{(n,k)}$) [20]. On the other hand, single-bounced rays have angles of arrival ($\alpha_R^{(m,l)}$ and $\beta_R^{(m,l)}$) that are dependent on the angles of departure ($\alpha_T^{(m,l)}$ and $\beta_T^{(m,l)}$) and angles of departure ($\alpha_T^{(n,k)}$ and $\beta_T^{(n,k)}$) that are dependent on the angles of arrival ($\alpha_R^{(n,k)}$ and $\beta_R^{(n,k)}$) [44].

- *Phases.* As in Section 3.1.5, we first note that the phase shift introduced by the (m,l)th scatterer located near the T_x is independent of the phase shift introduced by the (n,k)th scatterer located near the R_x. Then, without loss of generality, we can assume that the phases $\phi_{pq,ml}^{SBT}$, $\phi_{pq,nk}^{SBR}$, and $\phi_{pq,mlnk}^{DB}$ are independent random variables. Here, it is assumed that they are uniformly distributed on the interval $[0,2\pi)$ and independent from any other random variable.
- *Path lengths.* From Figure 3.10, the path lengths of the respective waves can be written as

$$d_{pq}^{LoS} = \varepsilon_{pq} \tag{3.165}$$

$$d_{pq,ml}^{SBT} = \varepsilon_{p,m,l} + \varepsilon_{m,l,q} \tag{3.166}$$

$$d_{pq,nk}^{SBR} = \varepsilon_{p,n,k} + \varepsilon_{n,k,q} \tag{3.167}$$

$$d_{pq,mlnk}^{DB} = \varepsilon_{p,m,l} + \varepsilon_{m,l,n,k} + \varepsilon_{n,k,q} \tag{3.168}$$

For outdoor propagation, which is typical for V-to-V communications, we can assume the radii R_{t2} and R_{r2} are much smaller than the distance D between the T_x and R_x (local scattering condition). Under this assumption, distances $\varepsilon_{p,q}$, $\varepsilon_{p,m,l}$, $\varepsilon_{m,l,q}$, $\varepsilon_{p,n,k}$, $\varepsilon_{n,k,q}$ and $\varepsilon_{m,l,n,k}$ can be approximated as follows:

$$\begin{aligned}\varepsilon_{p,q} \approx D - \frac{L_t+1-2p}{2} d_T \cos\psi_T \cos\theta_T \\ - \frac{L_r+1-2q}{2} d_R \cos\psi_R \cos\left(\alpha_{Rq}^{LoS} - \theta_R\right)\end{aligned} \tag{3.169}$$

$$\begin{aligned}\varepsilon_{p,m,l,} \approx R_t^{(l)} - \frac{L_t+1-2p}{2} \\ \times\left[d_{T_x} \cos\alpha_T^{(m,l)} \cos\beta_T^{(m,l)} + d_{Ty} \sin\alpha_T^{(m,l)} \cos\beta_T^{(m,l)} + d_{Tz} \sin\beta_T^{(m,l)}\right]\end{aligned} \tag{3.170}$$

$$\varepsilon_{m,l,q} \approx D - \frac{L_r + 1 - 2q}{2} d_R \cos\psi_R \left[\Delta_T^{(l)} \sin\theta_R \sin\alpha_T^{(m,l)} - \cos\theta_R \right] \tag{3.171}$$

$$\varepsilon_{p,n,k} \approx D - \frac{L_t + 1 - 2p}{2} d_T \cos\psi_T \left[\Delta_R^{(k)} \sin\theta_T \sin\alpha_R^{(n,k)} + \cos\theta_T \right] \tag{3.172}$$

$$\varepsilon_{n,k,q} \approx R_r^{(k)} - \frac{L_r + 1 - 2q}{2} \times \left[d_{Rx} \cos\alpha_R^{(n,k)} \cos\beta_R^{(n,k)} + d_{Ry} \sin\alpha_R^{(n,k)} \cos\beta_R^{(n,k)} + d_{Rz} \sin\beta_R^{(n,k)} \right] \tag{3.173}$$

$$\varepsilon_{m,l,n,k} \approx D \tag{3.174}$$

where p and q take values from the sets $p \in \{1,\ldots,L_t\}$ and $q \in \{1,\ldots,L_r\}$, respectively, $d_{Tx} = d_T \cos\psi_T \cos\theta_T$, $d_{Ty} = d_T \cos\psi_T \sin\theta_T$, $d_{Tz} = d_T \sin\psi_T$, $d_{Rx} = d_R \cos\psi_R \cos\theta_R$, $d_{Ry} = d_R \cos\psi_R \sin\theta_R$, $d_{Rz} = d_R \sin\psi_R$.

- *Multipath time delays*. The time delays of the multipath components can be inferred from the traversed distances and are given by the following expressions:

$$\tau_{pq}^{LoS} = \frac{\sqrt{D^2 + \Delta_H^2}}{c_0} \tag{3.175}$$

$$\tau_{m,l}^{SBT} = \frac{D + R_t^{(l)} \left(1 - \cos\alpha_T^{(m,l)}\right)}{c_0 \cos\beta_T^{(m,l)}} \tag{3.176}$$

$$\tau_{n,k}^{SBR} = \frac{D + R_r^{(k)} \left(1 + \cos\alpha_R^{(n,k)}\right)}{c_0 \cos\beta_R^{(n,k)}} \tag{3.177}$$

$$\tau_{m,l,n,k}^{DB} = \frac{D}{c_0} + \frac{R_t^{(l)} \left(1 - \cos\alpha_T^{(m,l)}\right)}{c_0 \cos\beta_T^{(m,l)}} + \frac{R_r^{(k)} \left(1 + \cos\alpha_R^{(n,k)}\right)}{c_0 \cos\beta_R^{(n,k)}} \tag{3.178}$$

where c_0 is the speed of light.

3.1.10 Spectral-Spatial-Temporal Fading Characteristics of V-to-V Frequency-Selective Multipath Fading Channels

Using the model described in Section 3.1.9, we can now derive the key temporal, spatial, and spectral characteristics of V-to-V frequency-selective multipath fading channels.

3.1.10.1 Space-Time-Frequency Correlation Function (STF-CF)

The normalized STF-CF between two time-variant transfer functions $T_{pq}(t,f)$ and $T_{\tilde{p}\tilde{q}}(t,f)$, which were defined in (3.152), is

$$R_{pq,\tilde{p}\tilde{q}}(\Delta t,\Delta f)=\frac{\mathrm{E}[T_{pq}(t,f)^{*}T_{\tilde{p}\tilde{q}}(t+\Delta t,f+\Delta f)]}{\sqrt{\mathrm{Var}\left[T_{pq}(t,f)^{*}\right]\mathrm{Var}\left[T_{\tilde{p}\tilde{q}}(t+\Delta t,f+\Delta f)\right]}} \tag{3.179}$$

where $(\cdot)^{*}$ is the complex conjugate operation, $\mathrm{E}[\cdot]$ is the statistical expectation operator, $\mathrm{Var}[\cdot]$ is the statistical variance operator, and indices $p,\tilde{p}\in\{1,\ldots,L_t\}$ and $q,\tilde{q}\in\{1,\ldots,L_t\}$. Since $T_{pq}^{SBT}(t,f)$ $T_{pq}^{SBR}(t,f)$, and $T_{pq}^{DB}(t,f)$ are independent zero-mean complex Gaussian random processes, (3.179) can be simplified to [44]

$$\begin{aligned}R_{pq,\tilde{p}\tilde{q}}(\Delta t,\Delta f)&=R_{pq,\tilde{p}\tilde{q}}^{LoS}(\Delta t,\Delta f)+R_{pq,\tilde{p}\tilde{q}}^{SBT}(\Delta t,\Delta f)\\&\quad+R_{pq,\tilde{p}\tilde{q}}^{SBR}(\Delta t,\Delta f)+R_{pq,\tilde{p}\tilde{q}}^{DB}(\Delta t,\Delta f)\end{aligned} \tag{3.180}$$

where $R_{pq,\tilde{p}\tilde{q}}^{LoS}(\Delta t,\Delta f)$, $R_{pq,\tilde{p}\tilde{q}}^{SBT}(\Delta t,\Delta f)$, $R_{pq,\tilde{p}\tilde{q}}^{SBR}(\Delta t,\Delta f)$, and $R_{pq,\tilde{p}\tilde{q}}^{DB}(\Delta t,\Delta f)$ are the normalized STF-CFs of the LoS, SBT, SBR, and DB components, respectively, and are defined as

$$R_{pq,\tilde{p}\tilde{q}}^{LoS}(\Delta t,\Delta f)=\frac{\mathrm{E}\left[T_{pq}^{LoS}(t,f)^{*}T_{\tilde{p}\tilde{q}}^{LoS}(t+\Delta t,f+\Delta f)\right]}{\Omega_{pq}^{2}} \tag{3.181}$$

$$R_{pq,\tilde{p}\tilde{q}}^{SBT}(\Delta t,\Delta f)=\frac{\mathrm{E}\left[T_{pq}^{SBT}(t,f)^{*}T_{\tilde{p}\tilde{q}}^{SBT}(t+\Delta t,f+\Delta f)\right]}{\Omega_{pq}^{2}} \tag{3.182}$$

$$R_{pq,\tilde{p}\tilde{q}}^{SBR}(\Delta t,\Delta f)=\frac{\mathrm{E}\left[T_{pq}^{SBR}(t,f)^{*}T_{\tilde{p}\tilde{q}}^{SBR}(t+\Delta t,f+\Delta f)\right]}{\Omega_{pq}^{2}} \tag{3.183}$$

$$R_{pq,\tilde{p}\tilde{q}}^{DB}(\Delta t,\Delta f)=\frac{\mathrm{E}\left[T_{pq}^{DB}(t,f)^{*}T_{\tilde{p}\tilde{q}}^{DB}(t+\Delta t,f+\Delta f)\right]}{\Omega_{pq}^{2}} \tag{3.184}$$

To characterize the $\mathrm{T_x}$ and $\mathrm{R_x}$ azimuth angles, we use the von Mises pdfs in (3.23) and (3.24), respectively. Prior work uses several different scatterer distributions, such as uniform, cosine, and Gaussian, to characterize the random elevation angles of departure and arrival, $\beta_T^{(m,l)}$ and $\beta_R^{(n,k)}$. We use the pdf [45]

$$p(\varphi)=\begin{cases}\dfrac{\pi}{4\,|\,\varphi_m\,|}\cos\left(\dfrac{\pi}{2}\dfrac{\varphi}{\varphi_m}\right), & |\varphi|\le|\varphi_m|\le\dfrac{\pi}{2}\\ 0 & ,\ \text{otherwise}\end{cases} \tag{3.185}$$

where φ_m is the maximum elevation angle and lies in the range $0^{\circ}\le|\varphi_m|\le 20^{\circ}$ [46]. Such maximum elevation angles have experimentally been found to be typical for the "street-canyon" type of propagation, which is prevalent in V-to-V communications where both the T_x and R_x are in motion and equipped with low elevation antennas. The pdfs for the T_x and R_x elevation angles are denoted as

$$p(\beta_T) = \pi \cos\left(\pi\beta_T / (2\beta_{T_m})\right) / \left(4\left|\beta_{T_m}\right|\right) \tag{3.186}$$

$$p(\beta_R) = \pi \cos\left(\pi\beta_R / (2\beta_{R_m})\right) / \left(4\left|\beta_{R_m}\right|\right) \tag{3.187}$$

respectively. Finally the radii $R_t^{(l)}$ and $R_r^{(k)}$ are characterized using the pdf $p(R) = 2R / \left(R_2^2 - R_1^2\right)$. Such a distribution implies that scatterers in the horizontal plane will have a uniform density between the cylinders R_{t1} and R_{t2} at the transmitter and cylinders R_{r1} and R_{r2} at the receiver, if scattering is isotropic in the horizontal plane. The pdfs for the T_x and R_x radii are denoted as

$$p(R_t) = \frac{2R_t}{R_{t2}^2 - R_{t1}^2} \tag{3.188}$$

$$p(R_r) = \frac{2R_r}{R_{r2}^2 - R_{r1}^2} \tag{3.189}$$

respectively.

It has been shown [44] that the STF-CFs of the LoS, SBT, SBR, and DB components can be closely approximated as

$$R_{pq,\tilde{p}\tilde{q}}^{LoS}(\Delta t, \Delta f) \approx \frac{K}{K+1} e^{j\frac{2\pi}{\lambda}[(p-\tilde{p})d_{Tx} - (q-\tilde{q})d_{Rx}] + j2\pi\Delta t[f_{T\max}\cos\gamma_T - f_{R\max}\cos\gamma_R]} \times e^{-j\frac{2\pi}{c_0}\Delta f\sqrt{D^2+\Delta_H^2}} \tag{3.190}$$

$$\begin{aligned} R_{pq,\tilde{p}\tilde{q}}^{SBT}(\Delta t, \Delta f) \approx & \frac{\cos\left(\frac{2\pi}{\lambda}\beta_{Tm}(p-\tilde{p})d_{Tz}\right)}{1-\left(\frac{4\beta_{Tm}(p-\tilde{p})d_{Tz}}{\lambda}\right)^2} e^{-j\frac{2\pi}{\lambda}(q-\tilde{q})d_{Rx} - j2\pi\Delta t f_{R\max}\cos\gamma_R} \\ & \times \frac{\eta_T}{I_0(k_T)(K+1)} \int_{R_{t1}}^{R_{t2}} \left(1-\gamma\frac{R_t}{D}\right) e^{-j\frac{2\pi}{c_0}\Delta f(D+R_t)} \\ & \times I_0\left(\sqrt{x_{SBT}^2 + y_{SBT}^2}\right) \frac{2R_t}{R_{t2}^2 - R_{t1}^2} dR_t \end{aligned} \tag{3.191}$$

$$R_{pq,\tilde{p}\tilde{q}}^{SBR}(\Delta t,\Delta f)\approx\frac{\cos\left(\frac{2\pi}{\lambda}\beta_{Rm}(q-\tilde{q})d_{Rz}\right)}{1-\left(\frac{4\beta_{Rm}(q-\tilde{q})d_{Rz}}{\lambda}\right)^2}e^{j\frac{2\pi}{\lambda}(p-\tilde{p})d_{Tx}+j2\pi\Delta t f_{T\max}\cos\gamma_T}$$
$$\times\frac{\eta_R}{I_0(k_R)(K+1)}\int_{R_{r1}}^{R_{r2}}\left(1-\gamma\frac{R_r}{D}\right)e^{-j\frac{2\pi}{c_0}\Delta f(D+R_r)}$$
$$\times I_0\left(\sqrt{x_{SBR}^2+y_{SBR}^2}\right)\frac{2R_r}{R_{r2}^2-R_{r1}^2}dR_r \quad (3.192)$$

$$R_{pq,\tilde{p}\tilde{q}}^{DB}(\Delta t,\Delta f)\approx A_{DB}\int_{R_{t1}}^{R_{t2}}2e^{-j\frac{2\pi}{c_0}\Delta fR_t}R_tI_0\left(\sqrt{x_{DB}^2+y_{DB}^2}\right)dR_t$$
$$\times\int_{R_{r1}}^{R_{r2}}e^{-j\frac{2\pi}{c_0}\Delta fR_r}R_rI_0\left(\sqrt{w_{DB}^2+z_{DB}^2}\right)\left(1-\gamma\frac{R_r}{D}\right)dR_r$$
$$+A_{DB}\int_{R_{r1}}^{R_{r2}}2e^{-j\frac{2\pi}{c_0}\Delta fR_r}R_rI_0\left(\sqrt{w_{DB}^2+z_{DB}^2}\right)dR_r$$
$$\times\int_{R_{t1}}^{R_{t2}}e^{-j\frac{2\pi}{c_0}\Delta fR_t}R_tI_0\left(\sqrt{x_{DB}^2+y_{DB}^2}\right)\left(1-\gamma\frac{R_t}{D}\right)dR_t \quad (3.193)$$

where parameters x_{SBT}, y_{SBT}, x_{SBR}, y_{SBR}, x_{DB}, y_{DB}, z_{DB}, w_{DB}, and A_{DB} are

$$x_{SBT}\approx j(2\pi/\lambda)(p-\tilde{p})d_{Tx}+j2\pi\Delta tf_{T\max}\cos\gamma_T$$
$$+j2\pi\Delta fR_t/c_0+k_T\cos\mu_T \quad (3.194)$$

$$y_{SBT}\approx j2\pi\left[(p-\tilde{p})d_{Ty}/\lambda+(q-\tilde{q})d_{Ry}\Delta_T/\lambda\right.$$
$$\left.+\Delta t\left(f_{T\max}\sin\gamma_T+f_{R\max}\Delta_T\sin\gamma_R\right)\right]+k_T\sin\mu_T \quad (3.195)$$

$$x_{SBR}\approx j(2\pi/\lambda)(q-\tilde{q})d_{Rx}+j2\pi\Delta tf_{R\max}\cos\gamma_R$$
$$-j2\pi\Delta fR_r/c_0+k_R\cos\mu_R \quad (3.196)$$

$$y_{SBR}\approx j2\pi\left[(q-\tilde{q})d_{Ry}/\lambda+(p-\tilde{p})d_{Ty}\Delta_R/\lambda\right.$$
$$\left.+\Delta t\left(f_{R\max}\sin\gamma_R+f_{T\max}\Delta_R\sin\gamma_T\right)\right]+k_R\sin\mu_R \quad (3.197)$$

$$x_{DB}\approx j(2\pi/\lambda)(p-\tilde{p})d_{Tx}+j2\pi\Delta tf_{T\max}\cos\gamma_T$$
$$+j2\pi\Delta fR_t/c_0+k_T\cos\mu_T \quad (3.198)$$

$$y_{DB}\approx j(2\pi/\lambda)(p-\tilde{p})d_{Ty}+j2\pi\Delta tf_{T\max}\sin\gamma_T+k_T\sin\mu_T \quad (3.199)$$

$$z_{DB}\approx j(2\pi/\lambda)(q-\tilde{q})d_{Rx}+j2\pi\Delta tf_{R\max}\cos\gamma_R$$
$$-j2\pi\Delta fR_r/c_0+k_R\cos\mu_R \quad (3.200)$$

$$w_{DB} \approx j(2\pi/\lambda)(q-\tilde{q})d_{Ry} + j2\pi\Delta t f_{R\max}\sin\gamma_R + k_R\sin\mu_R \tag{3.201}$$

$$A_{DB} = \frac{\eta_{TR}}{I_0(k_T)I_0(k_R)} \frac{\cos\left(\frac{2\pi}{\lambda}\beta_{Tm}(p-\tilde{p})d_{Tz}\right)}{1-\left(\frac{4\beta_{Tm}(p-\tilde{p})d_{Tz}}{\lambda}\right)^2} \times \frac{\cos\left(\frac{2\pi}{\lambda}\beta_{Rm}(q-\tilde{q})d_{Rz}\right)}{1-\left(\frac{4\beta_{Rm}(q-\tilde{q})d_{Rz}}{\lambda}\right)^2} \frac{e^{-j2\pi\Delta f D/c_0}}{\left(R_{t2}^2-R_{t1}^2\right)\left(R_{r2}^2-R_{r1}^2\right)} \tag{3.202}$$

Note that the STF-CFs for the SBT, SBR, and DB components of the time-variant transfer function must be evaluated numerically because the integrals in (3.191)–(3.193) do not have closed-form solutions. However, if we assume that the time and frequency dispersion are statistically independent, closed-form expressions for these STF-CFs can be obtained. For this reason, in the literature it is often assumed that the time dispersion (i.e., the time delays) and the frequency dispersion (i.e., the Doppler spreads) are statistically independent because the time delays depend on the relative locations of random scatterers (i.e., the angles of arrival, the angles of departure, and the distances among the T_x, R_x, and scatterers), whereas the Doppler spreads depend on the motions of the T_x and R_x [47, 48]. However, the Doppler spread depends not only on the speeds of the T_x and R_x, but also on the angles of departure and the angles of arrival. Hence, both the time delays and the Doppler spreads depend on the relative location of the random scatterers, which implies that they are not statistically independent. Since the angles of departure and the angles of arrival cannot be neglected in the Doppler spreads, another way to obtain the independence between the time delays and the Doppler spreads is to assume that the waves scattered from the relatively closely spaced scatterers have equal time delays. Then, the influence of the angles of arrival and the angles of departure on the time delays can be neglected, making the time delays and the Doppler spreads statistically independent. In 2-D geometry-based statistical models, this is often achieved by the independence between the time delays and the Doppler spreads, assuming that all waves scattered from the scatterers lying on the same circle have equal time delays [9]. These time delays are equal to the average of all time delays obtained from the scatterers lying on the same circle. The equivalent assumption in the 3-D concentric cylinders model is that all waves scattered from the scatterers lying on the same cylindric surface have equal (averaged) time delays. Using this approach, the time delays in (3.176)–(3.178) can be approximated as

$$\tau_{m,l} \approx \tau_l = \frac{D+R_t^{(l)}}{c_0} \tag{3.203}$$

$$\tau_{n,k} \approx \tau_k = \frac{D+R_r^{(k)}}{c_0} \tag{3.204}$$

$$\tau_{m,l,n,k} \approx \tau_{l,k} = \frac{D + R_t^{(l)} + R_r^{(k)}}{c_0} \tag{3.205}$$

Under these assumptions, the space-time-frequency in (3.180) can be factored as [48]

$$R_{pq,\tilde{p}\tilde{q}}(\Delta t, \Delta f) = R^{SBT}_{pq,\tilde{p}\tilde{q}}(\Delta t) R^{SBT}_{pq,\tilde{p}\tilde{q}}(\Delta f) + R^{SBR}_{pq,\tilde{p}\tilde{q}}(\Delta t) R^{SBR}_{pq,\tilde{p}\tilde{q}}(\Delta f)$$
$$+ R^{DB}_{pq,\tilde{p}\tilde{q}}(\Delta t) R^{DB}_{pq,\tilde{p}\tilde{q}}(\Delta f) + R^{LoS}_{pq,\tilde{p}\tilde{q}}(\Delta t) R^{LoS}_{pq,\tilde{p}\tilde{q}}(\Delta f) \tag{3.206}$$

where the closed-form expressions for the LoS, SBT, SBR, and DB components are given in [44].

Figures 3.12–3.14 compare the space-time correlation functions, the frequency correlation functions, and the space-time-frequency correlation functions in (3.180) and (3.206). The parameters used to obtain the results shown in Figures 3.12–3.14 are D = 500m, λ = 0.3m, $d_T = 0.5\lambda$, $d_R = 0.5\lambda$, $\theta_T = \pi/4$, $\theta_R = \pi/4$, $\psi_T = \pi/6$, $\psi_R = \pi/6$, $\gamma_T = 40°$, $\gamma_R = 20°$, $\Delta_H = 0$, $f_{T\max}$ = 200 MHz, $f_{R\max}$ = 200 MHz, $\gamma = 4$, $K = 0$, $\mu_T = \pi/2$, $\mu_R = 3\pi/2$, $\beta_{Tm} = 15°$, $\beta_{Rm} = 15°$, R_{t1} = 10m, R_{t2} = 100m, R_{r1} = 10m, R_{r2} = 100m, $\eta_T = 1/3$, $\eta_R = 1/3$, and $\eta_{TR} = 1/3$. Figure 3.12 shows excellent agreement between space-time correlation functions from (3.180) and from (3.206). This is an expected result, because the approximations in (3.203)–(3.205) affect only time delays, which are set to zero in both the closed-form and numerically obtained space-time correlation functions. Figure 3.13 shows relatively good agreement between the frequency correlation functions. Note that the frequency correlation functions are obtained from the space-time-frequency correlation functions by eliminating the space and time components (i.e., by setting $d_T = d_R = \Delta t = 0$). Hence, this result shows that the approximated

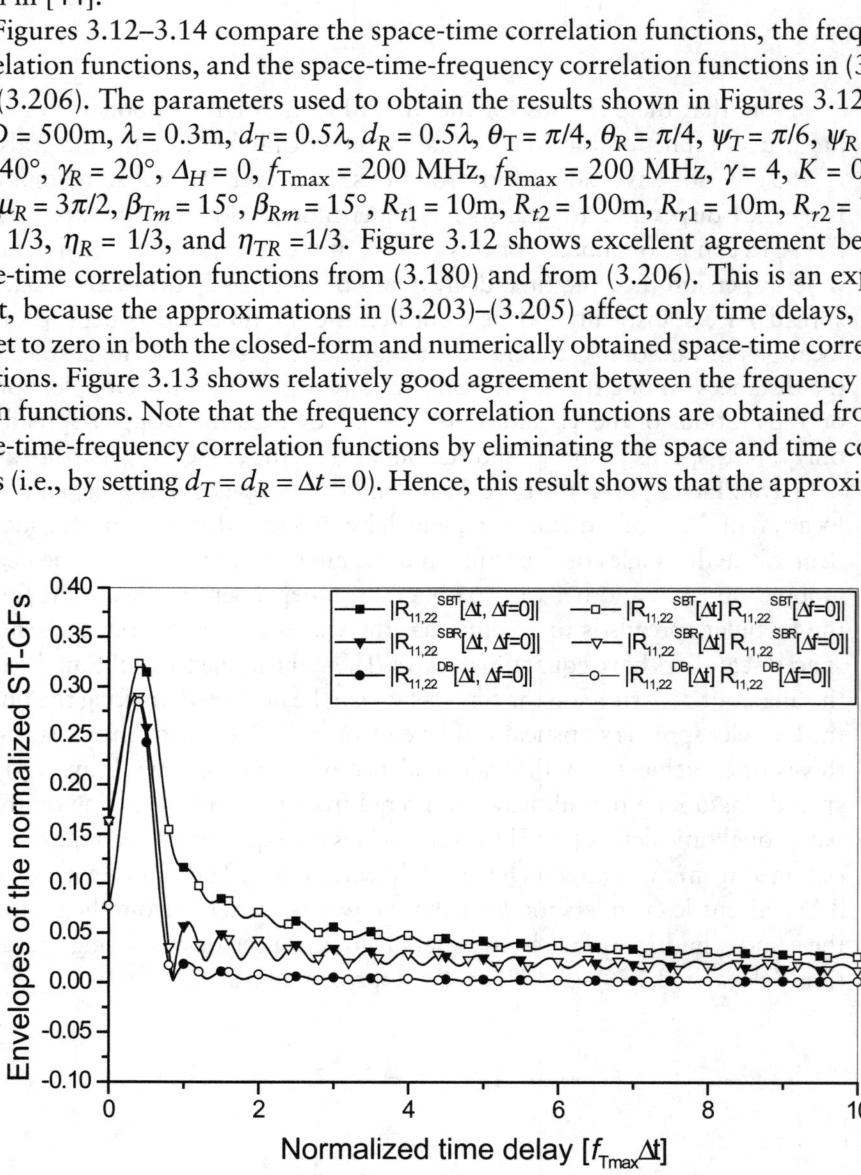

Figure 3.12 Comparison of the envelopes of normalized space-time correlation functions in (3.180) and (3.206).

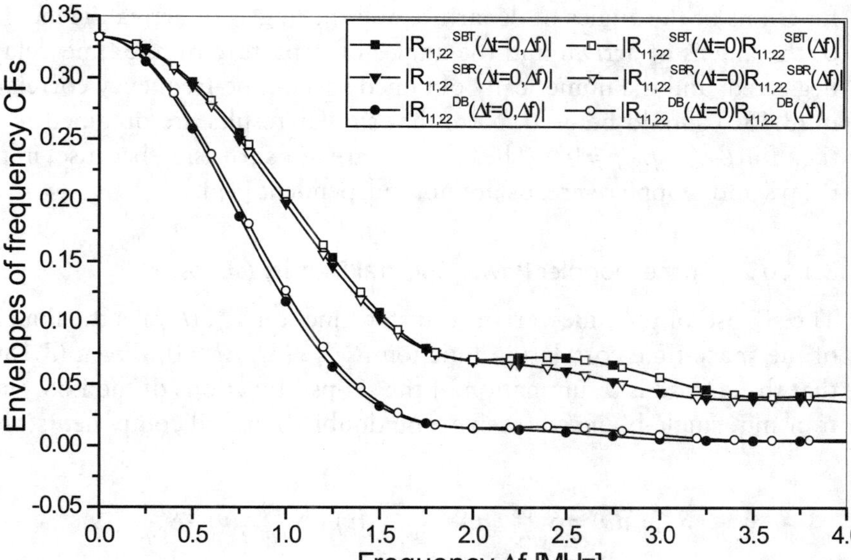

Figure 3.13 Comparison of the envelopes of frequency correlation functions in (3.180) and (3.206).

and exact time delays have similar values (i.e., verifies the approximations in (3.203)–(3.205)).

However, Figure 3.14 shows that STF-CF in (3.206) underestimates the space-time-frequency correlation function. This discrepancy is due to the fact that the time delays in closed-form correlations are not functions of the angles of departure or the angles of arrival, while the time delays in the numerically obtained correlations *are*

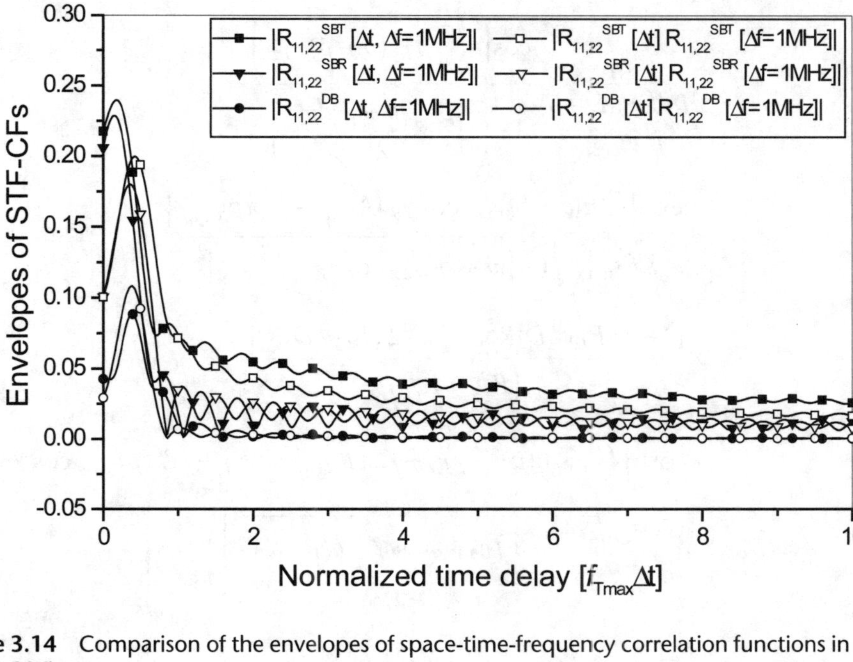

Figure 3.14 Comparison of the envelopes of space-time-frequency correlation functions in (3.180) and (3.206).

functions of the angles of departure and the angles of arrival. Hence, the influence of the angles of arrival and the angles of departure on the time delays cannot be neglected, and the numerically obtained space-time-frequency correlation function in (3.180) should be used. Note that similar results are obtained for 2-D models (i.e., for $\beta_{Tm} = \beta_{Rm} = 0°$). These results are not surprising because, in practice, time delays and Doppler spreads are not independent [49].

3.1.10.2 Space-Doppler Power Spectral Density (sD-psd)

The sD-psd of the time-variant transfer function $T_{pq}(t, f)$ is the Fourier transform of the space-time correlation function $R_{pq,\tilde{p}\tilde{q}}(\Delta t, \Delta f = 0)$. From (3.180), it follows that the sD-psd is a summation of the sD-psd functions of the LoS, single-bounced transmit, single-bounced receive, and double-bounced components; that is,

$$S_{pq,\tilde{p}\tilde{q}}(\nu) = S^{LoS}_{pq,\tilde{p}\tilde{q}}(\nu) + S^{SBT}_{pq,\tilde{p}\tilde{q}}(\nu) + S^{SBR}_{pq,\tilde{p}\tilde{q}}(\nu) + S^{DB}_{pq,\tilde{p}\tilde{q}}(\nu) \tag{3.207}$$

It has been shown [44] that the sD-psds of the LoS, single-bounced transmit, single-bounced receive, and double-bounced components are, respectively,

$$\begin{aligned} S^{LoS}_{pq,\tilde{p}\tilde{q}}(\nu) &= \mathcal{F}_{\Delta t}\left\{R^{LoS}_{pq,\tilde{p}\tilde{q}}(\Delta t, \Delta f = 0)\right\} = \frac{K}{K+1} e^{j\frac{2\pi}{\lambda}[(p-\tilde{p})d_{Tx} - (q-\tilde{q})d_{Rx}]} \\ &\times \delta(\nu + f_{T\max}\cos\gamma_T - f_{R\max}\cos\gamma_R) \end{aligned} \tag{3.208}$$

$$\begin{aligned} S^{SBT}_{pq,\tilde{p}\tilde{q}}(\nu) &= \mathcal{F}_{\Delta t}\left\{R^{SBT}_{pq,\tilde{p}\tilde{q}}(\Delta t, \Delta f = 0)\right\} \\ &= \frac{\eta_T}{I_0(k_T)(K+1)} \frac{\cos\left(\frac{2\pi}{\lambda}\beta_{T_m}(p-\tilde{p})d_{Tz}\right)}{1-\left(\frac{4\beta_{T_m}(p-\tilde{p})d_{Tz}}{\lambda}\right)^2} \\ &\times \frac{\exp\{j2\pi(\nu + f_{R\max}\cos\gamma_R)A_{SBT} - j2\pi p_{x_{SBR}}\}}{\pi f_{T\max}\sqrt{1-[(\nu + f_{R\max}\cos\gamma_R)/f_{T\max}]^2}} \\ &\times \frac{(3-2\gamma R_{t2}/D)R_{t2}^2 - (3-2\gamma R_{t1}/D)R_{t1}^2}{3\left(R_{t2}^2 - R_{t1}^2\right)} \\ &\times \cosh\Big[\big(k_T\sin(\mu_T-\gamma_T) + j2\pi p_{x_{SBT}}\sin\gamma_T - j2\pi p_{y_{SBT}}\cos\gamma_T\big) \\ &\times \sqrt{1-[(\nu + f_{R\max}\cos\gamma_R)/f_{T\max}]^2}\Big] \end{aligned} \tag{3.209}$$

$$
\begin{aligned}
S_{pq,\tilde{p}\tilde{q}}^{SBR}(v) &= \mathcal{F}_{\Delta t}\left\{R_{pq,\tilde{p}\tilde{q}}^{SBR}(\Delta t, \Delta f = 0)\right\} \\
&= \frac{\eta_R}{I_0(k_R)(K+1)} \frac{\cos\left(\dfrac{2\pi}{\lambda}\beta_{R_m}(q-\tilde{q})d_{Rz}\right)}{1-\left(\dfrac{4\beta_{R_m}(q-\tilde{q})d_{Rz}}{\lambda}\right)^2} \\
&\times \frac{\exp\left\{j2\pi(v-f_{T\max}\cos\gamma_T)A_{SBR} + j2\pi p_{x_{SBT}}\right\}}{\pi f_{R\max}\sqrt{1-\left[(v-f_{T\max}\cos\gamma_T)/f_{R\max}\right]^2}} \\
&\times \frac{(3-2\gamma R_{r2}/D)R_{r2}^2 - (3-2\gamma R_{r1}/D)R_{r1}^2}{3\left(R_{r2}^2 - R_{r1}^2\right)} \\
&\times \cosh\left[\left(k_R\sin(\mu_R-\gamma_R) + j2\pi p_{x_{SBR}}\sin\gamma_R - j2\pi p_{y_{SBR}}\cos\gamma_R\right)\right. \\
&\left.\times\sqrt{1-\left[(v-f_{T\max}\cos\gamma_T)/f_{R\max}\right]^2}\right]
\end{aligned}
\tag{3.210}
$$

$$
\begin{aligned}
S_{pq,\tilde{p}\tilde{q}}^{DB}(v) &= \mathcal{F}_{\Delta t}\left\{R_{pq,\tilde{p}\tilde{q}}^{DB}(\Delta t, \Delta f = 0)\right\} \\
&= \frac{\eta_{TR} I_{DB}}{I_0(k_T)I_0(k_R)(K+1)} \frac{\cos\left(\dfrac{2\pi}{\lambda}\beta_{T_m}(p-\tilde{p})d_{Tz}\right)}{1-\left(\dfrac{4\beta_{T_m}(p-\tilde{p})d_{Tz}}{\lambda}\right)^2} \frac{\cos\left(\dfrac{2\pi}{\lambda}\beta_{R_m}(q-\tilde{q})d_{Rz}\right)}{1-\left(\dfrac{4\beta_{R_m}(q-\tilde{q})d_{Rz}}{\lambda}\right)^2} \\
&\times \exp\left\{j\left(2\pi p_{x_{DB}}\cos\gamma_T + 2\pi p_{y_{DB}}\sin\gamma_T - jk_T\cos(\mu_T-\gamma_T)\right)v/f_{T\max}\right\} \\
&\times \cosh\left[\left(k_T\sin(\mu_T-\gamma_T) + j2\pi p_{x_{DB}}\sin\gamma_T - j2\pi p_{y_{DB}}\cos\gamma_T\right)\right. \\
&\qquad \left.\times\sqrt{1-(v/f_{T\max})^2}\right] \Big/ \left(\pi f_{T\max}\sqrt{1-(v/f_{T\max})^2}\right) \\
&\odot \exp\left\{j\left(2\pi p_{z_{DB}}\cos\gamma_R + 2\pi p_{w_{DB}}\sin\gamma_R - jk_R\cos(\mu_R-\gamma_R)\right)v/f_{R\max}\right\} \\
&\times \cosh\left[\left(k_R\sin(\mu_R-\gamma_R) + j2\pi p_{z_{DB}}\sin\gamma_R - j2\pi p_{w_{DB}}\cos\gamma_R\right)\right. \\
&\qquad \left.\times\sqrt{1-(v/f_{R\max})^2}\right] \Big/ \left(\pi f_{R\max}\sqrt{1-(v/f_{R\max})^2}\right)
\end{aligned}
\tag{3.211}
$$

where $\mathcal{F}\{\cdot\}$ is the Fourier transform, $\cosh(\cdot)$ is the hyperbolic cosine, $\delta(\cdot)$ is the Dirac delta function, $\odot$ is convolution, $|v+f_{R\max}\cos\gamma_R| \le f_{T\max}$ and $|v-f_{T\max}\cos\gamma_T| \le f_{R\max}$

for the single-bounced sD-psds, $|\nu| \leq f_{T\max} + f_{R\max}$ for the double-bounced sD-psd, and

$$A_{SBT} = (2\pi p_{x_{SBT}} \cos\gamma_T + 2\pi p_{y_{SBT}} \sin\gamma_T - w_{x_{SBT}}) / 2\pi f_{T\max} \tag{3.212}$$

$$p_{x_{SBT}} = (p - \tilde{p}) d_{Tx} / \lambda \tag{3.213}$$

$$w_{x_{SBT}} = jk_T \cos(\gamma_T - \mu_T) \tag{3.214}$$

$$p_{y_{SBT}} = \left[(p - \tilde{p}) d_{Ty} + (q - \tilde{q}) d_{Ry} (R_{t1} + 0.5(R_{t_2} - R_{t1})) / D \right] / \lambda \tag{3.215}$$

$$A_{SBR} = (2\pi p_{x_{SBR}} \cos\gamma_R + 2\pi p_{y_{SBR}} \sin\gamma_R + w_{x_{SBR}}) / 2\pi f_{R\max} \tag{3.216}$$

$$p_{x_{SBR}} = (q - \tilde{q}) d_{Rx} / \lambda \tag{3.217}$$

$$w_{x_{SBR}} = -jk_R \cos(\gamma_R - \mu_R) \tag{3.218}$$

$$p_{y_{SBR}} = \left[(q - \tilde{q}) d_{Ry} + (p - \tilde{p}) d_{Ty} \left(R_{r1} + 0.5 \left(R_{r2} - R_{r1} \right) \right) / D \right] / \lambda \tag{3.219}$$

$$p_{x_{DB}} = (p - \tilde{p}) d_{Tx} / \lambda \tag{3.220}$$

$$p_{y_{DB}} = (p - \tilde{p}) d_{Ty} / \lambda \tag{3.221}$$

$$p_{z_{DB}} = (q - \tilde{q}) d_{Rx} / \lambda \tag{3.222}$$

$$p_{w_{DB}} = (q - \tilde{q}) d_{Ry} / \lambda \tag{3.223}$$

$$\begin{aligned} I_{DB} = {} & \left(R_{t2}^2 - R_{t1}^2 \right) \left(0.5 R_{r2}^2 - \gamma R_{r2}^3 / (3D) - 0.5 R_{r1}^2 + \gamma R_{r1}^3 / (3D) \right) \\ & + \left(R_{r2}^2 - R_{r1}^2 \right) \left(0.5 R_{t2}^2 - \gamma R_{t2}^3 / (3D) - 0.5 R_{t1}^2 + \gamma R_{t1}^3 / (3D) \right) \end{aligned} \tag{3.224}$$

Figure 3.15 plots several SISO and MIMO sD-psds, assuming 3-D nonisotropic scattering and LoS conditions between the T_x and R_x. These sD-psds are typical of urban street environments. First, we compare the analytical SISO sD-psd in

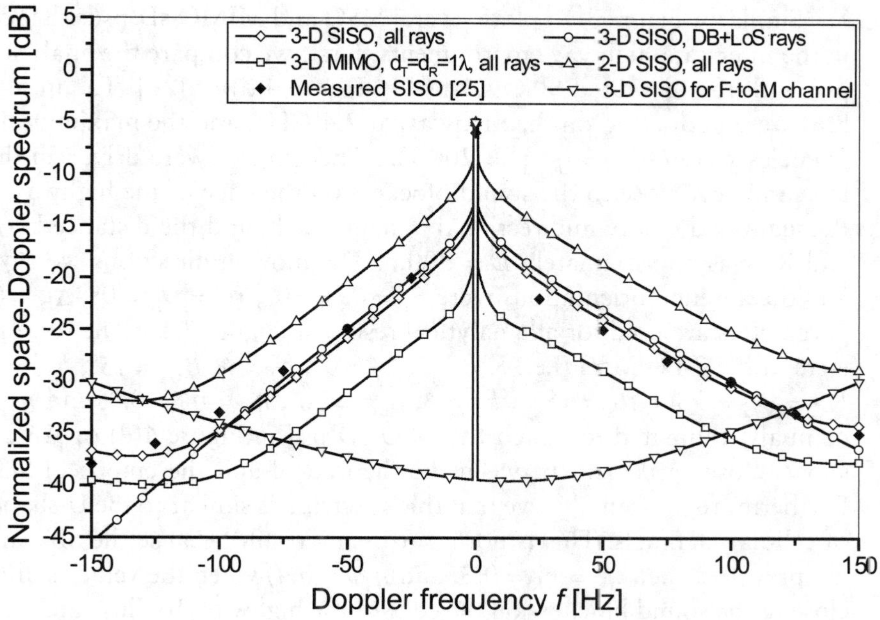

Figure 3.15 The normalized theoretical and measured sD-psd in an urban street environment.

(3.207) with the measured SISO sD-psd in Figure 8(a) of [25]. The measurements in [25] were performed in the urban street area, at 5.2GHz, and the maximum Doppler frequencies were $f_{T\max} = f_{R\max} = 75$ Hz. The T_x and R_x were equipped with one omnidirectional antenna each. The distance between the T_x and R_x was approximately $D = 300$m. The moving directions were $\gamma_T = \gamma_R = 90°$, the antenna orientations were $\theta_T = \theta_R = \psi_T = \psi_R = 0°$, and the difference in the T_x and R_x antenna heights was $\Delta_H = 0$. The same parameters are used to obtain all analytical results in Figure 3.15. The rest of the parameters in the 3-D model (i.e., $K = 4$, $\mu_T = 70°$, $k_T = 5$, $\beta_{T_m} = 20°$, $R_{t1} = R_{r1} = 9$m, $R_{t2} = R_{r2} = 90$m, $\mu_R = 250°$, $k_R = 3.3$, $\beta_{R_m} = 20°$, $\eta_T = \eta_R = 0.1$, and $\eta_{TR} = 0.8$), are estimated to fit the SISO sD-psd reported in [25]. Figure 3. 15 shows the close agreement between the theoretical and the empirical SISO sD-psds. Furthermore, we note that the V-to-V sD-psd in an urban area differs from the U-shaped sD-psd of F-to-M cellular radio channels. As noted in Section 3.1.6.2, this is because the DB rays are more dominant than the SBT and SBR rays (i.e., $\eta_T = \eta_R = 0.1$, and $\eta_{TR} = 0.8$) in an urban street area, whereas F-to-M cellular channels have only SBR and LoS rays (i.e., $\eta_T = 0$, $\eta_R = 1$, $\eta_{TR} = 0$). To illustrate the importance of combining the DB, SBT, and SBR rays, Figure 3.15 also plots the SISO sD-psd with only DB and LoS rays. The results show that for higher frequencies, the model with only DB and LoS rays overestimates the sD-psd. Furthermore, Figure 3.15 shows that the 2-D model (i.e., $\beta_{Tm} = \beta_{Rm} = 0°$) underestimates the sD-psd (i.e., a 3-D model should be used in urban environments). Finally, Figure 3.15 plots the MIMO sD-psd, $S_{11,22}(\nu)$, with $d_T = d_R = 1\lambda$. The results show that the sD-psd decreases with increased antenna element separation.

Similarly, Figure 3.16 plots several SISO and MIMO sD-psds that can be found in the interstate-highway environments. First, we compare the analytical SISO sD-psd with the measured SISO sD-psd in Figure 4(d) of [26]. The measurements in [26] were performed on the highway, at 2.4 GHz, and the maximum Doppler frequencies were $f_{T\max} = f_{R\max} = 200$ Hz. The vehicles were driven in the rightmost lane and very close to the sound blockers on the edge of the highway. The T_x and R_x again had one omnidirectional antenna each, and the distance between the T_x and R_x was approximately $D = 300$m. The moving directions were $\gamma_T = \gamma_R = 0°$, and the antenna orientations were $\theta_T = \theta_R = 90°$, $\psi_T = \psi_R = 0°$, $\Delta_H = 0$. The same parameters are used for all analytical results in Figure 3.16. The rest of the parameters in the 3-D model (i.e., $K = 1.5$, $\mu_T = 0°$, $k_T = 4$, $\beta_{T_m} = 15°$, $R_{t1} = R_{r1} = 9.6$m, $R_{t2} = R_{r2} = 96$m, $\mu_R = 180°$, $k_R = 4$, $\beta_{R_m} = 20°$, $\eta_T = \eta_R = 0.3$, and $\eta_{TR} = 0.4$), are manually estimated to match the SISO sD-psd in Figure 4(d) of [26]. Figure 3.16 shows close agreement between the theoretical and the empirical SISO sD-psds. Furthermore, we can observe that this spectrum is similar to the U-shaped spectrum of cellular channels. This is not a surprising result because the SBT and SBR rays are prevalent (i.e., $\eta_T = \eta_R = 0.3$, and $\eta_{TR} = 0.4$) when the vehicles are driven very close to the sound blockers on the edge of the highway. To illustrate the importance of combining the DB, SBT, and SBR rays, Figure 3.16 also plots the SISO sD-psd with only DB and LoS rays. The results show that the model with only DB and LoS rays has a sD-psd with a significantly different shape as compared to the model with SBT, SBR, DB, and LoS rays. Furthermore, Figure 3.16 shows that the 2-D model (i.e., $\beta_{T_m} = \beta_{R_m} = 0°$) underestimates the sD-psd (i.e., a 3-D model should be used in urban environments). Finally, Figure 3.16 plots the MIMO sD-psd, $S_{11,22}(\nu)$, for the antenna elements separation of $d_T = d_R = 1\lambda$ and again shows that the sD-psd decreases with increased antenna element separation.

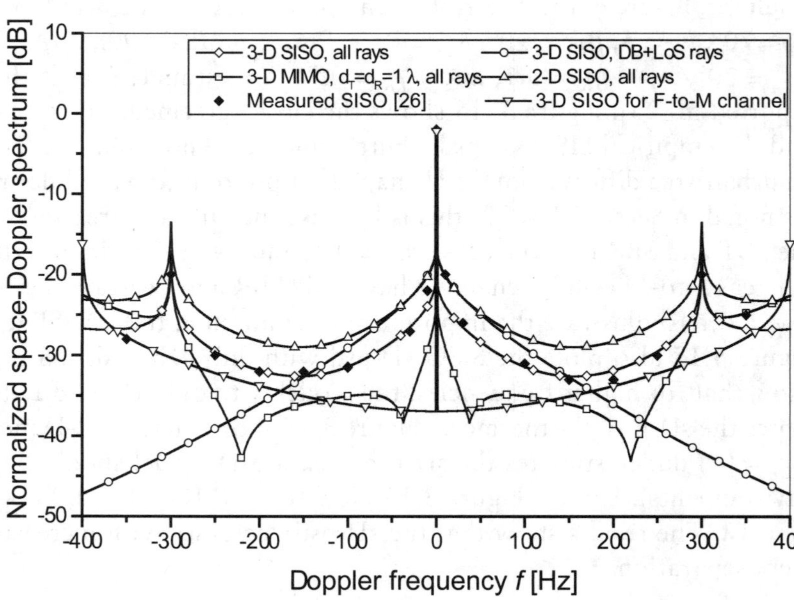

Figure 3.16 The normalized theoretical and measured sD-psd in an interstate-highway environment.

3.1.10.3 Power Space-Delay Spectral Density (psds)

The psds of the time-variant transfer function is the inverse Fourier transform of the space-frequency correlation function $R_{pq,\tilde{p}\tilde{q}}(\Delta t = 0, \Delta f)$. From (3.180), it follows that the psds is a summation of the psds functions of the LoS, SBT, SBR, and DB components, which have been shown [44] to be

$$P^{LoS}_{pq,\tilde{p}\tilde{q}}(\tau_{\text{rel}}) = \mathcal{F}^{-1}_{\Delta f}\left\{R^{LoS}_{pq,\tilde{p}\tilde{q}}(\Delta t = 0, \Delta f)\right\}$$

$$= \frac{K}{K+1} e^{j\frac{2\pi}{\lambda}\left[(p-\tilde{p})d_{T_x} - (q-\tilde{q})d_{R_x}\right]} \delta\left(\tau - \frac{\sqrt{D^2 + \Delta_H^2}}{c_0}\right) \tag{3.225}$$

$$P^{SBT}_{pq,\tilde{p}\tilde{q}}(\tau_{\text{rel}}) = \mathcal{F}^{-1}_{\Delta f}\left\{R^{SBT}_{pq,\tilde{p}\tilde{q}}(\Delta t = 0, \Delta f)\right\}$$

$$= \frac{\eta_T}{I_0(k_T)(K+1)} \frac{\cos\left(\frac{2\pi}{\lambda}\beta_{T_m}(p-\tilde{p})d_{T_z}\right)}{1 - \left(\frac{4\beta_{T_m}(p-\tilde{p})d_{T_z}}{\lambda}\right)^2} e^{-j\frac{2\pi}{\lambda}(q-\tilde{q})d_{Rx}}$$

$$\times \frac{2}{R_{t2}^2 - R_{t1}^2} \int_{R_{ta}}^{R_{tb}} \left(1 - \gamma\frac{R_t}{D}\right) R_t e^{j\frac{2\pi}{c_0} R_t B_{SBT}(c_0\tau_{\text{rel}}/R_t - 1)}$$

$$\times \frac{\cosh\left(2jC_{SBT}\sqrt{\frac{c_0\tau_{\text{rel}}}{2R_t}\left(1 - \frac{c_0\tau_{\text{rel}}}{2R_t}\right)}\right)}{\frac{2\pi}{c_0} R_t \sqrt{\frac{c_0\tau_{\text{rel}}}{2R_t}\left(1 - \frac{c_0\tau_{\text{rel}}}{2R_t}\right)}} dR_t \tag{3.226}$$

$$P^{SBR}_{pq,\tilde{p}\tilde{q}}(\tau_{\text{rel}}) = \mathcal{F}^{-1}_{\Delta f}\left\{R^{SBR}_{pq,\tilde{p}\tilde{q}}(\Delta t = 0, \Delta f)\right\}$$

$$= \frac{\eta_R}{I_0(k_R)(K+1)} \frac{\cos\left(\frac{2\pi}{\lambda}\beta_{R_m}(q-\tilde{q})d_{Rz}\right)}{1 - \left(\frac{4\beta_{R_m}(q-\tilde{q})d_{Rz}}{\lambda}\right)^2} e^{j\frac{2\pi}{\lambda}(p-\tilde{p})d_{Tx}}$$

$$\times \frac{2}{R_{r2}^2 - R_{r1}^2} \int_{R_{ra}}^{R_{rb}} \left(1 - \gamma\frac{R_r}{D}\right) R_r e^{j\frac{2\pi}{c_0} R_r B_{SBR}(c_0\tau_{\text{rel}}/R_r - 1)}$$

$$\times \frac{\cosh\left(2jC_{SBR}\sqrt{\frac{c_0\tau_{\text{rel}}}{2R_r}\left(1 - \frac{c_0\tau_{\text{rel}}}{2R_r}\right)}\right)}{\frac{2\pi}{c_0} R_r \sqrt{\frac{c_0\tau_{\text{rel}}}{2R_r}\left(1 - \frac{c_0\tau_{\text{rel}}}{2R_r}\right)}} dR_r \tag{3.227}$$

$$
\begin{aligned}
P^{DB}_{pq,\tilde{p}\tilde{q}}(\tau_{\mathrm{rel}}) &= \mathcal{F}^{-1}_{\Delta f}\left\{R^{DB}_{pq,\tilde{p}\tilde{q}}(\Delta t = 0, \Delta f)\right\} \\
&= \frac{\eta_{TR}}{I_0(k_T)I_0(k_R)(K+1)} \\
&\times \frac{\cos\left(\frac{2\pi}{\lambda}\beta_{T_m}(p-\tilde{p})d_{Tz}\right)}{1-\left(\frac{4\beta_{T_m}(p-\tilde{p})d_{Tz}}{\lambda}\right)^2}\frac{\cos\left(\frac{2\pi}{\lambda}\beta_{R_m}(q-\tilde{q})d_{Rz}\right)}{1-\left(\frac{4\beta_{R_m}(q-\tilde{q})d_{Rz}}{\lambda}\right)^2} \\
&\times \frac{2}{(R_{t2}^2 - R_{t1}^2)(R_{r2}^2 - R_{r1}^2)}\left[\int_{R_{ta}}^{R_{tb}} R_t e^{j\frac{2\pi}{c_0}R_t B_{DB}(c_0\tau_{\mathrm{rel}}/R_t - 1)}\right. \\
&\times \frac{\cosh\left(2jC_{DB}\sqrt{\frac{c_0\tau_{\mathrm{rel}}}{2R_t}\left(1-\frac{c_0\tau_{\mathrm{rel}}}{2R_t}\right)}\right)}{\frac{2\pi}{c_0}R_t\sqrt{\frac{c_0\tau_{\mathrm{rel}}}{2R_t}\left(1-\frac{c_0\tau_{\mathrm{rel}}}{2R_t}\right)}}dR_t \\
&\odot \int_{R_{ra}}^{R_{rb}}\left(1-\gamma\frac{R_r}{D}\right)R_r e^{j\frac{2\pi}{c_0}R_r D_{DB}(c_0\tau_{\mathrm{rel}}/R_r - 1)} \\
&\times \frac{\cosh\left(2jE_{DB}\sqrt{\frac{c_0\tau_{\mathrm{rel}}}{2R_r}\left(1-\frac{c_0\tau_{\mathrm{rel}}}{2R_r}\right)}\right)}{\frac{2\pi}{c_0}R_r\sqrt{\frac{c_0\tau_{\mathrm{rel}}}{2R_r}\left(1-\frac{c_0\tau_{\mathrm{rel}}}{2R_r}\right)}}dR_r \\
&+ \int_{R_{ta}}^{R_{tb}}\left(1-\gamma\frac{R_t}{D}\right)R_t e^{j\frac{2\pi}{c_0}R_t B_{DB}(c_0\tau_{\mathrm{rel}}/R_t - 1)} \\
&\times \frac{\cosh\left(2jC_{DB}\sqrt{\frac{c_0\tau_{\mathrm{rel}}}{2R_t}\left(1-\frac{c_0\tau_{\mathrm{rel}}}{2R_t}\right)}\right)}{\frac{2\pi}{c_0}R_t\sqrt{\frac{c_0\tau_{\mathrm{rel}}}{2R_t}\left(1-\frac{c_0\tau_{\mathrm{rel}}}{2R_t}\right)}}dR_t \\
&\odot \int_{R_{ra}}^{R_{rb}} R_r e^{j\frac{2\pi}{c_0}R_r D_{DB}(c_0\tau_{\mathrm{rel}}/R_r - 1)} \\
&\left.\times \frac{\cosh\left(2jE_{DB}\sqrt{\frac{c_0\tau_{\mathrm{rel}}}{2R_r}\left(1-\frac{c_0\tau_{\mathrm{rel}}}{2R_r}\right)}\right)}{\frac{2\pi}{c_0}R_r\sqrt{\frac{c_0\tau_{\mathrm{rel}}}{2R_r}\left(1-\frac{c_0\tau_{\mathrm{rel}}}{2R_r}\right)}}dR_r\right]
\end{aligned}
\tag{3.228}
$$

where cosh(·) is the hyperbolic cosine, ⊙ denotes convolution, $\delta(\cdot)$ denotes the Dirac delta function, and $\tau_{rel} = \tau - D/c_0$. The integrals in (3.226)–(3.228) need to be numerically evaluated over the range of possible radii R_t and R_r. For the range $0 \le \tau_{rel} \le 2R_{t1}/c_0$, the integration limits are $R_{ta} = R_{t1}$, $R_{tb} = R_{t2}$, $R_{ra} = R_{r1}$, and $R_{rb} = R_{r2}$. On the other hand, when $2R_{t1}/c_0 \le \tau_{rel} \le 2R_{t2}/c_0$, the integration limits are $R_{ta} = c_0\tau_{rel}/2$, $R_{tb} = R_{t2}$, $R_{ra} = c_0\tau_{rel}/2$, and $R_{rb} = R_{r2}$. Finally, the parameters in (3.226)–(3.228) are defined as

$$B_{SBT} = (2\pi(p-\tilde{p})d_{Tx}/\lambda - jk_T\cos\mu_T)c_0/(2R_t\pi) \tag{3.229}$$

$$C_{SBT} = (2\pi(p-\tilde{p})d_{Ty}/\lambda + 2\pi(q-\tilde{q})d_{Ry}(R_{t1}+0.5(R_{t2}-R_{t1}))/(D\lambda) - jk_T\sin\mu_T) \tag{3.230}$$

$$B_{SBR} = (2\pi(q-\tilde{q})d_{Rx}/\lambda - jk_R\cos\mu_R)c_0/(2R_r\pi) \tag{3.231}$$

$$C_{SBR} = (2\pi(q-\tilde{q})d_{Ry}/\lambda + 2\pi(q-\tilde{q})d_{Ty}(R_{r1}+0.5(R_{r2}-R_{r1}))/(D\lambda) - jk_R\sin\mu_R) \tag{3.232}$$

$$B_{DB} = (2\pi(p-\tilde{p})d_{Tx}/\lambda - jk_T\cos\mu_T)c_0/(2R_t\pi) \tag{3.233}$$

$$C_{DB} = 2\pi(p-\tilde{p})d_{Ty}/\lambda - jk_T\sin\mu_T \tag{3.234}$$

$$D_{DB} = (2\pi(q-\tilde{q})d_{Rx}/\lambda - jk_R\cos\mu_R)c_0/(2R_r\pi) \tag{3.235}$$

$$E_{DB} = 2\pi(q-\tilde{q})d_{Ry}/\lambda - jk_R\sin\mu_R \tag{3.236}$$

Note that power delay spectra that were derived assuming "one-circular-ring" model [50] are special cases of (3.227). For example, the 2-D nonisotropic power delay spectrum in [50] is obtained for $\beta_{R_m} = 0$, $\psi_R = 0$, and $d_T = d_R = 0$.

Figures 3.17 and 3.18 show several SISO and MIMO psds that can be found in highway environments and compares them with the measured SISO psds in [51]. The measurements in [51] were performed on the highway, at 2.4 GHz, and with maximum Doppler frequencies $f_{T\max} = f_{R\max} = 200$ Hz. The vehicles were driven in the rightmost lane of the highway, and the sound blockers on the edge of the highway were periodically present. The T_x and R_x were equipped with one omnidirectional antenna each. The distance between the T_x and R_x was approximately $D = 300$m. The moving directions were $\gamma_T = \gamma_R = 90°$, the antenna orientations were $\theta_T = \theta_R = \psi_T = \psi_R = 0°$, and the relative antenna elevation was $\Delta_H = 0$. The same parameters are used to obtain all analytical results in Figures 3.17 and 3.18. The measured SISO psds in Figure 3.17 is obtained on the part of the highway without sound blockers. The estimated parameters for the analytical SISO psds shown in Figure 3.17 are $K = 0$, $\gamma = 4$, $\mu_T = 90°$, $k_T = 9.1$, $\beta_{T_m} = 15°$, $R_{t1} = R_{r1} = 13$m, $R_{t2} = R_{r2} = 130$m, $\mu_R = 270°$, $k_R = 9.1$, $\beta_{R_m} = 15°$, $\eta_T = \eta_R = 0.2$, and $\eta_{TR} = 0.6$. Figure 3.17 shows the close agreement between the theoretical and empirical SISO psds.

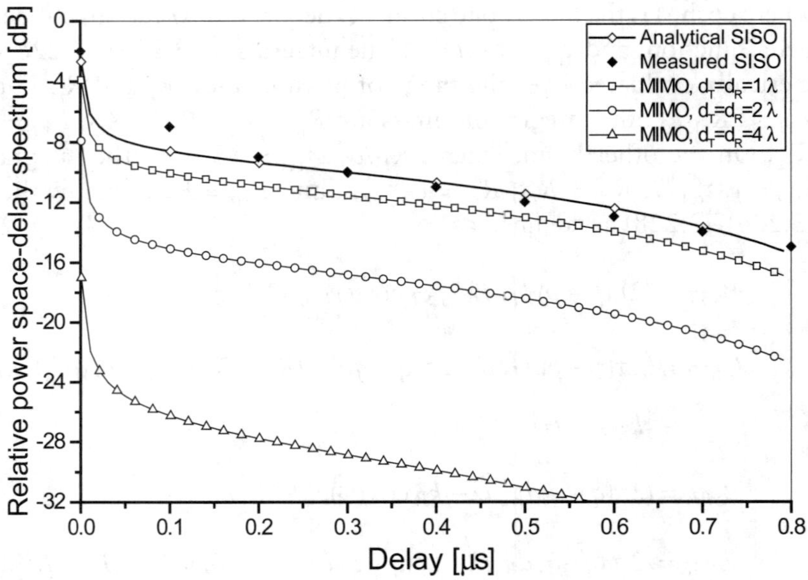

Figure 3.17 The analytical and measured psds in the rightmost lane of an interstate highway *without* sound blockers.

The results show that DB rays are prevalent on the part of the highway without sound blockers (i.e., $\eta_T = \eta_R = 0.2$, and $\eta_{TR} = 0.6$). In this case, the psds closely follows the one-sided exponential function and dies out after 0.8 μs.

The measured SISO psds in Figure 3.18 is obtained on the part of the highway with sound blockers. The estimated parameters for the analytical SISO psds shown in Figure 3.18 are $K = 0$, $\gamma = 4$, $\mu_T = 90°$, $k_T = 9.4$, $\beta_{T_m} = 15°$, $R_{t1} = R_{r1} = 12$m, $R_{t2} = R_{r2} = 120$m, $\mu_R = 270°$, $k_R = 9.4$, $\beta_{R_m} = 15°$, $\eta_T = \eta_R = 0.45$, and $\eta_{TR} = 0.1$.

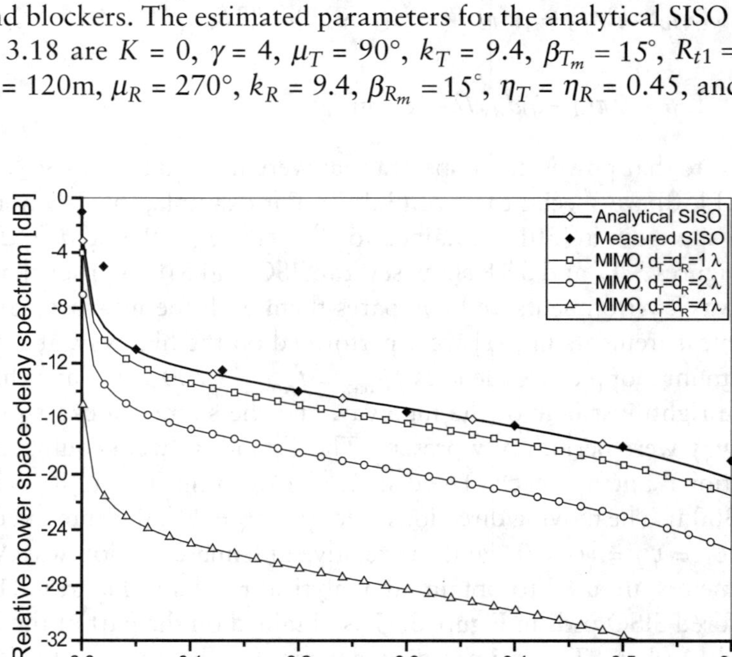

Figures 3.18 The analytical and measured psds in the rightmost lane of an interstate highway *with* sound blockers.

Figure 3.18 shows the close agreement between the theoretical and empirical SISO psds. The results show that SBT and SBR rays are prevalent on the part of the highway with sound blockers (i.e., $\eta_T = \eta_R = 0.45$, and $\eta_{TR} = 0.1$). We can observe that this psds does not follow the one-sided exponential function, which is characteristic for F-to-M cellular channels. The V-to-V psds has two distinct slopes and dies out after 0.6 μs. Figures 3.17 and 3.18 also show several MIMO psds, $P_{11,22}(\tau_{\text{rel}})$, with $d_T = d_R = \{1\lambda, 2\lambda, 4\lambda\}$. The results show that the psds decreases with increased antenna element separation.

3.1.10.4 Level Crossing Rate (LCR)

The level crossing rate of the channel envelope can be calculated using similar reasoning as in Section 3.1.6.3. The LCR for LoS conditions (Ricean fading, correlated in-phase and quadrature components) is given in (3.106) and the parameters b_0, b_1, and b_2 are defined in (3.107)–(3.109), respectively. When 3-D nonisotropic scattering is characterized by pdfs given in (3.23), (3.24), and (3.186)–(3.189), the parameters b_0, b_1, and b_2 can be written as [11]

$$b_n = b_n^{SBT} + b_n^{SBR} + b_n^{DB} \tag{3.237}$$

where $n \in \{0,1,2\}$ and parameters b_0^{SBT}, b_0^{SBR}, b_0^{DB}, b_1^{SBT}, b_1^{SBR}, b_1^{DB}, b_2^{SBT}, b_2^{SBR}, b_2^{DB}, and b_2^{DB} are

$$b_0^{SBT} = \frac{\eta_T}{2(K+1)} \tag{3.238}$$

$$b_0^{SBR} = \frac{\eta_R}{2(K+1)} \tag{3.239}$$

$$b_0^{DB} = \frac{\eta_{TR}}{2(K+1)} \tag{3.240}$$

$$\begin{aligned} b_1^{SBT} = 2\pi b_0^{SBT} \Bigg\{ & f_{T\max} \cos(\mu_T - \gamma_T) \frac{I_1(k_T)}{I_0(k_T)} \frac{\pi^2 \cos\beta_{T_m}}{\pi^2 - 4\beta_{T_m}^2} \\ & + \int_{R_{t1}}^{R_{t2}} \left[\frac{R_t}{D} \sin\gamma_R \sin\mu_T \frac{I_1(k_T)}{I_0(k_T)} + \cos\gamma_R \right] f_{R\max} \frac{2R_t}{R_{t2}^2 - R_{t1}^2} \\ & \times \frac{\pi^2 \left[\cos\left(\frac{R_t}{D}\beta_{T_m} + \frac{\Delta_H}{D} \right) + \cos\left(\frac{R_t}{D}\beta_{T_m} - \frac{\Delta_H}{D} \right) \right]}{2\left[\pi^2 - 4\left(\frac{R_t}{D} \right)^2 \beta_{T_m}^2 \right]} dR_t \Bigg\} \end{aligned} \tag{3.241}$$

$$b_1^{SBR} = 2\pi b_0^{SBR} \Bigg\{ f_{R\max} \cos(\mu_R - \gamma_R) \frac{I_1(k_R)}{I_0(k_R)} \frac{\pi^2 \cos\beta_{R_m}}{\pi^2 - 4\beta_{R_m}^2}$$

$$+ \int_{R_{r1}}^{R_{r2}} \left[\cos\gamma_T - \frac{R_r}{D} \sin\gamma_T \sin\mu_R \frac{I_1(k_R)}{I_0(k_R)} \right] f_{T\max} \frac{2R_r}{R_{r2}^2 - R_{r1}^2}$$

$$\times \frac{\pi^2 \left[\cos\left(\frac{R_r}{D} \beta_{R_m} + \frac{\Delta_H}{D} \right) + \cos\left(\frac{R_r}{D} \beta_{R_m} - \frac{\Delta_H}{D} \right) \right]}{2\left[\pi^2 - 4\left(\frac{R_r}{D} \right)^2 \beta_{R_m}^2 \right]} dR_r \Bigg\} \tag{3.242}$$

$$b_1^{DB} = 2\pi b_0^{DB} \Bigg\{ f_{T\max} \cos(\mu_T - \gamma_T) \frac{I_1(k_T)}{I_0(k_T)} \frac{\pi^2 \cos\beta_{T_m}}{\pi^2 - 4\beta_{T_m}^2}$$

$$+ f_{R\max} \cos(\mu_R - \gamma_R) \frac{I_1(k_R)}{I_0(k_R)} \frac{\pi^2 \cos\beta_{R_m}}{\pi^2 - 4\beta_{R_m}^2} \Bigg\} \tag{3.243}$$

$$b_2^{SBT} = (2\pi)^2 b_0^{SBT} \Bigg\{ f_{T\max}^2 \frac{\pi^2 \cos(2\beta_{T_m}) + \pi^2 - 16\beta_{T_m}^2}{2\left(\pi^2 - 16\beta_{T_m}^2\right)} \frac{I_0(k_T) + \cos\left(2(\mu_T - \gamma_T)\right) I_2(k_T)}{2I_0(k_T)}$$

$$+ \int_{R_{t1}}^{R_{t2}} \frac{4\beta_{T_m}^2 + 8\frac{R_t}{D}\beta_{T_m}^2 - \pi^2 + 4\left(\frac{R_t}{D}\right)^2 \beta_{T_m}^2}{4\left(8\beta_{T_m}^2 \pi^2 - 16\beta_{T_m}^4 - \pi^4 - 16\left(\frac{R_t}{D}\right)^4 \beta_{T_m}^4 + 8\pi^2 \left(\frac{R_t}{D}\right)^2 \beta_{T_m}^2 + 32\left(\frac{R_t}{D}\right)^2 \beta_{T_m}^4 \right)}$$

$$\times \left[\cos\left(\left(\frac{R_t}{D} - 1 \right) \beta_{T_m} + \frac{\Delta_H}{D} \right) + \cos\left(\left(\frac{R_t}{D} + 1 \right) \beta_{T_m} + \frac{\Delta_H}{D} \right) \right.$$

$$\left. + \cos\left(\left(\frac{R_t}{D} - 1 \right) \beta_{T_m} - \frac{\Delta_H}{D} \right) + \cos\left(\left(\frac{R_t}{D} + 1 \right) \beta_{T_m} - \frac{\Delta_H}{D} \right) \right]$$

$$\times \left[f_{T\max} f_{R\max} \frac{R_t}{D} \sin\gamma_R \left(\frac{I_2(k_T)}{I_0(k_T)} \sin(2\mu_T - \gamma_T) + \sin\gamma_T \right) \right.$$

$$\left. + 2 f_{T\max} f_{R\max} \cos\gamma_R \cos(\mu_T - \gamma_T) \frac{I_1(k_T)}{I_0(k_T)} \right]$$

$$+\left[\frac{\cos\left[2\left(\frac{R_t}{D}\beta_{T_m}+\frac{\Delta_H}{D}\right)\right]\pi^2+2\pi^2-32\left(\frac{R_t}{D}\right)^2\beta_{T_m}^2}{4\left[\pi^2-16\left(\frac{R_t}{D}\right)^2\beta_{T_m}\right]}\right.$$

$$\left.+\frac{\cos\left[2\left(\frac{R_t}{D}\beta_{T_m}-\frac{\Delta_H}{D}\right)\right]\pi^2}{4\left[\pi^2-16\left(\frac{R_t}{D}\right)^2\beta_{T_m}\right]}\right]$$

$$\times\left[f_{R\max}^2\cos\gamma_R^2+f_{R\max}^2\left(\frac{R_t}{D}\right)^2\sin^2\gamma_R\frac{I_0(k_T)-\cos(2\mu_T)I_2(k_T)}{2I_0(k_T)}\right.$$

$$\left.+f_{R\max}^2\frac{R_t}{D}\sin(2\gamma_R)\sin\mu_T\frac{I_1(k_T)}{I_0(k_T)}\right]\frac{2R_t}{R_{t2}^2-R_{t1}^2}dR_t\Bigg\} \quad (3.244)$$

$$b_2^{SBR}=(2\pi)^2 b_0^{SBR}\Bigg\{f_{R\max}^2\frac{\pi^2\cos(2\beta_{R_m})+\pi^2-16\beta_{R_m}^2}{2\left(\pi^2-16\beta_{R_m}^2\right)}\frac{I_0(k_R)+\cos(2(\mu_R-\gamma_R))I_2(k_R)}{2I_0(k_R)}$$

$$+\int_{R_{r1}}^{R_{r2}}\frac{4\beta_{R_m}^2+8\frac{R_r}{D}\beta_{R_m}^2-\pi^2+4\left(\frac{R_r}{D}\right)^2\beta_{R_m}^2}{4\left(8\beta_{R_m}^2\pi^2-16\beta_{R_m}^4-\pi^4-16\left(\frac{R_r}{D}\right)^4\beta_{R_m}^4+8\pi^2\left(\frac{R_r}{D}\right)^2\beta_{R_m}^2+32\left(\frac{R_r}{D}\right)^2\beta_{R_m}^4\right)}$$

$$\times\left[\cos\left(\left(\frac{R_r}{D}-1\right)\beta_{R_m}+\frac{\Delta_H}{D}\right)+\cos\left(\left(\frac{R_r}{D}+1\right)\beta_{R_m}+\frac{\Delta_H}{D}\right)\right.$$

$$\left.+\cos\left(\left(\frac{R_r}{D}-1\right)\beta_{R_m}-\frac{\Delta_H}{D}\right)+\cos\left(\left(\frac{R_r}{D}+1\right)\beta_{R_m}-\frac{\Delta_H}{D}\right)\right]$$

$$\times\left[f_{R\max}f_{T\max}\frac{R_r}{D}\sin\gamma_T\left(\frac{I_2(k_R)}{I_0(k_R)}\sin(2\mu_R-\gamma_R)+\sin\gamma_R\right)\right.$$

$$\left.-2f_{R\max}f_{T\max}\cos\gamma_T\cos(\mu_R-\gamma_R)\frac{I_1(k_R)}{I_0(k_R)}\right]$$

$$+\left[\frac{\cos\left[2\left(\frac{R_r}{D}\beta_{R_m}+\frac{\Delta_H}{D}\right)\right]\pi^2+2\pi^2-32\left(\frac{R_r}{D}\right)^2\beta_{R_m}^2}{4\left[\pi^2-16\left(\frac{R_r}{D}\right)^2\beta_{R_m}\right]}\right.$$

$$\left.+\frac{\cos\left[2\left(\frac{R_r}{D}\beta_{R_m}-\frac{\Delta_H}{D}\right)\right]\pi^2}{4\left[\pi^2-16\left(\frac{R_r}{D}\right)^2\beta_{R_m}\right]}\right]$$

$$\times\left[f_{T\max}^2\cos\gamma_T^2+f_{T\max}^2\left(\frac{R_r}{D}\right)^2\sin\gamma_T^2\frac{I_0(k_R)-\cos(2\mu_R)I_2(k_R)}{2I_0(k_R)}\right.$$

$$\left.-f_{T\max}^2\frac{R_r}{D}\sin(2\gamma_T)\sin\mu_R\frac{I_1(k_R)}{I_0(k_R)}\right]\frac{2R_r}{R_{r2}^2-R_{r1}^2}dR_r\Bigg\} \tag{3.245}$$

$$b_2^{DB}=(2\pi)^2 b_0^{DB}\left\{f_{T\max}^2\left[\frac{\pi^2\cos(2\beta_{T_m})}{2\left(\pi^2-16\beta_{T_m}^2\right)}+\frac{1}{2}\right]\frac{I_0(k_T)+\cos(2(\mu_T-\gamma_T))I_2(k_T)}{2I_0(k_T)}\right.$$

$$+2f_{T\max}f_{R\max}\frac{\pi^2\cos\beta_{T_m}}{\pi^2-4\beta_{T_m}^2}\frac{\pi^2\cos\beta_{R_m}}{\pi^2-4\beta_{R_m}^2}$$

$$\times\cos(\mu_T-\gamma_T)\frac{I_1(k_T)}{I_0(k_T)}\cos(\mu_R-\gamma_R)\frac{I_1(k_R)}{I_0(k_R)}$$

$$\left.+f_{R\max}^2\left[\frac{\pi^2\cos(2\beta_{R_m})}{2\left(\pi^2-16\beta_{R_m}^2\right)}+\frac{1}{2}\right]\frac{I_0(k_R)+\cos(2(\mu_R-\gamma_R))I_2(k_R)}{2I_0(k_R)}\right\} \tag{3.246}$$

Figure 3.19 compares the analytical LCR with the measured LCR in an urban environment. The measured data is from Figure 8(a) of [25]. The narrowband measurements were performed at 5.2 GHz, and the maximum Doppler frequencies were $f_{T\max}=f_{R\max}=$ 500 Hz. Both, the T_x and R_x were equipped with omnidirectional antennas. The distance between the T_x and R_x was approximately D = 300m. The moving directions and the antenna elevations are $\gamma_T=\gamma_R=\Delta_H=0$. The analytical LCR is obtained using the parameters K = 3.43, μ_T = 12.5°, k_T = 15.2, β_{Tm} = 10°, $\Delta_T=\Delta_R$ = 0.6, μ_R = 143.4°, k_R = 12.5, β_{Rm} = 5°, η_T =0.053, η_R = 0.057, and η_{TR} = 0.9377.

3.2 Impact of Antenna Coupling on MIMO V-to-V Channels

The V-to-V channel models discussed in previous sections neglect mutual coupling among T_x and R_x antenna elements by using wide spacing between antenna elements (i.e., several wavelengths λ). However, in V-to-V communications, it is often desirable to use robust compact antennas with reduced visual presence and low aerodynamic drag. For an example, a car owner would not be happy with an antenna array that occupies the entire roof of the car. Hence, T_x and R_x antennas with wide interelement spacings are usually not practical, and close antenna element spacing inevitably leads to mutual coupling [52] that must be accounted for. Hence, this section focuses on the impact of mutual coupling on MIMO V-to-V systems.

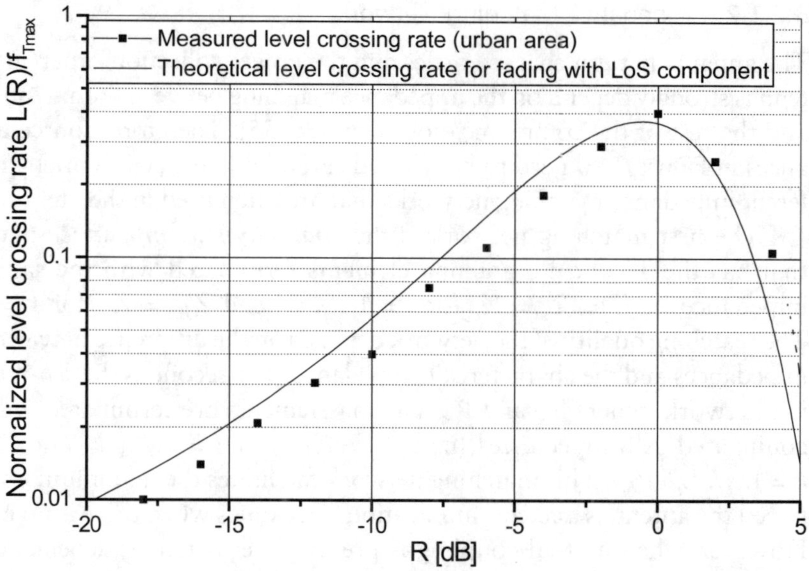

Figure 3.19 Comparison of the analytical LCR and measured LCR in an urban environment.

3.2.1 MIMO V-to-V System Model

Here, we consider a wideband MIMO V-to-V communication system with L_t transmit and L_r receive omnidirectional antenna elements, as shown in Figure 3.20. It is assumed that both T_x and R_x are in motion and equipped with vertically polarized half-wavelength electric dipole antenna elements. We also assume here that the propagation occurs in an outdoor metropolitan environment characterized by 3-D nonisotropic scattering with either LoS or NLoS conditions between the T_x and R_x. Finally, we assume that the T_x and R_x antennas and the scatterers are in the far-field of one another.

3.2.1.1 Transmit and Receive Antenna Subsystems

The voltage sources V_{si} in Figure 3.20, for $i = 1,\ldots, L_t$ have impedances Z_{si} and feed L_t antenna elements at the T_x. The mutual coupling among the T_x antenna elements in Figure 3.20 is characterized by an $L_t \times L_t$ impedance matrix $\mathbf{Z}_{TT}$. The vector of voltages at the entrance of T_x coupling network, $\mathbf{V}_T$ is defined as [53–54]

$$\mathbf{V}_T = \mathbf{V}_S - \mathbf{Z}_S \mathbf{i}_T \tag{3.247}$$

where $\mathbf{V}_S = [V_{s1}, V_{s2},\ldots, V_{sL_t}]^T$ is the vector of source voltages, $\mathbf{i}_T = [i_{t1}, i_{t2},\ldots i_{tL_t}]^T$ is the vector of T_x currents, $\mathbf{Z}_S$ is the diagonal matrix of source impedances, and $[\cdot]^T$ denotes transpose operation. Figure 3.20 also shows the R_x that consists of L_r antenna elements and the load impedances Z_{Li}, for $i = 1, \ldots, L_r$. The mutual coupling among the R_x antenna elements is characterized by an $L_r \times L_r$ impedance matrix $\mathbf{Z}_{RR}$. The vector of voltages at the exit of the R_x coupling network, $\mathbf{V}_R$, is defined as

$$\mathbf{V}_R = -\mathbf{Z}_L \mathbf{i}_R \tag{3.248}$$

where $\mathbf{i}_R = [i_{r1}, i_{r2},\ldots i_{rL_r}]^T$ is the vector of R_x currents, and $\mathbf{Z}_L$ is the diagonal matrix of load impedances.

3.2.1.2 Impedance Matching Networks

The antenna pattern characteristics and the power collection capability of coupled antennas strongly depend on the impedance matching between the antenna inputs/outputs and the rest of the communication network [55]. Therefore, source and load impedances in Figure 3.20 need to be selected carefully. This section briefly reviews four different impedance matching networks that are often used in the literature [55, 56].

The first matching network is the *characteristic impedance* matching network that has the T_x and R_x antenna elements terminated with the same characteristic impedance Z_c (i.e., $Z_{si} = Z_c$ for $i = 1,\ldots, L_t$ and $Z_{Lk} = Z_c$, for $k = 1,\ldots, L_r$ [56]). The matching quality of the network depends on the difference between actual antenna impedances and the characteristic impedance. The second is the *self-impedance* matching network, where T_x and R_x antenna elements are terminated with the complex-conjugated self-impedances (i.e., $Z_{si} = Z^*_{TT_{ii}}$ for $i = 1,\ldots, L_t$ and $Z_{Lk} = Z^*_{RR_{kk}}$, for $k = 1,\ldots, L_r$ [56]). This matching network facilitates the maximum power transfer between the antennas and communication electronics when there is no mutual coupling. However, when mutual coupling is present, the matching depends on the behavior of antenna mutual impedances. The third is the *input-impedance* matching network, where T_x and R_x antenna elements are terminated with the complex-conjugate input impedances (i.e., $Z_{si} = Z^{\text{in}^*}_{Ti}$ for $i = 1,\ldots, L_t$ and $Z_{Lk} = Z^{\text{in}^*}_{Rk}$ for $k = 1,\ldots, L_r$, where Z^{in}_{Ti} and Z^{in}_{Rk} are calculated using (10) in [56]). Similar to the self-impedance matching network, the input-impedance matching network has a single-branch matching. However, while the self-impedance matching network takes into account only the self-impedances of antenna elements, the input-impedance matching network takes

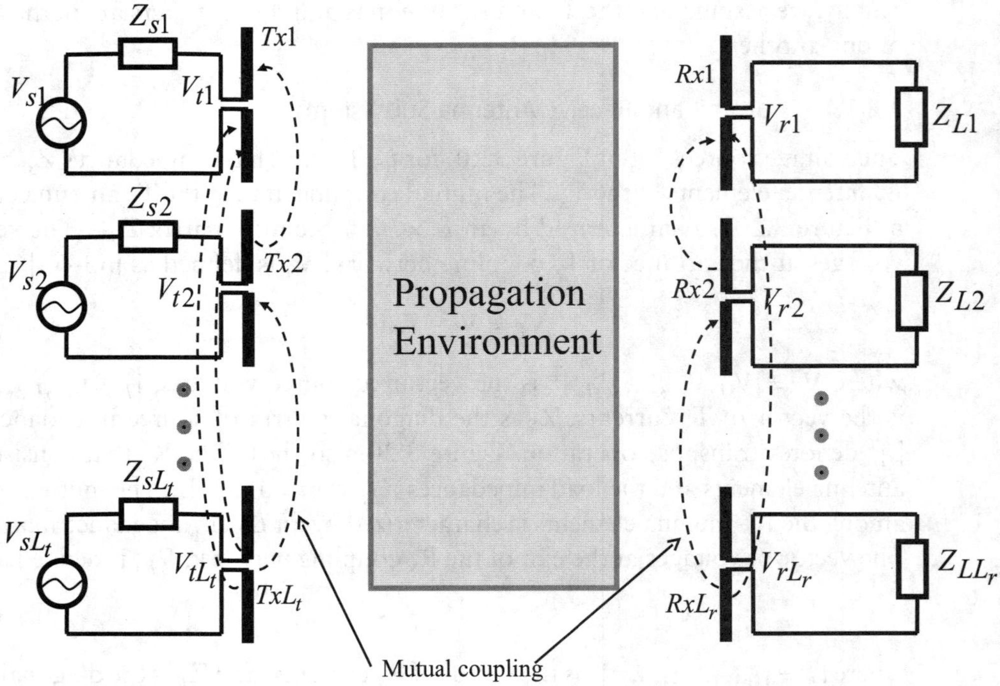

Figure 3.20 An $L_t \times L_r$ MIMO V-to-V system with mutual coupling.

into account both the self and mutual impedances. Finally, the fourth is the *multiport conjugate* matching network [55] that, similarly to the input-impedance matching network, takes into account mutual coupling among the antenna elements [55]. However, unlike the input-impedance match, it allows interconnections among the ports. Although this matching network offers both zero return loss and zero correlation for uniformly distributed angles of departure and arrival, it has very narrow bandwidth and high complexity, which prevents practical implementation. In this section, we will focus only on the first three matching networks.

3.2.2 Modeling of MIMO V-to-V Channels with Mutual Coupling

To evaluate the channel with mutual coupling, we model the MIMO V-to-V system in Figure 3.20 as a multiport network with the voltages and currents related as follows:

$$\begin{bmatrix} \mathbf{V}_R \\ \mathbf{V}_T \end{bmatrix} = \begin{bmatrix} \mathbf{Z}_{RR} & \mathbf{Z}_{RT} \\ \mathbf{Z}_{TR} & \mathbf{Z}_{TT} \end{bmatrix} \begin{bmatrix} \mathbf{i}_R \\ \mathbf{i}_T \end{bmatrix} \tag{3.249}$$

The subblock matrices $\mathbf{Z}_{TT}$ and $\mathbf{Z}_{RR}$ in (3.249) denote mutual coupling among T_x and R_x antenna elements, respectively, the subblock matrix $\mathbf{Z}_{RT}$ represents propagation form the T_x to the R_x, and the subblock matrix $\mathbf{Z}_{TR}$ represents propagation form the R_x to the T_x. In the analysis that follows, we assume that backscattering from the R_x to the T_x is negligible, (i.e., $\mathbf{Z}_{TR} = 0$).

3.2.2.1 MIMO V-to-V Channel without Mutual Coupling

The MIMO V-to-V channel without mutual coupling can be described by an $L_r \times L_t$ matrix $\mathbf{H} = [h_{ij}(t=0, \tau=0)]_{L_r \times L_t}$ of input delay-spread functions that is defined as

$$\mathbf{V}_{Ro} = \mathbf{H}\,\mathbf{V}_{To} \tag{3.250}$$

where $\mathbf{V}_{Ro} = [V_{ro1}, V_{ro2}, \ldots, V_{roL_r}]^T$ and $\mathbf{V}_{To} = [V_{to1}, V_{to2}, \ldots, V_{toL_t}]^T$ vectors contain open circuited voltages for the R_x and T_x, respectively. To characterize input delay-spread functions of matrix $\mathbf{H}$, we use the 3-D MIMO V-to-V channel model described in Section 3.1.9.

3.2.2.2 MIMO V-to-V Channel with Mutual Coupling

To include the effects of mutual coupling into MIMO V-to-V channel matrix $\mathbf{H}$, we use an $L_r \times L_t$ channel matrix $\mathbf{H}^c$, defined as [53]

$$\mathbf{V}_R = \mathbf{H}^c \mathbf{V}_S \tag{3.251}$$

It has been shown [54] that the channel matrix $\mathbf{H}^c$ can be written as

$$\mathbf{H}^c = \frac{\mathbf{Z}_L}{(\mathbf{Z}_L + \mathbf{Z}_{RR})}\mathbf{H}\frac{\mathbf{Z}_{TT}}{(\mathbf{Z}_S + \mathbf{Z}_{TT})} = \mathbf{Z}_R \mathbf{H} \mathbf{Z}_T \tag{3.252}$$

where $\mathbf{Z}_R = \mathbf{Z}_L(\mathbf{Z}_L + \mathbf{Z}_{RR})^{-1}$ and $\mathbf{Z}_T = \mathbf{Z}_{TT}(\mathbf{Z}_S + \mathbf{Z}_{TT})^{-1}$. Then, the input delay-spread functions of the MIMO V-to-V channel matrix with mutual coupling, $\mathbf{H}^c$, can be written as

$$h_{pq}^c = \sum_{m=1}^{L_t} Z_{Tmq} \sum_{n=1}^{L_r} Z_{Rpn} h_{nm} \tag{3.253}$$

3.2.3 Analysis of MIMO V-to-V Channel with Mutual Coupling

This section discusses the impact of mutual coupling on T_x and R_x antenna patterns, spatial correlation, received power, and capacity of MIMO V-to-V systems. The T_x and R_x are equipped with two identical vertically polarized half-wavelength ($\lambda/2$) electric dipole antennas with radii $\lambda/600$, where λ = 123.3 m. The antenna dipoles are modeled using a software package based on the method of moments [57]. Characteristic impedances of T_x and R_x antenna elements, Z_c, are chosen to be 50Ω, an assumption commonly used in practice.

3.2.3.1 Antenna Patterns and Matching Networks

To illustrate the effect of mutual coupling on the dipole antenna pattern, Figure 3.21 plots the radiation pattern of one of the T_x dipoles for several different antenna spacings, $d_T \in \{0, 0.1, 0.3, 0.5, 1\}$. The other T_x dipole is terminated with the characteristic impedance Z_c. The R_x antenna elements have negligible effect on the T_x antenna patterns because they are in the far-field of T_x antenna elements. From Figure 3.21, we can observe that the isolated dipole antenna (i.e., $d_T = 0$) has the omnidirectional pattern. When the second antenna element is introduced and their spacing is very small (i.e., $d_T = 0.1\lambda$), the pattern remains omnidirectional, but the radiated power is significantly reduced. For the distances 0.3λ–0.5λ, mutual coupling noticeably changes the radiating pattern and the radiated power. The antenna pattern does not remain omnidirectional, and the radiated power becomes a function of the angles of departure and arrival. Finally, for antenna spacings greater than or equal to λ, mutual coupling has less impact on the antenna radiation and the pattern resembles that of an isolated dipole antenna.

To characterize the influence of different matching networks on the dipole antenna pattern, Figure 3.22 plots the radiation pattern of one of the T_x dipoles, when all antenna elements are terminated with characteristic impedance, self-impedance, or input-impedance matching networks. For comparison, we also plot the antenna pattern of an isolated dipole antenna. The coupling impedance matrices $\mathbf{Z}_{TT}$ and $\mathbf{Z}_{RR}$ are obtained using antenna-analysis software [57], the input impedances are calculated using the results in [56], and the spacing between antenna elements is $d_T = 0.3\lambda$. The results in Figure 3.22 show that mutual coupling noticeably impacts the radiating pattern and the radiated power, even in the presence of a matching network. The coupled dipole antenna pattern is not omnidirectional, and the radiated power strongly depends on the angles of departure and arrival. For example, the radiated power of a coupled dipole antenna is larger than that of an isolated dipole for angles of departure/arrival between 120° and 240°. However, for the rest of angles of departure/arrival, the radiated power of a coupled dipole

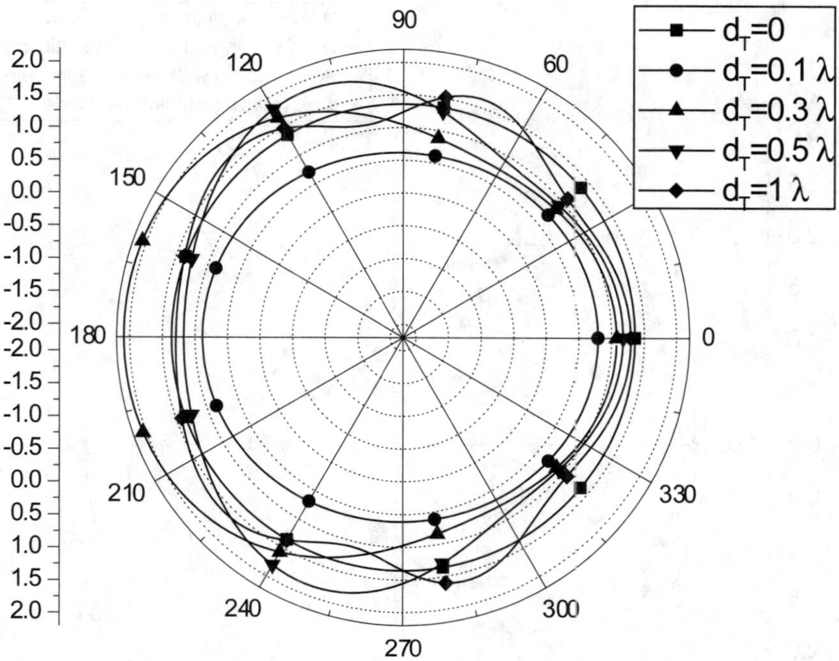

Figure 3.21 The coupled dipole antenna pattern in the azimuthal plane as a function of antenna element spacings. The second dipole antenna is terminated with $Z_c = 50\Omega$.

antenna is smaller than that of an isolated dipole. Finally, Figure 3.22 shows that input-impedance and self-impedance matching networks significantly improve the radiated power of coupled antennas, while the characteristic impedance matching network is less effective.

3.2.3.2 Spatial Correlation and Received Power

The normalized spatial correlation function between two input delay-spread functions with mutual coupling defined in (3.253), i.e., h^c_{pq} and $h^c_{\tilde{p}\tilde{q}}$ is defined as

$$R^c_{pq,\tilde{p}\tilde{q}} = \frac{\mathrm{E}\left[h^{c*}_{pq} h^c_{\tilde{p}\tilde{q}}\right]}{\sqrt{\mathrm{Var}\left[h^{c*}_{pq}\right]\mathrm{Var}\left[h^c_{\tilde{p}\tilde{q}}\right]}} \tag{3.254}$$

where E[·] is the statistical expectation operator, Var[·] is the statistical variance operator, $p, \tilde{p}, \in \{1, ..., L_t\}$, and $q, \tilde{q}, \in \{1, ..., L_r\}$. Substituting (3.253) into (3.254), the normalized spatial correlation with mutual coupling becomes

$$R^c_{pq,\tilde{p}\tilde{q}} = \sum_{m=1}^{L_t} \sum_{n=1}^{L_t} \frac{Z_{Tmq}}{C_T} \frac{Z^*_{Tn\tilde{q}}}{C^*_T} \sum_{k=1}^{L_r} \sum_{l=1}^{L_r} \frac{Z_{Rpk}}{C_R} \frac{Z^*_{R\tilde{p}l}}{C^*_R} R_{pq,\tilde{p}\tilde{q}} \tag{3.255}$$

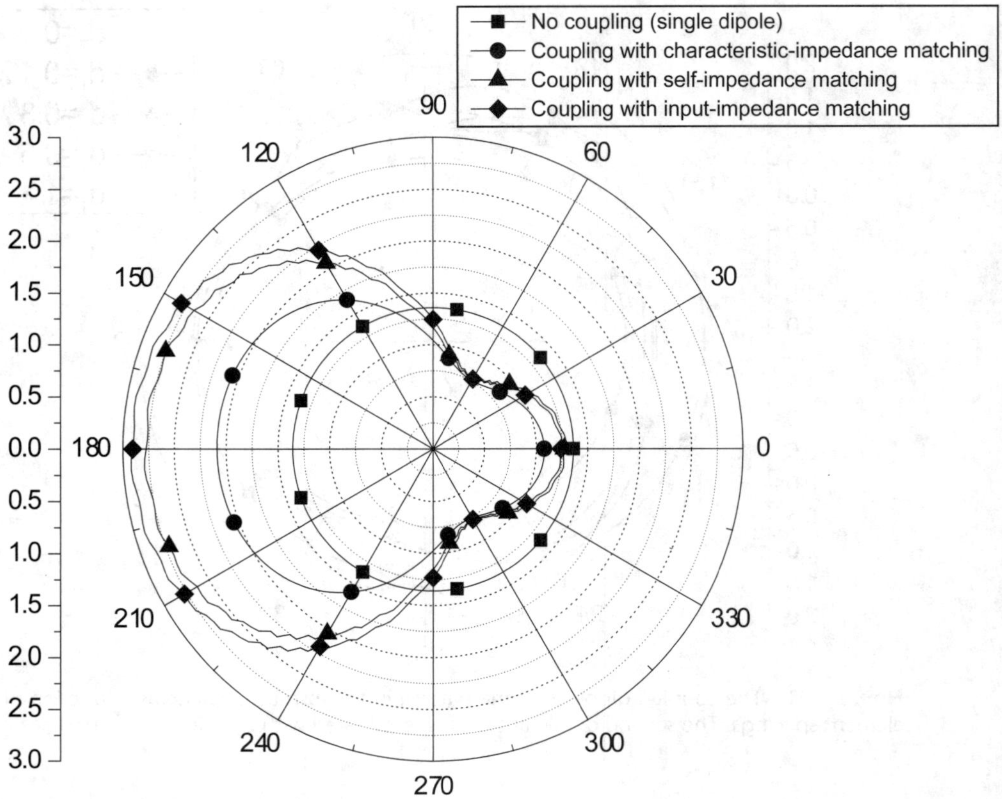

Figure 3.22 The coupled dipole antenna patterns in the azimuthal plane for antenna elements terminated with the characteristic impedance, self-impedance, and input-impedance matching networks.

where $R_{pq,\tilde{p}\tilde{q}}$ denotes the normalized spatial correlation function defined in (3.180), and C_T and C_R are normalization factors that lead to $\mathbf{V}_T = \mathbf{V}_S$ and $\mathbf{V}_R = \mathbf{V}_{Ro}$ for zero mutual coupling at the T_x and R_x, respectively.

Figure 3.23 shows the effect of mutual coupling on the spatial correlation. The coupled T_x and R_x antennas are terminated with the characteristic, self-, or input-impedance matching networks. We assume 3-D nonisotropic radio propagation ($k_T = k_R = 3$, $\beta_{Tm} = \beta_{Rm} = 10°$) in an urban street environment ($\eta_T = \eta_R = 0.05$, $\eta_{TR} = 0.9$). The curves are obtained for parameters $\mu_T = \mu_R = 0°$, $R_{t1} = R_{r1} = 3$m, $R_{t2} = R_{r2} = 30$m, and $K = 0$. The distance between the T_x and R_x is $D = 300$m and the antenna array elements at both vehicles have azimuth and elevation angles $\theta_T = \theta_R = 0°$ and $\psi_T = \psi_R = 0°$, respectively. All antenna elements have the same height, $\Delta_H = 0$, and the pathloss exponent γ is set to 4. The results show that, for antenna element spacings 0λ to 0.4λ, the system with the input-impedance matching network has the smallest spatial correlation, while the spatial correlations of systems with self- and characteristic impedance matching networks are slightly smaller or equal to that of the system without mutual coupling. The system with the input-impedance matching network has the smallest spatial correlation because it has the largest dissimilarity in radiation patterns (see Figure 3.22). On the other hand, for distances

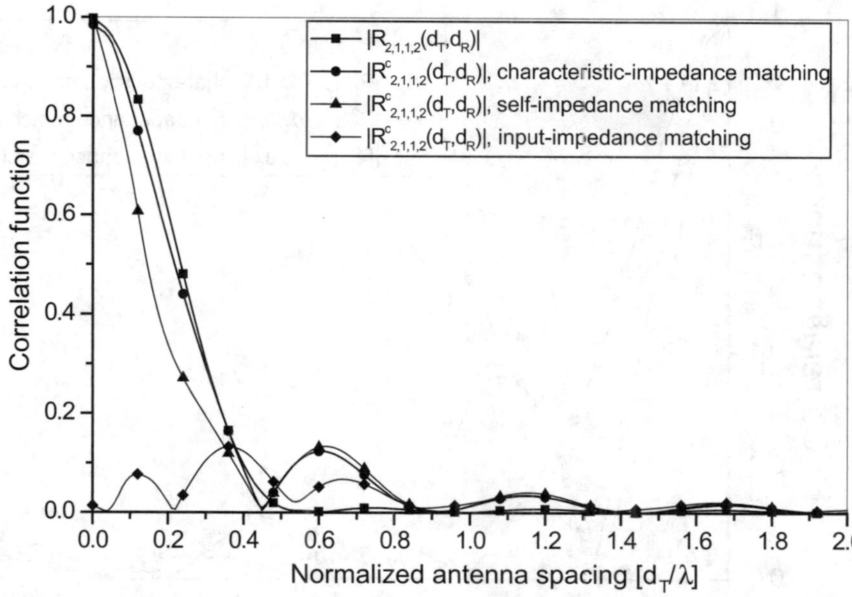

Figure 3.23 The spatial correlation function with and without mutual coupling in urban street environments.

larger than 0.5λ, dissimilarity in antenna patterns is less dominant and the system without mutual coupling has the smallest spatial correlation.

Figure 3.24 shows the effect of mutual coupling on the spatial correlation in interstate-highway (IH) environments. We assume 3-D nonisotropic radio propagation ($k_T = k_R = 3$, $\beta_{T_m} = \beta_{R_m} = 10°$) in IH environments ($\eta_T = \eta_R = 0.35$, $\eta_{TR} = 0.3$). The coupled T_x and R_x antennas are terminated with characteristic, self-, or input-impedance matching networks. The rest of the parameters in Figure 3.24 are chosen as in Figure 3.23. The results show that for the antenna element spacings 0λ to 0.3λ, the system with the input-impedance matching network has the smallest spatial correlation, while spatial correlations of systems with self- and characteristic impedance matching network are smaller than that of the system without mutual coupling. On the other hand, for distances larger than 0.3λ, the system without mutual coupling has the smallest correlation. Figures 3.23 and 3.24 show that the correlation of system with mutual coupling depends on the type of propagation environment and the type of matching network.

Low correlation between coupled antenna elements does not necessarily lead to maximum diversity gain. Since mutual coupling may cause unequal power distribution in T_x and R_x branches, power collection capabilities of MIMO V-to-V system are also important. To assess this property of the system, we analyze the received power of the coupled-dipole antenna elements normalized to the received power of the single-dipole antenna. Figure 3.25 compares the normalized received power of systems without and with mutual coupling. The parameters used to obtain curves in Figure 3.25 are the same as those in Figure 3.23, and the system with mutual coupling is terminated with an input-impedance matching network. The results show that, for the antenna element spacings 0λ to 0.3λ, the received power of a system with mutual coupling is significantly smaller than that of the system without

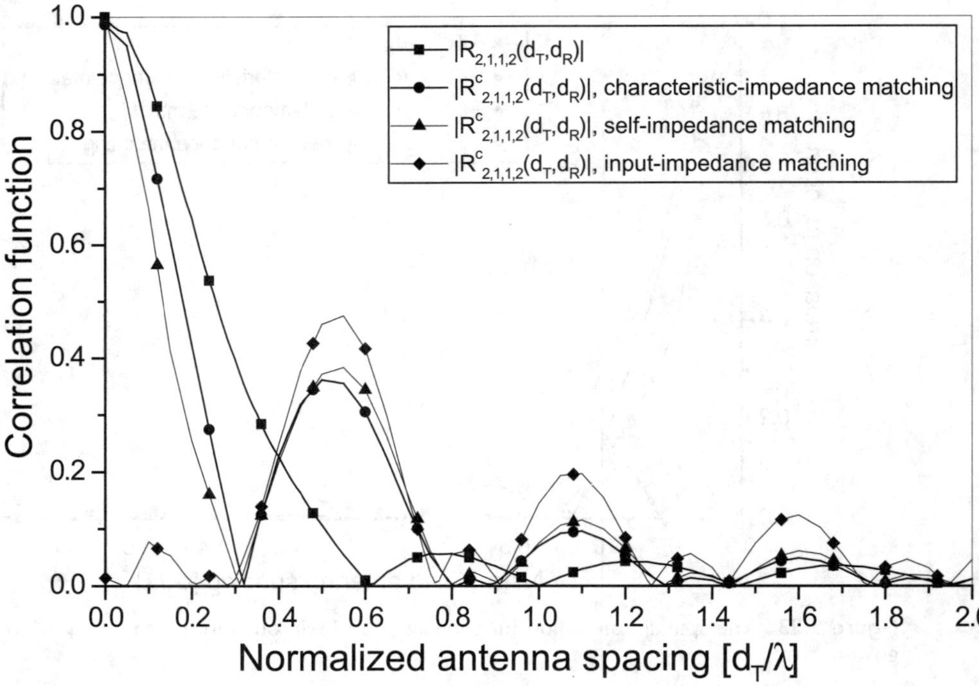

Figure 3.24 The spatial correlation function with and without mutual coupling in interstate highway environments.

mutual coupling, even though the input-impedance matching network increases decorrelation of the antenna elements. The results also show that two coupled antenna elements with spacings between 0.3λ–0.6λ can collect more power than two widely separated ones (i.e., the normalized received power is greater than two). One possible explanation for this increased effectiveness of coupled dipoles is that a portion of the power is scattered by one receiving antenna and can be recaptured by the other receiving antenna. The received power of systems with characteristic and self-impedance matching networks are similar to that of the system with input-impedance match and hence omitted in Figure 3.25.

Figure 3.26 shows the normalized received power of a system with input-impedance matching as a function of the orientations of local scatterers in the azimuthal plane (i.e., μ_T and μ_R) around the T_x and R_x, respectively. The parameters used to obtain the curve in Figure 3.26 are the same as those in Figure 3.23. The results show that the received power with coupled antenna elements strongly depend on the position of local scatterers and how they interact with directional radiating patterns caused by coupling. Finally, Figure 3.26 shows that available diversity may vary between one and two antenna elements.

3.2.3.3 Channel Capacity

The instantaneous channel capacity (in bit/s/Hz) of a stochastic MIMO V-to-V channel *with* mutual coupling, under an average transmit power constraint is [16]

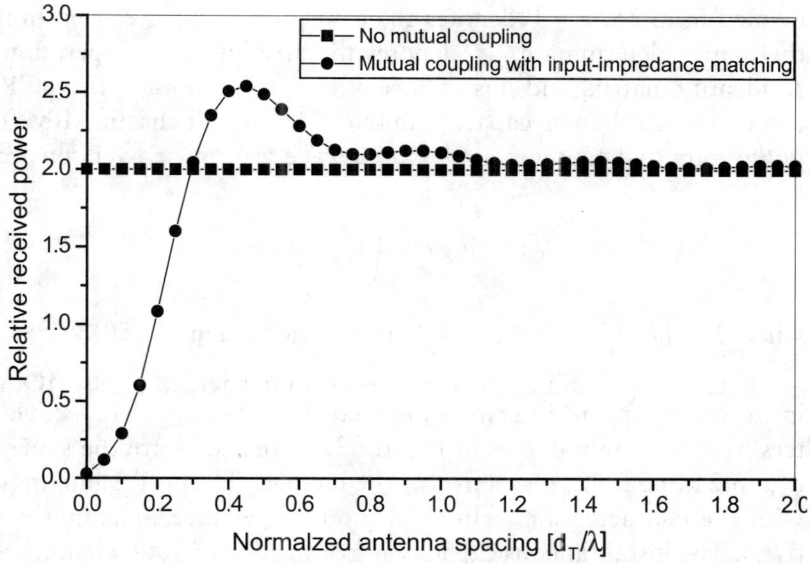

Figure 3.25 The normalized received power with and without mutual coupling as a function of the normalized antenna spacing.

$$C_c = \log_2 \det\left(\mathbf{I}_{Lr} + \frac{\rho}{L_t} \frac{\mathbf{H}^c \mathbf{H}^{c^H}}{|C_T|^2 |C_R|^2} \right) \tag{3.256}$$

under the assumptions that $L_t \geq L_r$, that T_x has no channel knowledge, and that R_x has perfect channel knowledge. In (3.256), $\mathbf{H}^c = \left[h_{ij}^c \right]_{L_r \times L_t}$ is the $L_r \times L_t$ matrix

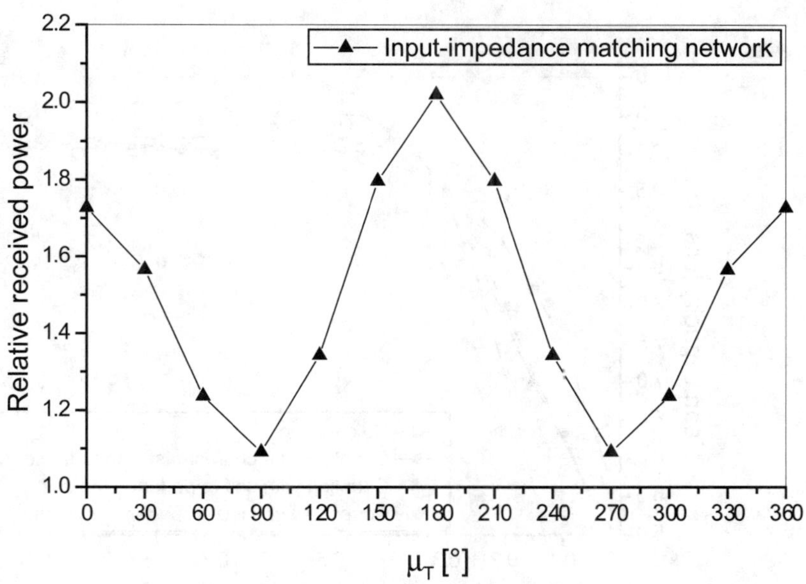

Figure 3.26 The normalized received power with mutual coupling as a function of the orientations of local scatterers in the azimuthal plane.

defined in (3.252), $(\cdot)^H$ denotes the transpose conjugate operation, $\det(\cdot)$ denotes the matrix determinant, $|\cdot|$ denotes the absolute value operation, $\mathbf{I}_{L_r}$ is the $L_r \times L_r$ identity matrix, and ρ is the average signal-to-noise ratio (SNR). Similarly, the instantaneous channel capacity (in bit/s/Hz) of a stochastic MIMO V-to-V channel *without* mutual coupling, under an average transmit power constraint is

$$C = \log_2 \det\left(\mathbf{I}_{L_r} + \frac{\rho}{L_t}\mathbf{H}\mathbf{H}^H\right) \tag{3.257}$$

where $\mathbf{H} = \left[h_{ij}\right]_{L_r \times L_t}$ is the $L_r \times L_t$ matrix defined in (3.250).

Figures 3.27 and 3.28 compare mean capacities of MIMO V-to-V channels in urban street and interstate highway environments, respectively. The parameters used to obtain curves in Figures 3.27 and 3.28 are the same as those in Figures 3.21 and 3.22, respectively. The coupled T_x and R_x antennas are terminated with the characteristic, self-, and input-impedance matching networks, respectively. The instantaneous capacity is averaged over 5000 channel realizations. The results in Figures 3.27 and 3.28 show that the capacities in urban street and interstate highway environments are relatively similar. This is in contrast to the results obtained for the correlation and received power, which are strongly dependent on the propagation environment. Figures 3.27 and 3.28 show that for antenna element spacings 0λ–0.3λ, the system with an input-impedance matching network provides the smallest capacity, while the capacity of systems with self- and Z_c impedance matching networks are slightly smaller than that of the system without mutual coupling. These results show that the MIMO V-to-V channel model without mutual coupling overestimates the available capacity, which illustrates the importance of incorporating the effect of mutual coupling into propagation models. Furthermore, these results differ from the results obtained for F-to-M channels,

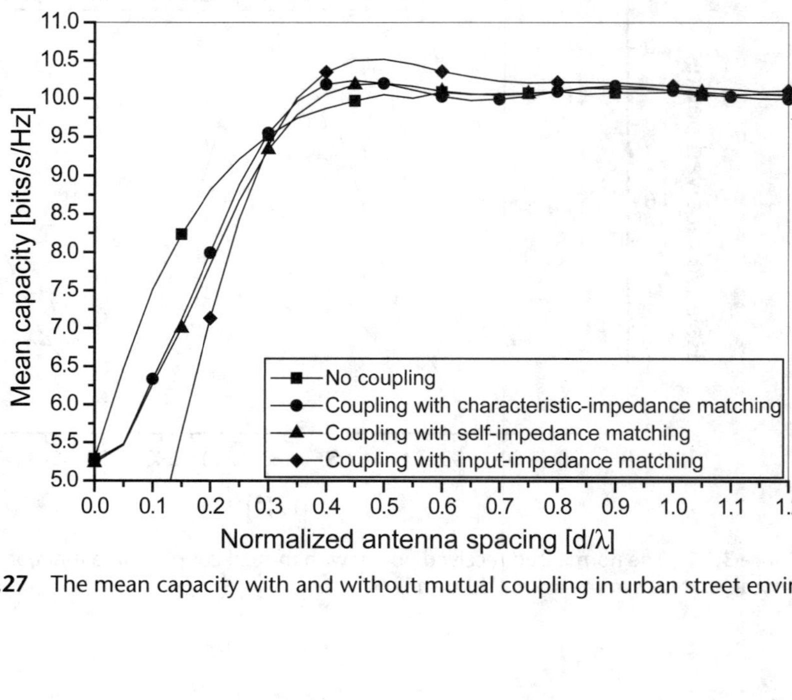

Figure 3.27 The mean capacity with and without mutual coupling in urban street environments.

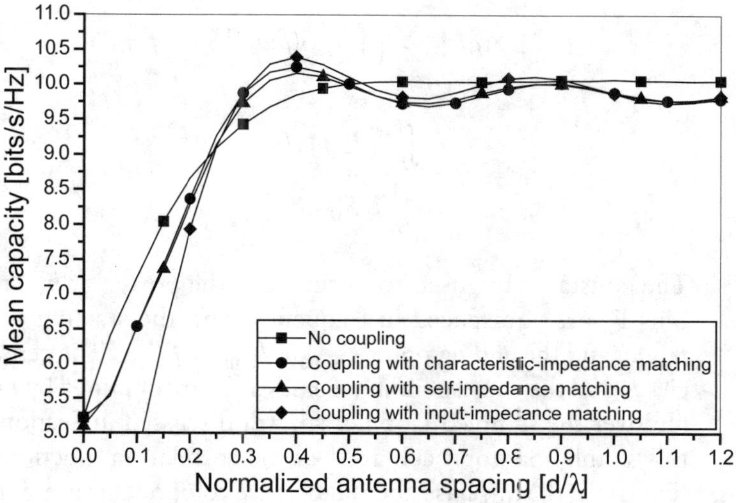

Figure 3.28 The mean capacity with and without mutual coupling in interstate-highway environments.

where the capacity with mutual coupling is larger than the capacity without mutual coupling, regardless of which antenna spacing is used [55]. On the other hand, for the antenna element spacings 0.3λ–0.5λ, the results show that all three systems with mutual coupling have similar capacities and outperform the system without mutual coupling. Finally, for the antenna spacings larger than 0.8λ, the capacities of systems with mutual coupling converge to that of the system without mutual coupling.

3.3 Simulators for Vehicle-to-Vehicle Radio Channels

Simulation of V-to-V channels is a crucial part of link layer simulations and of the overall characterization of a V-to-V system performance. The channel models detailed in Section 3.1 are very useful mathematical tools for understanding propagation mechanisms and for statistical characterization of channels. However, these models assume an infinite number of scatterers, making these models impractical for simulators. We now describe simulation models with finite numbers of scatterers that still match the statistical properties of reference models.

The following approaches are available when simulating the V-to-V fading channels:

- *Discrete line spectrum method*. This method generates the channel impulse response by discretizing the power spectral density prior to transforming it back into the time domain. Wang and Cox [59] introduced a model that approximates the continuous Doppler spectrum by a discrete line spectrum. This model can be written as follows:

$$h(t) = \frac{1}{A}\sum_{n=1}^{N}\sqrt{S(f_n)}e^{-j(2\pi f_n t+\phi_n)} \tag{3.258}$$

where

$$S(f_n) = \sum_{n=1}^{N}\left(\int_{F_n}^{F_{n+1}} S(f)df\right)\delta(f - f_n) \tag{3.259}$$

$$f_n = \frac{\int_{F_n}^{F_{n+1}} fS(f)df}{\int_{F_n}^{F_{n+1}} S(f)df} \tag{3.260}$$

The constant A is used to normalized the power of $h(t)$ to unity. Frequency bins F_n are equispaced in frequency with the starting point $F_1 = -(f_{\text{Tmax}} + f_{\text{Rmax}})$ and the ending point $F_{N+1} = f_{\text{Tmax}} + f_{\text{Rmax}}$. The center-of-mass frequencies f_n and the corresponding masses $S(f_n)$ are obtained by numerically integrating over the sampled Doppler spectra discussed in Section 3.1 Drawbacks of this simulation approach are that the correlation functions vary with simulation trials, the in-phase and quadrature components are statistically correlated, the autocorrelation function exhibits periodicities, and the need for numerical integration [60].

- *Sum-of-sinusoids method.* SoS models generate the channel waveform by superimposing a finite number of properly selected sinusoids. These models have low complexity, produce channel waveforms that have high accuracy and can be easily extended to develop space-time-frequency correlated channel models for MIMO antenna systems. Due to these advantages, this chapter focuses on SoS models.

  As pointed out in Section 2.4, SoS models can be classified as either deterministic or statistical. Deterministic SoS models have random phases that are fixed in all simulation runs, deterministic amplitudes, and deterministic Doppler frequencies. In contrast, statistical SoS models leave at least one of the parameter sets (amplitudes, phases, or Doppler frequencies) as a random variable that varies with each simulation trial. The statistical properties of statistical SoS models also vary for each simulation trial, but converge to the desired properties when averaged over a sufficient number of simulation trials.

3.3.1 Simulators for V-to-V Frequency-Flat Multipath Fading Channels

3.1.1.1 SoS Simulation Models for SISO V-to-V Channels

Patel et al. [60] were the first to propose SoS simulation models for SISO V-to-V channels. They used the two-ring concept [19] to derive their models. Using this model, the received complex faded envelope under narrowband, frequency flat fading, and NLoS propagation assumptions is defined as [60]

$$h(t) = \sqrt{\frac{2}{NM}} \sum_{n=1}^{N}\sum_{m=1}^{M} \exp\{j[\omega_1 t\cos(\alpha_n) + \omega_2 t\cos(\beta_m) + \phi_{nm}]\} \tag{3.261}$$

where NM is the number of propagation paths, ω_1 and ω_2 are the maximum angular Doppler frequencies, ϕ_{mn} denotes the phase, and α_n and β_m denote the angles of departure and arrival, respectively.

The first model proposed by Patel et al. [60] is a *deterministic* statistical SoS model. By choosing only the phases to be random variables, the statistical properties of this model converge to the desired properties in a single simulation trial. We refer to this model as SISO V-to-V Model I.

SISO V-to-V Model I The complex faded envelope is $h(t) = h_i(t) + jh_q(t)$, where

$$h_i(t) = \sqrt{\frac{2}{N_i M_i}} \sum_{m=1}^{M_i} \sum_{n=1}^{N_i} \cos(\omega_1 t \cos(\alpha_n^i) + \omega_2 t \cos(\beta_m^i) + \phi_{nm}^i) \tag{3.262}$$

$$h_q(t) = \sqrt{\frac{2}{N_q M_q}} \sum_{m=1}^{M_q} \sum_{n=1}^{N_q} \cos(\omega_1 t \cos(\alpha_n^q) + \omega_2 t \cos(\beta_m^q) + \phi_{nm}^q) \tag{3.263}$$

and where ϕ_{nm}^i and ϕ_{nm}^q are independent random phases uniformly distributed on the interval $[-\pi, \pi)$. The nth angle of departure and the mth angle of arrival are

$$\alpha_n^{i/q} = \frac{\pi(n-0.5)}{(2N_{i/q})} \tag{3.264}$$

$$\beta_m^{i/q} = \frac{\pi(m-0.5)}{(M_{i/q})} \tag{3.265}$$

for $n = 1, 2, \ldots, N_{i/q}$ and $m = 1, 2, \ldots, M_{i/q}$, respectively.

This model has the disadvantage that its statistical properties match those of the reference model only for a small range of normalized time delays ($0 \leq f_1 T_s \leq 3$, where T_s denotes the sampling period). To improve these statistical properties, Patel et al. [60] also proposed a *statistical* SoS model. We refer to this model as SISO V-to-V Model II.

SISO V-to-V Model II The complex faded envelope is defined as $h(t) = h_i(t) + jh_q(t)$, where

$$h_i(t) = \sqrt{\frac{2}{NM}} \sum_{n=1}^{N} \sum_{m=1}^{M} \cos(\omega_1 t \cos(\alpha_n) + \omega_2 t \cos(\beta_m) + \phi_{nm}^i) \tag{3.266}$$

$$h_q(t) = \sqrt{\frac{2}{NM}} \sum_{n=1}^{N} \sum_{m=1}^{M} \cos(\omega_1 t \sin(\alpha_n) + \omega_2 t \cos(\beta_m) + \phi_{nm}^q) \tag{3.267}$$

and where ϕ_{nm}^i and ϕ_{nm}^q are independent random phases uniformly distributed on the interval $[-\pi,\pi)$. Model II assumes N scatterers located on the T_x ring and M scatterers located on the R_x ring. The nth angle of departure and the mth angle of arrival are

$$\alpha_n = \frac{2\pi n - \pi + \theta}{4N} \tag{3.268}$$

$$\beta_m = \frac{2\pi m - \pi + \psi}{2M} \tag{3.269}$$

where θ and ψ are independent random variables uniformly distributed in the interval $[-\pi, \pi)$.

SISO V-to-V Model II obtains better statistical properties than SISO V-to-V Model I. However, this model requires a large number of simulation trials (at least 50) to obtain adequate ensemble averaged statistical properties. Furthermore, Models I and II have a noted difficulty in producing time averaged auto- and cross-correlation functions that match those of the reference model. The solution to this problem is offered by Zajić et. al in [61] by using a two-ring concept and by choosing orthogonal functions for the I and Q components of the complex faded envelope. Also, this model is designed to directly generate multiple uncorrelated complex envelopes. We refer to this model as SISO V-to-V Model III.

SISO V-to-V Model III The kth complex faded envelope is $h_k(t) = h_{ik}(t) + jh_{qk}(t)$, where

$$h_{ik}(t) = \frac{2}{\sqrt{N_0 M}} \sum_{n=1}^{N_0} \sum_{m=1}^{M} \cos(\omega_2 t \cos\beta_{mk}) \cos(\omega_1 t \cos\alpha_{nk} + \phi_{nmk}) \tag{3.270}$$

$$h_{qk}(t) = \frac{2}{\sqrt{N_0 M}} \sum_{n=1}^{N_0} \sum_{m=1}^{M} \sin(\omega_2 t \cos\beta_{mk}) \sin(\omega_1 t \sin\alpha_{nk} + \phi_{nmk}) \tag{3.271}$$

It assumed that P independent complex envelopes are desired ($k = 0, \ldots, P-1$) each having MN_0 sinusoidal terms in the I and Q components. The angles of departures and the angles of arrivals are chosen as follows:

$$\alpha_{nk} = \frac{2\pi n}{4N_0} + \frac{2\pi k}{4PN_0} + \frac{\theta - \pi}{4N_0} \tag{3.272}$$

$$\beta_{mk} = 0.5\left(\frac{2\pi m}{M} + \frac{2\pi k}{PM} + \frac{\psi - \pi}{M}\right) \tag{3.273}$$

for $n = 1, \ldots, N_0$, $m = 1, \ldots, M$, $k = 0, \ldots, P-1$. The angles of departures and the angles of arrivals in the kth complex faded envelope are obtained by rotating the angles of departures and the angles of arrivals in the $(k-1)$th complex envelope by $(2\pi)/(4PN_0)$ and $(2\pi/2PM)$, respectively. The parameters ϕ_{nmk}, θ, and ψ are independent random variables uniformly distributed on the interval $[-\pi, \pi)$.

To compare complexity of the SISO V-to-V Models I–III, Table 3.3 summarizes the number of simulation trials required to obtain desired statistical properties, the number of operations needed to generate one sample of the complex envelope, and the relative simulation times needed to generate a sample of the complex faded envelope with desired statistical properties, in Matlab on a Pentium III processor. Here, we count only the frequently executed operations and the number of random variables. Table 3.3 shows that Model III has slightly higher complexity compared to Models I and II.

Figures 3.29 and 3.30 compare the variances of the auto-correlation functions of the in-phase components averaged over 10^3 simulation trials for the SISO V-to-V

Table 3.3 Complexity of Different SISO V-to-V Models

Simulation Models	*Number of Simulation Trials*	*Estimated Number of Computations Needed to Generate One Sample of $h_k(t)$*			*Relative Simulation Time for One Simulation Trial*
		Cosine	*Addition*	*Random Variable*	
SISO V-to-V Model I	1	6*MN*	4*MN*	2*MN*	*Tx*
SISO V-to-V Model II	50	6*MN*	4*MN*	2*MN* + 2	1.25*Tx*
SISO V-to-V Model III	30	8*MN*	2*MN*	*MN* + 2	1.65*Tx*

Aki and Haber's theoretical model and Models I–III. For the statistical models, the variance is calculated as [62]

$$\mathrm{Var}[R(\cdot)] = E[|\,\widehat{R}(\cdot) - \lim_{N\to\infty} R(\cdot)\,|^2] \tag{3.274}$$

where $\widehat{R}(\cdot)$ denotes time-averaged correlations and $R(\cdot)$ denotes the statistical correlation function. For the deterministic model, the equivalent quantity is the squared error $|\,\widehat{R}(\cdot) - R(\cdot)\,|^2$. The variance for the Akki and Haber's theoretical model described in Section 3.1.3 is obtained using [61]

$$\begin{aligned}\mathrm{Var}\left[R_{h_i,h_i}(\tau)\right] &= \mathrm{Var}\left[R_{h_q,h_q}(\tau)\right] \\ &= \left[1 + J_0(2\omega_1\tau)J_0(2\omega_2\tau) - 2J_0^2(\omega_1\tau)J_0^2(\omega_2\tau)\right]/(2N)\end{aligned} \tag{3.275}$$

The variance provides a measure of the usefulness of the model in simulating the desired channel with a finite *N*. A lower variance means that a smaller number of simulation trials are needed to achieve the desired statistical properties and, hence, the corresponding model is better. Since the theoretical model does not exploit the

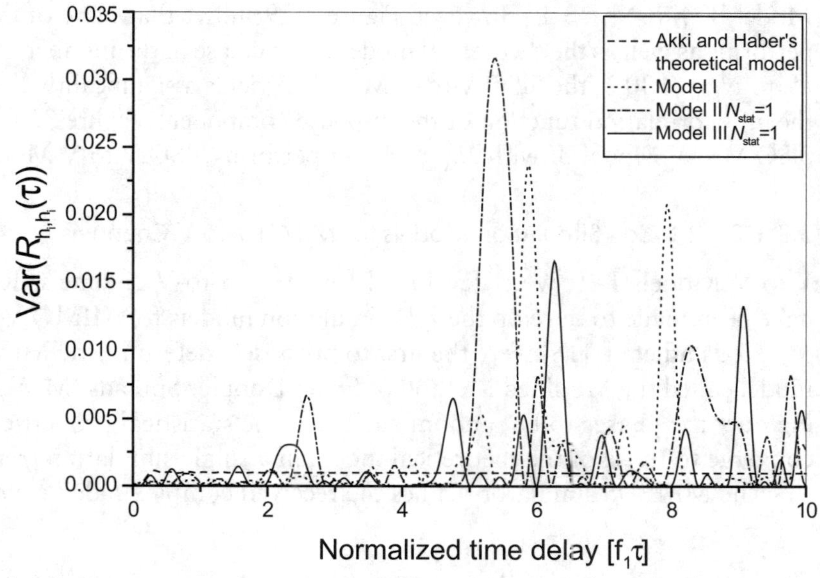

Figure 3.29 The variances of the auto-correlation functions of the in-phase components for the SISO V-to-V Aki and Haber's theoretical model and Models I–III and $N_{\text{stat}} = 1$.

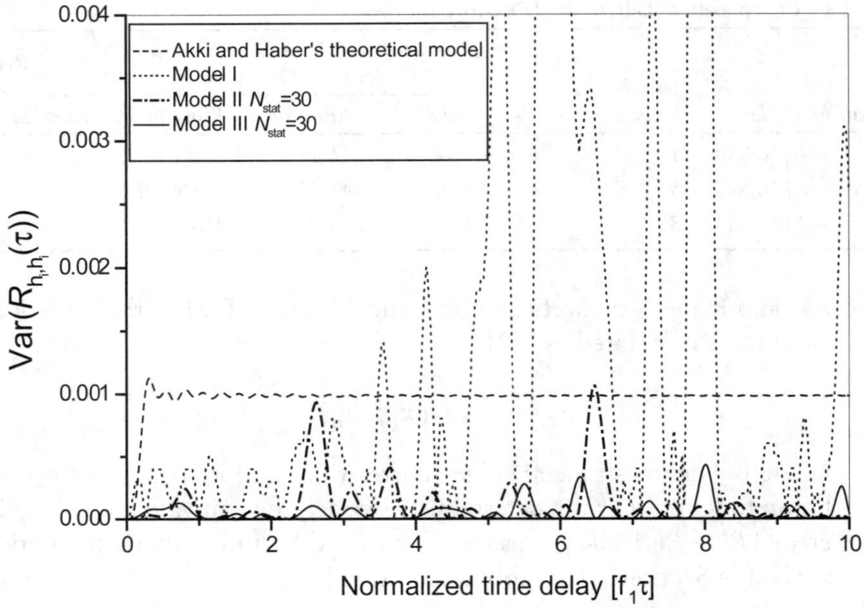

Figure 3.30 The variances of the auto-correlation functions of the in-phase components for the SISO V-to-V Aki and Haber's theoretical model, Model I, and Models II and III with $N_{stat} = 30$.

symmetry of the double ring model, for fair comparison, we use $N = 4N_0 \times 2M = 512$ sinusoids for simulation of the theoretical model. In all simulations, we use a normalized sampling period $f_1T_s = 0.01(f_1 = f_2$ are the maximum Doppler frequencies and T_s is the sampling period) and $M = N_0 = P = 8$. For Model II, we use $N_i = M_i = 8$ and $N_q = M_q = 9$ to obtain a complex envelope with uncorrelated quadrature components.

From Figure 3.29, we can observe that SISO V-to-V Model III with $N_{stat} = 1$ has a similar variance to the SISO V-to-V Model I and a lower variance than SISO V-to-V Model II with $N_{stat} = 1$. However, Figure 3.29 shows that none of the three models performs as well as the theoretical model. An increase in the number of simulation trials to $N_{stat} = 30$ in the SISO V-to-V Model III yields a significantly lower variance of the autocorrelation function of the in-phase component. Figure 3.30 shows that the SISO V-to-V Model III with $N_{stat} = 30$ outperforms SISO V-to-V Models I and II.

3.3.1.2 2-D SoS Simulation Models for MIMO V-to-V Channels

V-to-V Models I–III were developed for SISO V-to-V channels. Recently, efforts have been made to develop the 2-D simulation models for MIMO V-to-V channels.

Hogstad et al. [16] were the first to propose a deterministic MIMO V-to-V SoS model called the Modified Method of Exact Doppler Spreads (MMEDS). By choosing only the phases to be random variables, the statistical properties of this model converge to those of the theoretical model in a single simulation trial.

The MMEDS simulator defines the received complex faded envelope as [16]

$$h_{pq}(t) = \sum_{m,n=1}^{M,N} \frac{a_{p,m}c_{m,n}b_{n,q}}{\sqrt{MN}} e^{j\left[2\pi f_{T\max} t\cos(\alpha_T^{(m)}-\gamma_T)+2\pi f_{R\max} t\cos(\alpha_R^{(n)}-\gamma_R)+\phi_{mn}\right]} \tag{3.276}$$

where

$$a_{p,m} = \exp\left(j\pi(L_t + 1 - 2p)\delta_T \cos(\alpha_T^{(m)} - \theta_T)/\lambda\right) \tag{3.277}$$

$$c_{m,n} = \exp\left(j2\pi(R_t \cos\alpha_T^{(m)} - R_r \cos\alpha_R^{(n)})/\lambda\right) \tag{3.278}$$

$$b_{n,q} = \exp\left(j\pi(L_r + 1 - 2q)\delta_R \cos(\alpha_R^{(n)} - \theta_R)/\lambda\right) \tag{3.279}$$

and parameters p and q take from the sets $p \in \{1, \ldots, L_t\}$ and $q \in \{1, \ldots, L_r\}$, respectively, where L_t and L_r are numbers of the T_x and R_x antenna elements, respectively. Parameters θ_T and θ_R are the T_x and R_x antenna array orientations, δ_T and δ_R are the T_x and R_x spacings among antenna elements, R_t and R_r are the radii of circles around the T_x and R_x, γ_T and γ_R are the directions of movement of the T_x and R_x, and $f_{T\max}$, and $f_{R\max}$ are the T_x and R_x maximum Doppler frequencies, respectively. With the MMEDS model, the phases ϕ_{mn} are independent random variables uniformly distributed on the interval $[0, 2\pi)$, and the angles of departure $\alpha_T^{(m)}$ and the angles of arrival $\alpha_R^{(n)}$ are chosen as follows:

$$\alpha_T^{(m)} = \frac{2\pi}{M}(m - 0.5) + \gamma_T \tag{3.280}$$

$$\alpha_R^{(n)} = \frac{2\pi}{N}(n - 0.5) + \gamma_R \tag{3.281}$$

for $m = 1, \ldots, M$, $n = 1, \ldots, N$, respectively.

Another deterministic and a statistical simulation model has been proposed in [18]. These models are based on the theoretical MIMO 2-D model detailed in Section 3.1.5. In these models, a received complex faded envelope is defined as $h_{pq}(t) = h_{pq}^{(I)}(t) + jh_{pq}^{(Q)}(t)$ where

$$\begin{aligned}
h_{pq}^{(I)}(t) = {} & \sum_{m=1}^{M} \frac{2P_T}{\sqrt{M}} \cos\left[\beta_m - K_q \cos\theta_R - 2\pi t f_{R\max} \cos\gamma_R\right] \\
& \times \cos\left[K_p \cos\left(\theta_T - \alpha_T^{(m)}\right) + 2\pi t f_{T\max} \cos\left(\alpha_T^{(m)} - \gamma_T\right)\right. \\
& \left. + K_q \Delta_T \sin\theta_R \sin\alpha_T^{(m)} + 2\pi t f_{R\max} \Delta_T \sin\gamma_R \sin\alpha_T^{(m)} + \phi_m\right] \\
& + \sum_{n=1}^{N} \frac{2P_R}{\sqrt{N}} \cos\left[\beta_n + K_p \cos\theta_T + 2\pi t f_{T\max} \cos\gamma_T\right] \\
& \times \cos\left[K_q \cos\left(\alpha_R^{(n)} - \theta_R\right) + 2\pi t f_{R\max} \cos\left(\alpha_R^{(n)} - \gamma_R\right)\right. \\
& \left. + K_p \Delta_R \sin\theta_T \sin\alpha_R^{(n)} + 2\pi t f_{T\max} \Delta_R \sin\gamma_T \sin\alpha_R^{(n)} + \phi_n\right]
\end{aligned}$$

$$+\sum_{m,n=1}^{M,N}\frac{2P_{TR}}{\sqrt{MN}}\cos\left[K_p\cos\left(\theta_T-\alpha_T^{(m)}\right)+2\pi t f_{T\max}\cos\left(\alpha_T^{(m)}-\gamma_T\right)\right]$$

$$\times\cos\left[K_q\cos\left(\alpha_R^{(n)}-\theta_R\right)+2\pi t f_{R\max}\cos\left(\alpha_R^{(n)}-\gamma_R\right)+\phi_{mn}\right]$$

$$+P_{LoS}\cos\left[K_p\cos\theta_T+K_q\cos\left(\alpha_{Rq}^{LoS}-\theta_R\right)+2\pi t\left(f_T^{LoS}+f_R^{LoS}\right)\right] \tag{3.282}$$

$$h_{pq}^{(Q)}(t)=\sum_{m=1}^{M}\frac{2P_T}{\sqrt{M}}\sin\left[\beta_m-K_q\cos\theta_R-2\pi t f_{R\max}\cos\gamma_R\right]$$

$$\times\cos\left[K_p\sin\left(\theta_T-\alpha_T^{(m)}\right)+2\pi t f_{T\max}\sin\left(\alpha_T^{(m)}-\gamma_T\right)\right.$$

$$\left.+K_q\Delta_T\sin\theta_R\cos\alpha_T^{(m)}+2\pi t f_{R\max}\Delta_T\sin\gamma_R\cos\alpha_T^{(m)}+\phi_m\right]$$

$$+\sum_{n=1}^{N}\frac{2P_R}{\sqrt{N}}\sin\left[\beta_n+K_p\cos\theta_T+2\pi t f_{T\max}\cos\gamma_T\right]$$

$$\times\cos\left[K_q\sin\left(\alpha_R^{(n)}-\theta_R\right)+2\pi t f_{R\max}\sin\left(\alpha_R^{(n)}-\gamma_R\right)\right.$$

$$\left.+K_p\Delta_R\sin\theta_T\cos\alpha_R^{(n)}+2\pi t f_{T\max}\Delta_R\sin\gamma_T\cos\alpha_R^{(n)}+\phi_n\right]$$

$$+\sum_{m,n=1}^{M,N}\frac{2P_{TR}}{\sqrt{MN}}\sin\left[K_p\cos\left(\theta_T-\alpha_T^{(m)}\right)+2\pi t f_{T\max}\cos\left(\alpha_T^{(m)}-\gamma_T\right)\right]$$

$$\times\sin\left[K_q\sin\left(\alpha_R^{(n)}-\theta_R\right)+2\pi t f_{R\max}\sin\left(\alpha_R^{(n)}-\gamma_R\right)+\phi_{mn}\right]$$

$$+P_{LoS}\sin\left[K_p\cos\theta_T+K_q\cos\left(\alpha_{Rq}^{LoS}-\theta_R\right)+2\pi t\left(f_T^{LoS}+f_R^{LoS}\right)\right] \tag{3.283}$$

are the in-phase and quadrature components of the complex faded envelope and β_m and β_n are the random path gains. The rest of the parameters are

$$P_T = \sqrt{\frac{\eta_T\Omega}{K+1}} \tag{3.284}$$

$$P_R = \sqrt{\frac{\eta_R\Omega}{K+1}} \tag{3.285}$$

$$P_{TR} = \sqrt{\frac{\eta_{TR}\Omega}{K+1}} \tag{3.286}$$

$$K_q = \frac{2\pi(0.5L_r+0.5-q)\delta_R}{\lambda} \tag{3.287}$$

$$P_{LoS} = \sqrt{\frac{K\Omega}{1+K}} \tag{3.288}$$

$$K_p = \frac{2\pi(0.5L_t + 0.5 - p)\delta_T}{\lambda} \tag{3.289}$$

$$f_T^{LoS} = f_{T\max}\cos\left(\pi - \alpha_{Rq}^{LoS} - \gamma_T\right) \tag{3.290}$$

$$f_R^{LoS} = f_{R\max}\cos\left(\alpha_{Rq}^{LoS} - \gamma_R\right) \tag{3.291}$$

The following two modeling methods are considered that trade simulation precision for simulation speed and complexity:

- *Modified MMEDS model.* The deterministic SoS model needs only a single simulation trial to obtain a channel impulse response with desired statistical properties. Here, only the phases ϕ_n, ϕ_m, and ϕ_{mn} are generated as independent random variables uniformly distributed on the interval $[-\pi, \pi)$. The AoDs, AoAs, and path gains are deterministically fixed to

$$\alpha_T^{(m)} = \frac{\pi}{M}(m - 0.5) + \gamma_T \tag{3.292}$$

$$\alpha_R^{(n)} = \frac{2\pi}{N}(n - 0.5) + \gamma_R \tag{3.293}$$

$$\beta_m = \frac{\pi(m - 0.5)}{M} \tag{3.294}$$

$$\beta_n = \frac{2\pi(n - 0.5)}{N} \tag{3.295}$$

  for $m = 1, \ldots, M$, $n = 1, \ldots, N$, respectively.
- *Statistical model.* The statistical SoS model generates channel realizations with statistical properties that vary with every simulation run. However, the statistical properties converge to the desired ones when averaged over a sufficient channel ensemble. The angles of departures and the angles of arrivals are chosen as follows:

$$\alpha_T^{(m)} = 0.5\left(\frac{2\pi m}{M} + \frac{\psi - \pi}{M}\right) \tag{3.296}$$

$$\alpha_R^{(n)} = \frac{2\pi n}{N} + \frac{\theta - \pi}{N} \tag{3.297}$$

  for $m = 1, \ldots, M$, $n = 1, \ldots, N$, respectively. The parameters ϕ_{mn}, β_m, β_n, θ, and ψ are independent random variables uniformly distributed on the interval $[-\pi, \pi)$.

Here we compare the performance and complexity of the modified MMEDS model, statistical model, and MMEDS model. For meaningful comparison, we remove the LoS and single-bounced components of the complex faded envelope from the first two models.

Figure 3.31 compares the temporal auto-correlation functions of the complex faded envelopes obtained by the modified MMEDS, the statistical, and the MMEDS model. For the MMEDS model and the modified MMEDS model, the temporal correlation functions are obtained using $M = 40$, $N = 38$ and $M = 22$, $N = 40$ scatterers, respectively. For the statistical model, the temporal correlation function is obtained using $N = M = 14$ scatterers and $N_{stat} = 50$ trials. In all simulations, we use a normalized sampling period $f_{T\max}T_s = 0.01$. ($f_{T\max} = f_{R\max}$ are the maximum Doppler frequencies, and T_s is the sampling period.) The orientations of the T_x and R_x antenna arrays are chosen to be $\theta_T = \theta_R = \pi/2$. The angles of motion for the T_x and R_x are chosen to be $\gamma_T = \pi/4$ and $\gamma_R = 0$, respectively. Figure 3.31 shows that the temporal correlation function of the MMEDS model approaches the theoretical one for normalized time delays in the range $0 \le f_{T\max}\, T_s \le 6$. The temporal correlation function of the modified MMEDS model approaches the theoretical one in the same range of the normalized time delays but requires a smaller number of scatterers. Finally, the temporal correlation function of the statistical model approaches the theoretical one in a wider range of normalized time delays (i.e., $0 \le f_{T\max}\, T_s \le 10$) than the MMEDS and the modified MMEDS model.

Figure 3.32 compares the space-time correlation functions of the complex faded envelopes for system with $L_t = L_r = 2$ antennas, obtained by the modified MMEDS, the statistical, and the MMEDS model. For the MMEDS model and the modified MMEDS model, the space-time correlation functions are obtained using the same number of scatterers as in Figure 3.31 and antenna distances $\delta_T = \delta_R = 1\lambda$. For the

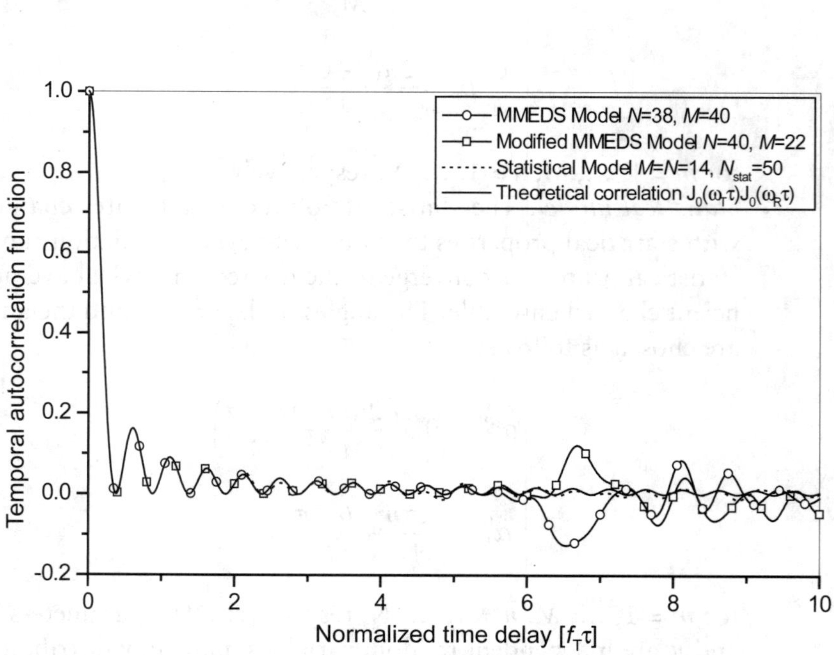

Figure 3.31 The normalized temporal autocorrelation function ($\delta_T = \delta_R = 0$) of the complex faded envelope of the statistical, modified MMEDS, MMEDS, and 2-D MIMO V-to-V theoretical model.

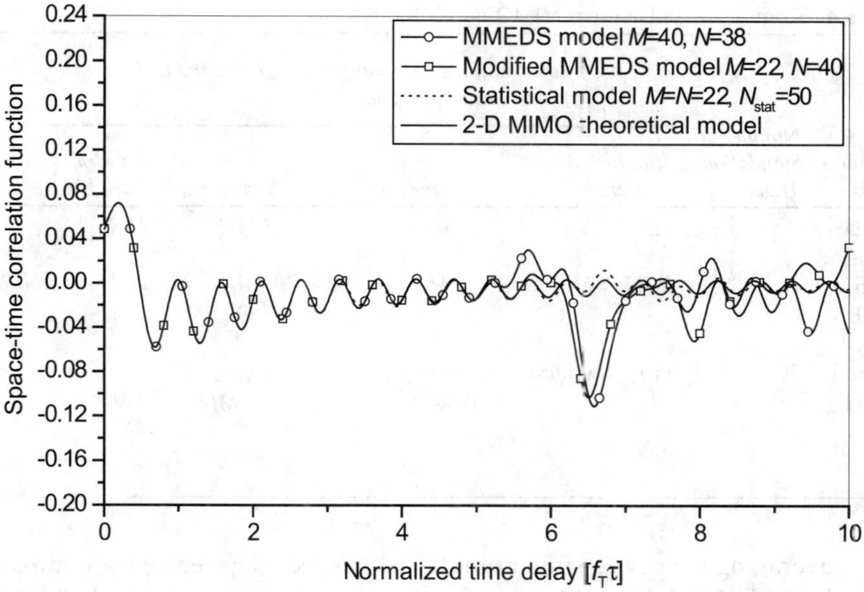

Figure 3.32 The normalized space-time correlation functions ($\delta_T = \delta_R = 1\lambda$) of the complex faded envelope of the statistical, modified MMEDS, MMEDS, and 2-D MIMO V-to-V theoretical model.

statistical model, the space-time correlation function is obtained using $N = M = 22$ scatterers, antenna distances $\delta_T = \delta_R = 1\lambda$, and $N_{stat} = 50$ trials. Figure 3.32 shows that the space-time correlation function of the MMEDS model approaches the theoretical one for normalized time delays in the range $0 \leq f_{T\max} T_s \leq 5$. The space-time correlation function of the modified MMEDS model approaches the theoretical one in a somewhat wider range of normalized time delays (i.e., $0 \leq f_{T\max} T_s \leq 6$), while requiring a smaller number of scatterers than the MMEDS model. The space-time correlation function of the statistical model approaches the theoretical one in the widest range of normalized time delays (i.e., $0 \leq f_{T\max} T_s \leq 10$).

Simulation results presented in Figure 3.32 indicate that larger distances between antenna elements require a larger number of scatterers to match theoretical statistics for the same normalized time delays as in Figure 3.31. For the statistical model, instead of using many scatterers (e.g., 40) in all simulations, we can adaptively select the number of scatterers depending on the distances between antenna elements. If we assume that a mean square error (MSE) of $\leq 10^{-3}$ is required for the simulated space-time correlation function of the complex faded envelope (relative to the theoretical one), we need at least $M = 14 + 2\lceil \delta_T/0.3\lambda \rceil$ and $N = 14 + 2\lceil \delta_R/0.3\lambda \rceil$ scatterers.[1] This methodology can also be applied to the modified MMEDS and MMEDS models. However, to obtain similar statistics as with the statistical model, the number of scatterers needs to be larger (i.e., $70 + 10\lceil \delta_{T/R}/0.3\lambda \rceil$).

To compare complexity of the various models, Table 3.4 summarizes the number of simulation trials and the number of scatterers required to obtain similar statistical properties, the number of operations in one simulation trial needed to generate the complex faded envelope, and the relative simulation times (including

1. Operation $\lceil \cdot \rceil$ denotes rounding up to the next integer.

Table 3.4 Complexity of Different MIMO V-to-V Models

Simulation Models	*Number of Simulation Trials*	*Estimated Number of Computations Needed to Generate $h_{pq}(t)$ (one simulation trial)*: *Number of Scatterers*	*Cosine*	*Addition*	*Random Variable*	*Relative Simulation Time Needed to Generate $h_{pq}(t)$*
MMEDS Model	1	MN	$7MN$	$10MN$	MN	T
Modified MMEDS Model	1	$M_dN_d = MN/2$	$12M_dN_d = 6MN$	$14M_dN_d = 7MN$	$M_dN_d = MN/2$	$0.24T$
Statistical Model	50	$M_sN_s \approx MN/25$	$12M_sN_s = 0.48MN$	$14M_sN_s = 0.56MN$	$M_sN_s + 2 = MN/25 + 2$	$0.95T$ (including averaging over 50 trials)

averaging over N_{stat} simulation trials) needed to generate the complex faded envelopes, in Matlab on a Pentium III laptop. Here, we count only the frequently executed operations and the number of random variables. Table 3.4 shows that the modified MMEDS model has a lower complexity then the MMEDS model, whereas the statistical model has the highest complexity. Note that the modified MMEDS and statistical model require a smaller number of scatterers than the MMEDS model, which leads to shorter simulation times.

From Table 3.4, Figure 3.31, and Figure 3.32 we can conclude that the modified MMEDS model performs similarly to the MMEDS model but has lower complexity and shorter simulation time. On the other hand, the statistical model, with a slight increase in complexity, outperforms both the modified MMEDS and the MMEDS models.

3.3.2 Simulators for V-to-V Frequency-Selective Multipath Fading Channels

The propagation model for the wideband MIMO V-to-V channel described in Section 3.1.9 assumes an infinite number of scatterers, which prevents practical implementation. It is desirable to design simulation model with a finite (preferably small) number of scatterers, while still matching the statistical properties of the reference model.

Assuming 3-D nonisotropic scattering and using the reference model described in Section 3.1.9, the following function is used as a time-variant transfer function [44]:

$T_{pq}(t,f) = T_{pq}^{(I)}(t,f) + jT_{pq}^{(Q)}(t,f)$ where

$$T_{pq}^{(I)}(t,f) = \frac{\rho_{SBT}}{\sqrt{M}} \sum_{l,m,i=1}^{L,M_A^{(l)},M_E^{(l)}} \left(1 - \frac{\gamma}{2}\frac{R_t^{(l)}}{D}\right) \cos\Big[K_p D_T^{SBT} + K_q D_R^{SBT}$$

$$+ 2\pi t f_{T\max} \cos\left(\alpha_T^{(m,l)} - \gamma_T\right)\cos\beta_T^{(i,l)} + \phi_{m,i,l}$$

$$+ 2\pi t f_{R\max}\left(\cos\gamma_R - \Delta_T^{(l)}\sin\gamma_R \sin\alpha_T^{(m,l)}\right)\cos\left(\Delta_T^{(l)}\beta_T^{(i,l)}\right)$$

$$
\begin{aligned}
& -\frac{2\pi}{c_0} f\left(D + R_t^{(l)}\left(1 - \cos\alpha_T^{(m,l)}\right)\right)\Big] \\
& + \frac{\rho_{SBR}}{\sqrt{N}} \sum_{k,n,g=1}^{F,N_A^{(k)},N_E^{(k)}} \left(1 - \frac{\gamma}{2}\frac{R_r^{(k)}}{D}\right) \cos\Big[K_p D_T^{SBR} + K_q D_R^{SBR} \\
& + 2\pi t f_{R\max} \cos\left(\alpha_R^{(n,k)} - \gamma_R\right) \cos\beta_R^{(g,k)} + \phi_{n,g,k} \\
& + 2\pi t f_{T\max} \left(\Delta_R^{(k)} \sin\gamma_T \sin\alpha_R^{(n,k)} + \cos\gamma_T\right) \\
& \times \cos\left(\Delta_R^{(k)} \beta_R^{(g,k)}\right) - \frac{2\pi}{c_0} f\left(D + R_r^{(k)}\left(1 + \cos\alpha_R^{(n,k)}\right)\right)\Big] \\
& + \frac{\rho_{DB}}{\sqrt{MN}} \sum_{l,m,i=1}^{L,M_A^{(l)},M_E^{(l)}} \sum_{k,n,g=1}^{F,N_A^{(k)},N_E^{(k)}} \left(1 - \frac{\gamma}{2}\frac{R_t^{(l)} + R_r^{(k)}}{2D}\right) \\
& \times \cos[K_p D_T^{DB} + K_q D_R^{DB} + 2\pi t f_{T\max} \cos\left(\alpha_T^{(m,l)} - \gamma_T\right) \cos\beta_T^{(i,l)} \\
& + 2\pi t f_{R\max} \cos\left(\alpha_R^{(n,k)} - \gamma_R\right) \cos\beta_R^{(g,k)} \\
& - \frac{2\pi f}{c_0}\left(D + R_t^{(l)}\left(1 - \cos\alpha_T^{(m,l)}\right) + R_r^{(k)}\left(1 + \cos\alpha_R^{(n,k)}\right)\right) + \phi_{m,i,l,n,g,k}\Big] \\
& + \rho_{LoS} \cos\Big[2\pi t\left(f_T^{LoS} + f_R^{LoS}\right) - \frac{2\pi}{c_0} f\sqrt{D^2 + \Delta_H{}^2} + K_p d_{Tx} \\
& + K_q d_R \cos\psi_R \cos\left(\alpha_{Rq}^{LoS} - \theta_R\right)\Big]
\end{aligned}
\tag{3.298}
$$

$$
\begin{aligned}
T_{pq}^{(Q)}(t,f) = {} & \frac{\rho_{SBT}}{\sqrt{M}} \sum_{l,m,i=1}^{L,M_A^{(l)},M_E^{(l)}} \left(1 - \frac{\gamma}{2}\frac{R_t^{(l)}}{D}\right) \sin\Big[K_p D_T^{SBT} + K_q D_R^{SBT} \\
& + 2\pi t f_{T\max} \cos\left(\alpha_T^{(m,l)} - \gamma_T\right) \cos\beta_T^{(i,l)} + \phi_{m,i,l} \\
& + 2\pi t f_{R\max} \left(\cos\gamma_R - \Delta_T^{(l)} \sin\gamma_R \sin\alpha_T^{(m,l)}\right) \cos\left(\Delta_T^{(l)} \beta_T^{(i,l)}\right) \\
& - \frac{2\pi}{c_0} f\left(D + R_t^{(l)}\left(1 - \cos\alpha_T^{(m,l)}\right)\right)\Big] \\
& + \frac{\rho_{SBR}}{\sqrt{N}} \sum_{k,n,g=1}^{F,N_A^{(k)},N_E^{(k)}} \left(1 - \frac{\gamma}{2}\frac{R_r^{(k)}}{D}\right) \sin\Big[K_p D_T^{SBR} + K_q D_R^{SBR} \\
& + 2\pi t f_{R\max} \cos\left(\alpha_R^{(n,k)} - \gamma_R\right) \cos\beta_R^{(g,k)} + \phi_{n,g,k}
\end{aligned}
$$

$$+ 2\pi t f_{T\max}\left(\Delta_R^{(k)}\sin\gamma_T\sin\alpha_R^{(n,k)} + \cos\gamma_T\right)\cos\left(\Delta_R^{(k)}\beta_R^{(g,k)}\right)$$

$$\left. - \frac{2\pi}{c_0} f\left(D + R_r^{(k)}\left(1 + \cos\alpha_R^{(n,k)}\right)\right)\right]$$

$$+ \frac{\rho_{DB}}{\sqrt{MN}} \sum_{l,m,i=1}^{L,M_A^{(l)},M_E^{(l)}} \sum_{k,n,g=1}^{F,N_A^{(k)},N_E^{(k)}} \left(1 - \frac{\gamma}{2}\frac{R_t^{(l)} + R_r^{(k)}}{2D}\right)$$

$$\times \sin\left[K_p D_T^{DB} + K_q D_R^{DB} + 2\pi t f_{T\max}\cos\left(\alpha_T^{(m,l)} - \gamma_T\right)\cos\beta_T^{(i,l)}\right.$$

$$+ 2\pi t f_{R\max}\cos\left(\alpha_R^{(n,k)} - \gamma_R\right)\cos\beta_R^{(g,k)}$$

$$\left. - \frac{2\pi f}{c_0}\left(D + R_t^{(l)}\left(1 - \cos\alpha_T^{(m,l)}\right) + R_r^{(k)}\left(1 + \cos\alpha_R^{(n,k)}\right)\right) + \phi_{m,i,l,n,g,k}\right]$$

$$+ \rho_{LoS}\sin\left[2\pi t\left(f_T^{LoS} + f_R^{LoS}\right) - \frac{2\pi}{c_0} f\sqrt{D^2 + {\Delta_H}^2} + K_p d_{Tx}\right.$$

$$\left. + K_q d_R \cos\psi_R \cos\left(\alpha_{R_q}^{LoS} - \theta_R\right)\right] \tag{3.299}$$

are the in-phase (I) and the quadrature (Q) components of the time-variant transfer function. The rest if the parameters are

$$M = \sum_{l=1}^{L} M^{(l)} = \sum_{l=1}^{L} M_A^{(l)} M_E^{(l)} \tag{3.300}$$

$$N = \sum_{k=1}^{F} N^{(k)} = \sum_{k=1}^{F} N_A^{(k)} N_E^{(k)} \tag{3.301}$$

$$\rho_{SBT} = \sqrt{\frac{\eta_T}{K+1}} \tag{3.302}$$

$$\rho_{SBR} = \sqrt{\frac{\eta_R}{K+1}} \tag{3.303}$$

$$\rho_{DB} = \sqrt{\frac{\eta_{TR}}{K+1}} \tag{3.304}$$

$$\rho_{LoS} = \sqrt{\frac{K}{1+K}} \tag{3.305}$$

$$K_p = \frac{\pi(L_t + 1 - 2p)}{\lambda} \tag{3.306}$$

$$K_q = \frac{\pi(L_r + 1 - 2q)}{\lambda} \quad (3.307)$$

$$D_T^{SBT} = d_{Tx} \cos\alpha_T^{(m,l)} + d_{Ty} \sin\alpha_T^{(m,l)} + d_{Tz} \sin\beta_T^{(i,l)} \quad (3.308)$$

$$D_R^{SBT} = d_{Ry}\Delta_T^{(l)} \sin\alpha_T^{(m,l)} - d_{Rx} \quad (3.309)$$

$$D_T^{SBR} = d_{Ty}\Delta_R^{(k)} \sin\alpha_R^{(n,k)} + d_{Tx} \quad (3.310)$$

$$D_R^{SBR} = d_{Rx} \cos\alpha_R^{(n,k)} + d_{Ry} \sin\alpha_R^{(n,k)} + d_{Rz} \sin\beta_R^{(g,k)} \quad (3.311)$$

$$D_T^{DB} = d_{Tx} \cos\alpha_T^{(m,l)} + d_{Ty} \sin\alpha_T^{(m,l)} + d_{Tz} \sin\beta_T^{(i,l)} \quad (3.312)$$

$$D_R^{DB} = d_{Rx} \cos\alpha_R^{(n,k)} + d_{Ry} \sin\alpha_R^{(n,k)} + d_{Rz} \sin\beta_R^{(g,k)} \quad (3.313)$$

$$f_T^{LoS} = f_{T\max} \cos(\pi - \alpha_{Rq}^{LoS} - \gamma_T) \quad (3.314)$$

$$f_R^{LoS} = f_{R\max} \cos(\alpha_{Rq}^{LoS} - \gamma_R) \quad (3.315)$$

$$d_{Tx} = d_T \cos\psi_T \cos\theta_T \quad (3.316)$$

$$d_{Ty} = d_T \cos\psi_T \sin\theta_T \quad (3.317)$$

$$d_{Tz} = d_T \sin\psi_T \quad (3.318)$$

$$d_{Rx} = d_R \cos\psi_R \cos\theta_R \quad (3.319)$$

$$d_{Ry} = d_R \cos\psi_R \sin\theta_R \quad (3.320)$$

$$d_{Rz} = d_R \sin\psi_R \quad (3.321)$$

As before, the following two modeling methods are considered. They trade simulation precision for simulation speed and complexity:

- *Deterministic MIMO 3-D wideband SoS model.* The deterministic SoS model needs only a single simulation trial to obtain a channel impulse response with desired statistical properties. The AAoDs, $\alpha_T^{(m,l)}$, and the AAoAs, $\alpha_R^{(n,k)}$, are modeled using the von Mises pdf and are generated as follows:

$$\alpha_T^{(m,l)} = F^{-1}\left(\frac{m-0.5}{M_A^{(l)}}\right) \tag{3.322}$$

$$\alpha_R^{(n,k)} = F^{-1}\left(\frac{n-0.5}{N_A^{(k)}}\right) \tag{3.323}$$

for $m = 1, \ldots, M_A^{(l)}$, $n = 1, \ldots, N_A^{(k)}$, where $F(\cdot)^{-1}$ denotes the inverse cumulative von Mises distribution function and is evaluated using the method from [63]. The EAoDs, $\beta_T^{(i,l)}$, and the EAoAs, $\beta_R^{(g,k)}$, are modeled using pdfs from (3.186) and (3.187), and are generated as follows:

$$\beta_T^{(i,l)} = \frac{2\beta_{Tm}}{\pi}\arcsin\left(\frac{2i-1}{M_E^{(l)}}-1\right) \tag{3.324}$$

$$\beta_R^{(g,k)} = \frac{2\beta_{Rm}}{\pi}\arcsin\left(\frac{2g-1}{N_E^{(k)}}-1\right) \tag{3.325}$$

for $i = 1,\ldots, M_E^{(l)}$, $g = 1,\ldots, N_E^{(k)}$. The radii $R_t^{(l)}$ and $R_r^{(k)}$ are modeled using pdfs (3.188) and (3.189), and are generated as follows:

$$R_t^{(l)} = \sqrt{\frac{(l-0.5)(R_{t2}^2 - R_{t1}^2)}{L} + R_{t1}^2} \tag{3.326}$$

$$R_r^{(k)} = \sqrt{\frac{(k-0.5)(R_{r2}^2 - R_{r1}^2)}{F} + R_{r1}^2} \tag{3.327}$$

for $l = 1, \ldots, L$, $k = 1, \ldots, F$. Phases $\phi_{m,i,l}$, $\phi_{n,g,k}$, and $\phi_{m,i,l,n,g,k}$ are generated as independent uniform random variables on the interval $[-\pi, \pi)$.

- *Statistical MIMO 3-D wideband SoS model.* The statistical SoS model generates channel realizations with statistical properties that vary with every simulation run but converge to the desired properties when averaged over a sufficient channel ensemble. The AAoDs, the AAoAs, the EAoDs, the EAoAs, and the radii are generated as follows:

$$\alpha_T^{(m,l)} = F^{-1}\left(\frac{m+\theta_A^{(l)}-1}{M_A^{(l)}}\right) \tag{3.328}$$

$$\alpha_R^{(n,k)} = F^{-1}\left(\frac{n+\psi_A^{(k)}-1}{N_A^{(k)}}\right) \tag{3.329}$$

$$\beta_T^{(i,l)} = \frac{2\beta_{Tm}}{\pi}\arcsin\left(\frac{2(i+\theta_E^{(l)}-1)}{M_E^{(l)}}-1\right) \tag{3.330}$$

$$\beta_R^{(g,k)} = \frac{2\beta_{Rm}}{\pi}\arcsin\left(\frac{2\left(g+\psi_E^{(k)}-1\right)}{N_E^{(k)}}-1\right) \quad (3.331)$$

$$R_t^{(l)} = \sqrt{\frac{(l+\sigma_T-1)(R_{t2}^2-R_{t1}^2)}{L}+R_{t1}^2} \quad (3.332)$$

$$R_r^{(k)} = \sqrt{\frac{(k+\sigma_R-1)(R_{r2}^2-R_{r1}^2)}{K}+R_{r1}^2} \quad (3.333)$$

for $m = 1, \ldots, M_A^{(l)}$, $n = 1, \ldots, N_A^{(k)}$, $i = 1, \ldots, M_E^{(l)}$, $g = 1, \ldots, N_E^{(k)}$, $l = 1, \ldots, L$, and $k = 1, \ldots, K$ respectively. The parameters $\theta_A^{(l)}$, $\psi_A^{(k)}$, $\theta_E^{(l)}$, $\psi_E^{(k)}$, σ_T, and σ_R are independent random variables uniformly distributed on the interval [0,1). The function $F^{-1}(\cdot)$ denotes the inverse von Mises cumulative distribution function.

Here we compare the deterministic and statistical simulation model with the theoretical model detailed in Section 3.1.9. In all simulations, we use a normalized sampling period $f_{T\max}T_s = 0.01$, ($f_{T\max} = f_{R\max}$ are the maximum Doppler frequencies, and T_s is the sampling period.) The number of transmit and receive antennas is set to $L_t = L_r = 2$. The parameters used to obtain curves in Figure 3.33 are $\beta_{Tm} = \beta_{Rm} = 15^\circ$, $\theta_T = \theta_R = \pi/4$, $\psi_T = \psi_R = \pi/3$, $\gamma_T = 0$, $\gamma_R = \pi/3$, $k_T = k_R = 0$, $\lambda = 0.3$ m, $R_{t1} = R_{r1} = 30$m, $R_{t2} = R_{r2} = 300$ m, $D = 5000$ m, and $\gamma = 4$.

Figure 3.33 compares the space-time-frequency correlation functions ($d_T = d_R = 0.5\lambda$, $\Delta f = 100$ Hz) of the theoretical, deterministic, and statistical models. The

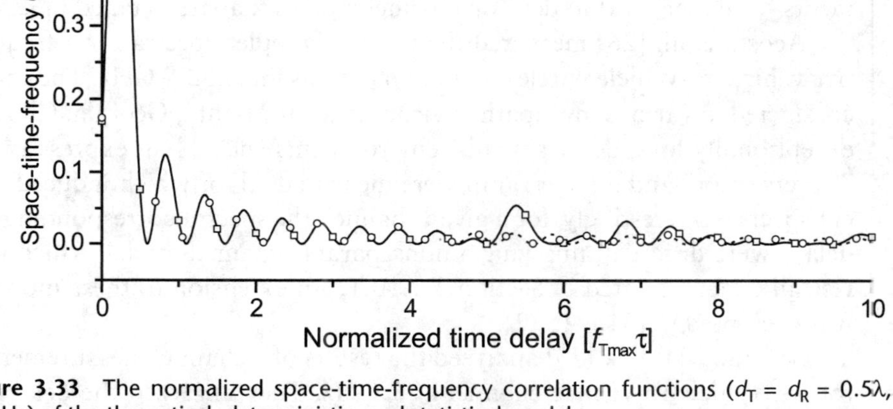

Figure 3.33 The normalized space-time-frequency correlation functions ($d_T = d_R = 0.5\lambda$, $\Delta f = 100$ Hz) of the theoretical, deterministic, and statistical models.

space-time-frequency correlation function of the deterministic model is obtained with $M_A^{(l)} = 32$, $M_E^{(l)} = 7$, $N_A^{(k)} = 32$, and $N_E^{(k)} = 7$ scatterers and six tap-delays ($L = 3$ and $K = 3$), whereas the space-time-frequency correlation function of the statistical model is obtained with $M_A^{(l)} = 12$, $M_E^{(l)} = 3$, $N_A^{(k)} = 12$, and $N_E^{(k)} = 3$ scatterers, six tap-delays ($L = 3$ and $K = 3$), and $N_{\text{stat}} = 10$ simulation trials. The results in Figure 3.33 show that the space-time-frequency correlation function of the deterministic model closely matches the theoretical one in the range of normalized time delays, $0 \leq f_{T\max} T_s \leq 4$, whereas the space-time correlation function of the statistical model approaches the theoretical one in a wider range of normalized time delays (i.e., $0 \leq f_{T\max} T_s \leq 10$).

3.4 Vehicle-to-Vehicle Measurement Campaigns and Modeling

A large number of papers have reported measurement results, and channel models based on these measurements, for vehicle-to-vehicle channels [64].

Harley [65] was one of the first to highlight that traditional F-to-M cellular propagation models do not apply to short-range communications systems with low antenna heights at both the transmitting and receiving ends. He also conducted measurements and extracted a suitable propagation model.

One of the first measurements conducted explicitly for the vehicle-to-vehicle scenario have been reported in [66]. The authors conducted an extensive measurement campaign at 900 MHz with the aim of determining the pathloss and delay spread probability distribution parameters, such as Ricean K.

Kovacs et al. [67] investigated the mobile-to-mobile radio channel in different suburban outdoor-to-indoor environments. As already discussed in Section 3.1, the measured signal envelope statistics were found to be a combination of single and double Rayleigh distributions, which generally leads to a system performance degradation. A through description of conducted measurement campaigns and derived propagation models are presented in [68].

Maurer et al. characterized the narrowband intervehicle transmission channel at 5.2 GHz. Based on realistic road traffic scenarios, measurements were obtained for four different classes of environments: urban, suburban, motorway, and highway. The measured channel was then characterized in terms of the CDF of the received signal, the average Doppler power density spectrum, and the LCR. These measurements were then used to develop channel models that are detailed in [25, 69, 70].

Acosta et al. [26] measured the per-tap Doppler spectra for a frequency selective vehicle-to-vehicle wireless communications link at 2.4 GHz. The campaign was conducted in various multipath environments in Atlanta, Georgia. Chosen for their exceptionally long delay spreads, environments such as an expressway, an urban T-intersection, and an exit ramp were measured, all of which induced fairly different spectra. Interestingly, for a given channel, the spectra corresponding to different delays were different, implying a nonseparable channel model, which is also theoretically demonstrated in Section 3.1.10.1. An extension to these modeling efforts was presented in [51, 71, 72].

Sen and Matolak [73] analyzed the results of a channel measurement campaign for the vehicle-to-vehicle channel at 5 GHz. They analyzed the delay spread and amplitude statistics as well as correlations. From these measurements, several statistical channel models were developed, and, using simulation results, they gauged tradeoffs between model implementation complexity and modeling precision. The results are presented in [74–76].

A comprehensive measurement and modeling campaign has also been conducted by the European IST WINNER and WINNER 2 projects [77]. Most notably, the publicly available Winner II [78] covers propagation scenarios that are pertinent to mobile-to-mobile communications.

The disadvantage of all channel models derived from measurements is that they are site specific (i.e., there is no guarantee that the model will be applicable to other environments). On the other hand, the theoretical models and simulators detailed in Sections 3.1 and 3.3 need to be validated against the measurements. To validate these theoretical and simulation models, an experimental MIMO V-to-V channel-sounding campaign was conducted for V-to-V vehicular communication with vehicles traveling along surface streets near to Georgia Tech campus and along the interstate highways near midtown in the Atlanta metropolitan area [11].

Furthermore, to compare the first- and second-order channel statistics obtained from the theoretical model described in Section 3.1.9 with those obtained from the empirical measurements, a maximum likelihood–based stochastic estimator was derived to extract the parameters needed for the theoretical model from the measured data [11]. Now, we present a comparison between the measured data and the model detailed in Section 3.1.9.

3.4.1 Validation of the 3-D Wideband Model for V-to-V Frequency-Selective Multipath Fading Channels

The wideband channel measurements collected in the urban street area were performed at 2.435 GHz, and the maximum Doppler frequencies were $f_{T\max} = f_{R\max} =$ 90.86 Hz. The distance between the T_x and R_x was approximately D = 300m and the moving directions were $\gamma_T = \gamma_R = 90^\circ$. Both, the T_x and R_x were equipped with two omnidirectional antenna elements with the azimuth and elevation angles $\theta_T = \theta_R = 0^\circ$ and $\psi_T = \psi_R = 0^\circ$, respectively. All antenna elements have the same height, $\Delta_H = 0$, and the distance between the antenna elements is $d_T = d_R = 2.943\lambda$. It is assumed that the pathloss exponent γ is 4, which is typical for radio propagation over a flat reflecting surface [9].

The theoretical model in Section 3.1.9 assumes an infinite number of scatterers around the T_x and R_x (i.e, $M, N \to \infty$). Using the Central Limit Theorem [9], we can conclude that the theoretical time-variant transfer function $T_{pq}(t, f)$ is a complex Gaussian random process with the mean $E[T_{pq}(t, f)] = T_{pq}^{LoS}(t, f)$ and the variance $\mathrm{Var}[T_{pq}(t, f)] = \Omega_{pq}/(K + 1)$. Before we compare the theoretical and measured sD-psd, psds, and LCR, we need to confirm that the measured time-variant transfer functions are also complex Gaussian random processes with the means and variances similar to those of the theoretical model. Figure 3.34 compares the cumulative distribution functions (CDFs) of the real and imaginary components of the measured time-variant transfer function, $\hat{T}_{11}(t, 180\text{ MHz})$, with the CDFs of the real and imaginary components of the complex Gaussian process. The complex Gaussian process is obtained with the mean $E[T_{11}(t, 180\text{ MHz})]$ and the variance

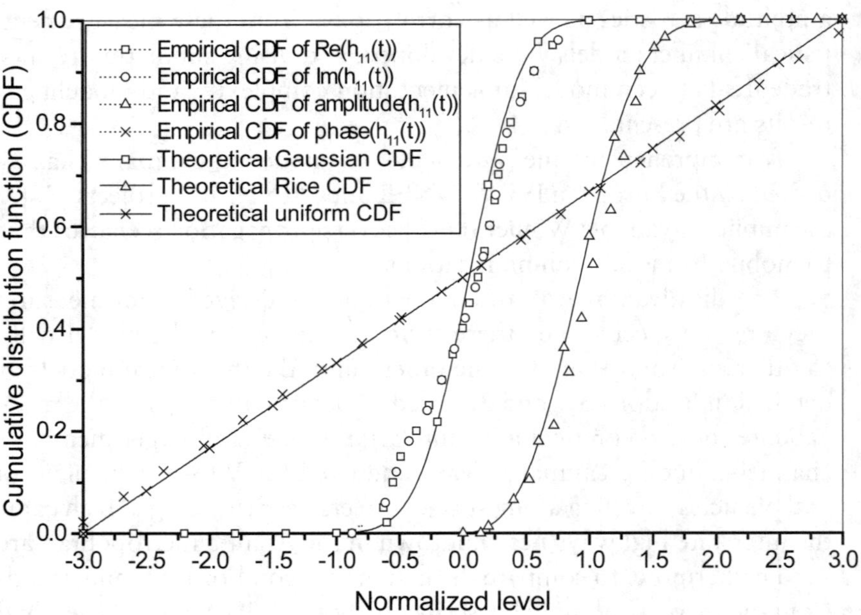

Figure 3.34 Theoretical and empirical distribution functions of h11(t,τ) in an urban street area.

Var[$T_{11}(t$,180 MHz)], respectively, where $T_{11}(t$,180 MHz) is the time-variant transfer function of the theoretical model. The results show that CDFs of the real and imaginary components of the measured time-variant transfer function, $\hat{T}_{11}(t$,180 MHz), can be closely approximated by the Gaussian distributions and that the theoretical model and measured data have the similar means and variances. Furthermore, the CDFs of the measured amplitude and phase can be closely approximated by a Rice and uniform distribution,[2] respectively.

Figures 3.35–3.37 compare theoretical and measured sD-psd, psds, and LCR for the urban street environment, respectively. The analytical curves are obtained with the parameters $\beta_{Tm} = 5.1^\circ$, $\beta_{Rm} = 10.2^\circ$, $\mu_T = 73.3^\circ$, $\mu_R = 264.7^\circ$, $k_T = 5.7$, $k_R = 6.4$, $\eta_T = 0.043$, $\eta_R = 0.137$, $\eta_{TR} = 0.82$, $R_{t1} = R_{r1} = 9.6$ m, $R_{t2} = R_{r2} = 96$ m, and $K = 2.41$. These parameters are estimated from the measured data using the parameter estimator in [11]. The parameters γ, D, γ_T, γ_R, θ_T, θ_R, ψ_T, ψ_R, d_T, d_R, Δ_H, $f_{T\max}$, and $f_{R\max}$ are selected to match the measurement conditions as described earlier. Figures 3.35–3.37 show close agreement between the theoretical and empirical curves. The results for urban street environments show that double-bounced rays bear more energy than the single-bounced rays. Note also that the estimated scattering radii $R_{t2} = R_{r2} = 96$ m are sufficiently small compared to the distance $D = 300$ m in which the local scattering condition is satisfied.

The wideband channel measurements, collected on an interstate highway, were performed at 2.435 GHz, and the maximum Doppler frequencies were $f_{T\max} = f_{R\max} =$ 181.72 Hz. The distance between T_x and R_x was approximately $D = 180$ m, and the moving directions were $\gamma_T = \gamma_R = 90^\circ$. The T_x was equipped with two and the R_x was equipped with four omnidirectional antenna elements. The antenna array elements

2. With nonisotropic scattering, the phase may not be uniformly distributed.

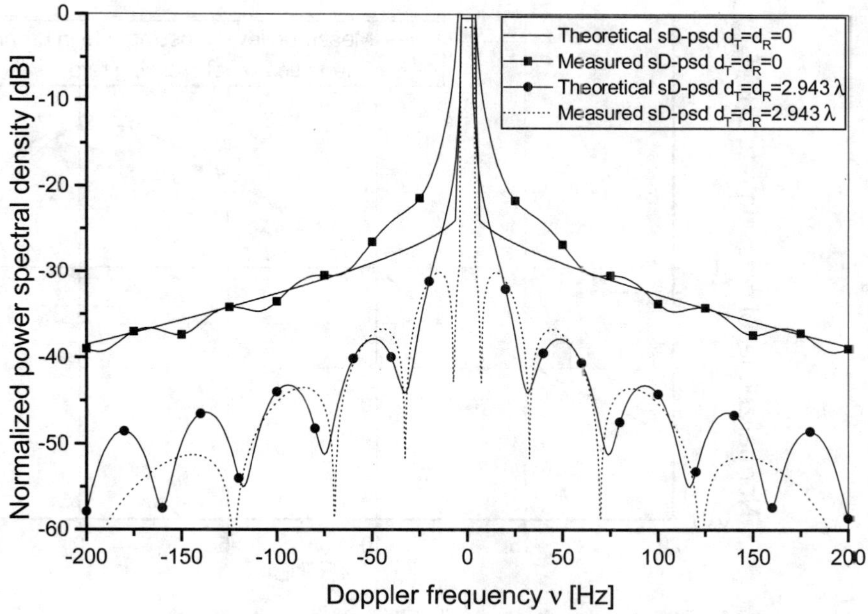

Figure 3.35 Theoretical and the measured sD-psd in an urban street area.

at both vehicles have the azimuth and elevation angles $\theta_T = \theta_R = 0^\circ$ and $\psi_T = \psi_R = 0^\circ$, respectively. All antenna elements have the same height, $\Delta_H = 0$, and the distance between the antenna elements is $d_T = d_R = 2.943\lambda$. The pathoss exponent γ is set to 4.

Figures 3.38–3.40 compare the theoretical and measured sD-psd, psds, and LCR, respectively, on the interstate highway. The analytical curves are obtained

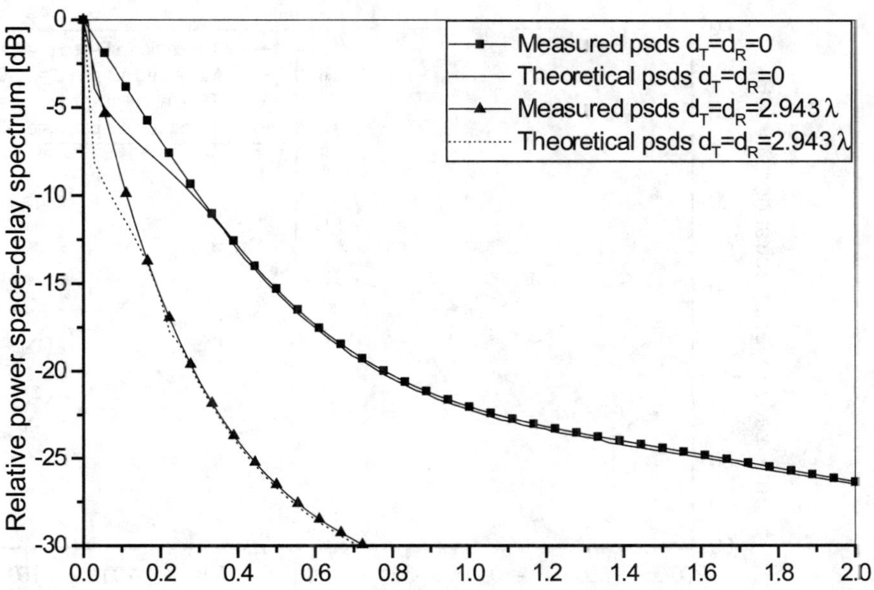

Figure 3.36 Theoretical and measured psds in an urban street area.

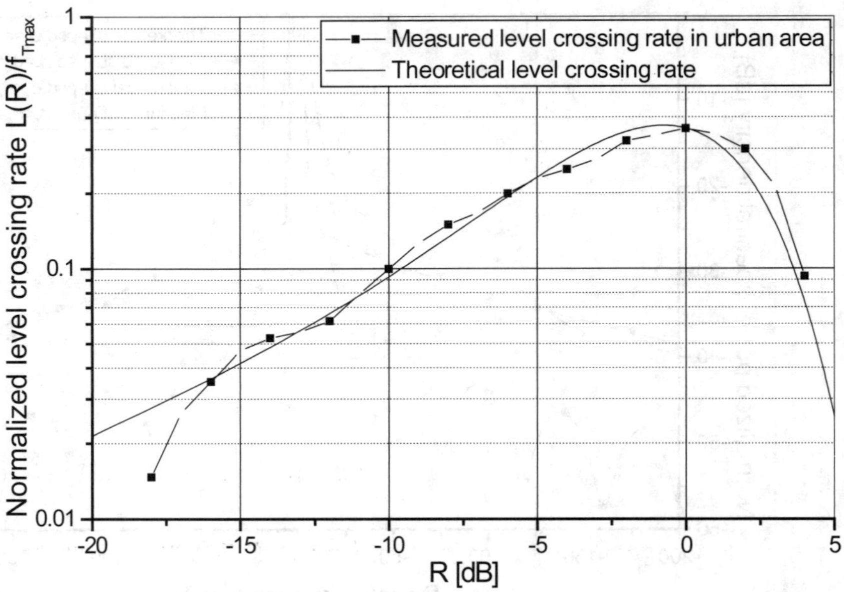

Figure 3.37 Theoretical and measured LCR in an urban street area.

with the parameters $\beta_{Tm} = 7.4^\circ$, $\beta_{Rm} = 8.3^\circ$, $\mu_T = 101.4^\circ$, $\mu_R = 281.5^\circ$, $k_T = 5.5$, $k_R = 5.2$, $\eta_T = 0.358$, $\eta_R = 0.288$, $\eta_{TR} = 0.354$, $R_{t1} = R_{r1} = 4.5$ m, $R_{t2} = R_{r2} = 45$ m, and $K = 1.29$. The remaining parameters D, γ, γ_T, γ_R, θ_T, θ_R, ψ_T, ψ_R, d_T, d_R, Δ_H, $f_{T\max}$, and $f_{R\max}$ are selected as in the highway measurement setup, described earlier. Figures 3.38–3.40 show close agreement between the theoretical and empirical curves. In this set of data, single-bounced rays are more dominant than double-

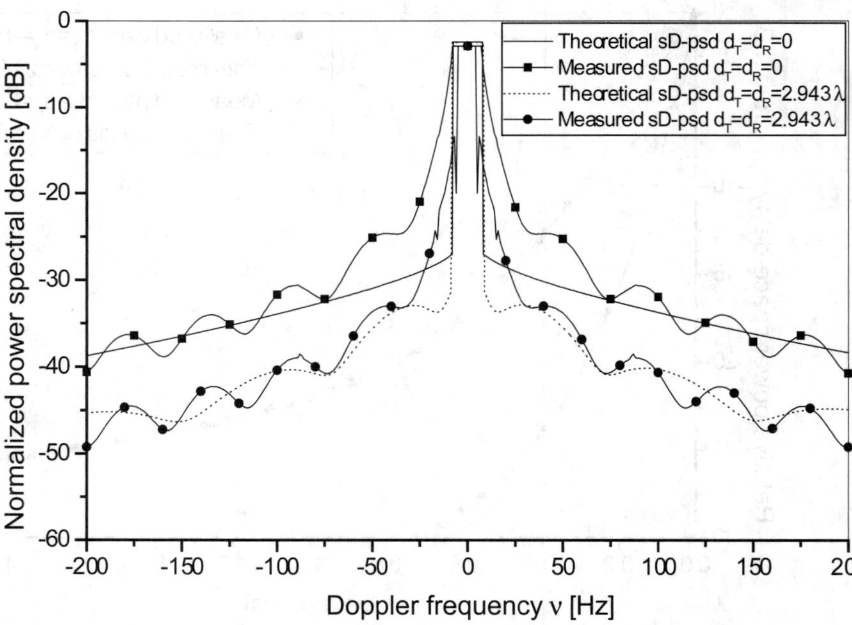

Figure 3.38 Theoretical and measured sD-psd on an interstate highway.

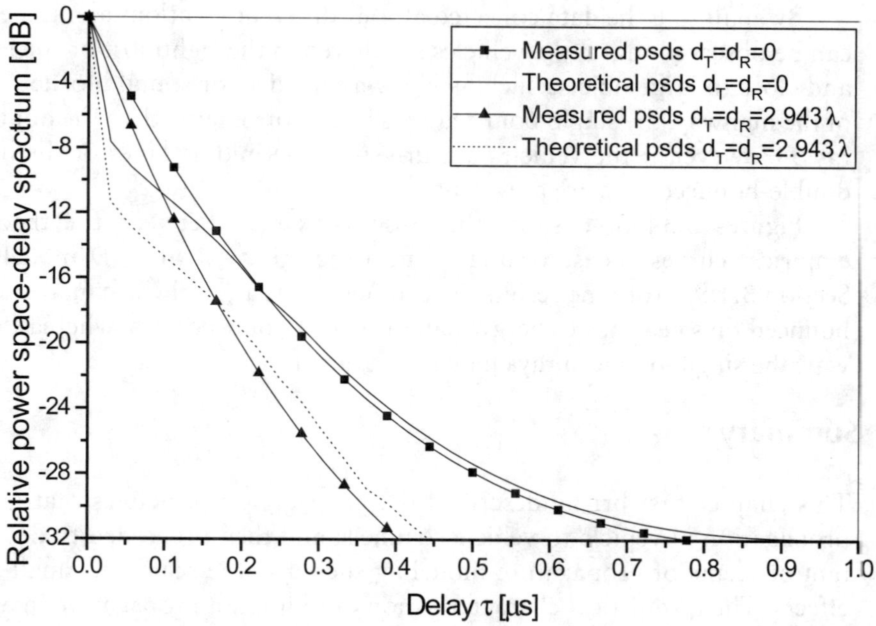

Figure 3.39 Theoretical and measured psds on an interstate highway.

bounced rays. This effect can be explained by the fact that both vans were in the rightmost lane and very close to the sound blockers on the edge of the highway (this was verified using a video recording of the campaign). Finally, the estimated scattering radii $R_{t2} = R_{r2} = 45$ m are sufficiently small compared to the distance D = 180 m in which the local scattering condition is satisfied.

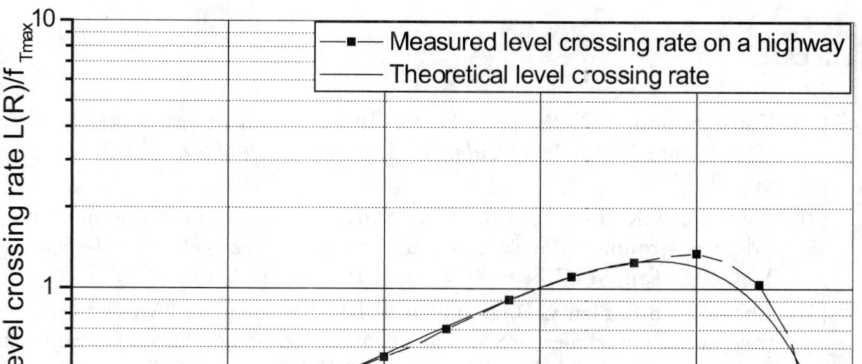

Figure 3.40 Theoretical and measured LCR on an interstate highway.

By analyzing the data collected at four different locations along the highway, it can be observed that if the vehicles are driven in the rightmost or the leftmost lane and close to large objects such as highway dividers or sound blockers on the edge of the highway, the single-bounced rays bear more energy than the double-bounced rays. However, if the vehicles are driven in the middle lanes of the highway, the double-bounced rays are prevalent.

Figures 3.34–3.40 show the close agreement between the theoretical and empirical curves. These results confirm the utility of the 3-D model detailed in Section 3.1.9. From the results we can observe that, in the urban area, the double-bounced rays bear more energy than the single-bounced rays, whereas, on the highway, the single-bounced rays may be prevalent.

3.5 Summary

This chapter first briefly described the propagation principles that govern signal propagation in vehicle-to-vehicle channels, and then proceeded to describe important concepts of propagation modeling such as large-scale and small-scale fading effects. Then, statistical characterization of multipath propagation in V-to-V channels was presented, and several 2-D and 3-D models and their statistics, such as correlation function, Doppler spectrum, power delay profile, and level crossing rate were detailed. These reference models are very useful mathematical tools needed for statistical characterization of channels. However, these models assume a large number of scatterers, making these models impractical for simulators, so simulators for 2-D frequency-flat and 3-D frequency-selective multipath fading were presented, with a small number of propagation rays that still match the statistical properties of the V-to-V channel. Finally, V-to-V measurement campaigns were reviewed, and validation the 3-D wideband model for V-to-V frequency-selective multipath fading channels against the measurements was presented.

References

[1] Konstantinou, K., S. Kang, C. Tzaras, "A measurement-based model for mobile-to-mobile UMTS links," *IEEE Vehicular Technology Conference, VTC2007-Spring*, Apr. 2007, pp. 529–533.

[2] Wang Z., E.K. Tameh, A.R. Nix, "Statistical peer-to-peer channel models for outdoor urban environments at 2 GHz and 5 GHz," *IEEE Vehicular Technology Conference, VTC2004-Fall*, Vol.7, Sept. 2004, pp. 5101–5105.

[3] Cheng L., B.E. Henty, D.D. Stancil, F. Bai, P. Mudalige, "Mobile vehicle-to-vehicle narrow-band channel measurement and characterization of the 5.9 GHz dedicated short range communication (DSRC) frequency band," *IEEE Journal on Selected Areas in Communications*, Vol. 25, Oct. 2007, pp. 1501–1516.

[4] Cheng L., B.E. Henty, F. Bai, D.D. Stancil, "Highway and rural propagation channel modeling for vehicle-to-vehicle communications at 5.9 GHz," *IEEE Antennas and Propagation Society International Symposium, AP-S 2008*, July 2008, pp. 1–4.

[5] Kunisch, J., J. Pamp, "Wideband Car-to-Car Radio Channel Measurements and Model at 5.9 GHz," *IEEE Vehicular Technology Conference, VTC 2008-Fall*, Sept. 2008, pp. 1–5.

[6] Karedal, J., N. Czink, A. Paier, F. Tufvesson, A.F. Molisch, "Pathloss modeling for vehicle-to-vehicle communications," *IEEE Transactions on Vehicular Technology*, Vol. 60, Jan. 2011, pp. 323–328.

[7] Paschalidis, P., K. Mahler, A. Kortke, M. Peter, W. Keusgen, "Pathloss and multipath power decay of the wideband car-to-car channel at 5.7 GHz," *IEEE Vehicular Technology Conference, VTC 2011-Spring*, May 2011, pp. 1–5.

[8] Wang Z., E.K. Tameh, A. R. Nix, "Joint shadowing process in urban peer-to-peer radio channels," *IEEE Transactions on Vehicular Technology*, Vol. 57, Jan. 2008, pp. 52–64.

[9] Stüber, G. L., *Principles of mobile communication*, 2nd ed., Norwell, MA: Kluwer, 2001.

[10] Akki A. S. and F. Haber, "A statistical model for mobile-to-mobile land communication channel," *IEEE Transactions on Vehicular Technology*, Vol. 35, February 1986, pp. 2–10.

[11] Zajić A. G., G. L. Stüber, T. G. Pratt, and S. Nguyen, "Wideband MIMO mobile-to-mobile channels: geometry-based statistical modeling with experimental verification," *IEEE Transactions on Vehicular Technology*, Vol. 58, Feb. 2009, pp. 517–534.

[12] Von Mises R. "Über die "Ganzzahligkeit" der Atomgewicht und verwandte Fragen," *Physikal. Z.*, Vol. 19, 1918, pp. 490–500.

[13] Gradshteyn I. S. and I. M. Ryzhik, *Table of Integrals, Series, and Products*, San Diego CA: Academic Press, 1980.

[14] Zajić A. G. and G. L. Stüber, "Space-time correlated MIMO mobile-to-mobile channels," *IEEE 17th International Symposium on Personal, Indoor and Mobile Radio Communications, PIMRC 2006*, Sept. 2006, pp. 1–5.

[15] Pätzold M, B.O. Hogstad, N. Youssef, and D. Kim, "A MIMO mobile-to-mobile channel model: part I-the reference model," *IEEE 16th International Symposium on Personal, Indoor and Mobile Radio Communications, PIMRC 2005*, Vol. 1, Sept. 2005, pp. 573–578.

[16] Hogstad B. O., M. Pätzold, N. Youssef and D. Kim, "A MIMO mobile-to-mobile channel model: part II-the simulation model," *IEEE 16th International Symposium on Personal, Indoor and Mobile Radio Communications, PIMRC 2005*, Vol. 1, Sept. 2005, pp. 562–567.

[17] Pätzold M, B.O. Hogstad, and N. Youssef, "Modeling, analysis, and simulation of MIMO mobile-to-mobile fading channels," *IEEE Transactions on Wireless Communications*, Vol. 7, Feb. 2008, pp. 510–520.

[18] Zajić A. G. and G. L. Stüber, "Space-time correlated mobile-to-mobile channels: modeling and simulation," *IEEE Transactions on Vehicular Technology*, Vol. 57, Mar. 2008, pp. 715–726.

[19] Byers G. J. and F. Takawira, "Spatially and temporally correlated MIMO channels: modeling and capacity analysis," *IEEE Transactions on Vehicular Technology*, Vol. 53, May 2004, pp. 634–643.

[20] Gesbert D. , H. Bölcskei, D.A. Gore, and A.J. Paulraj, "Outdoor MIMO wireless channels: models and performance prediction," *IEEE Transactions on Communications*, Vol. 50, Dec. 2002, pp. 1926–1934.

[21] Clarke, R. H., "A statistical theory of mobile-radio reception," *Bell Systems Technical Journal*, July 1968, pp. 957–1000.

[22] Shiu D., G. J. Foschini, M. J. Gans, and J. M. Khan, "Fading correlation and its effect on the capacity of multielement antenna systems," *IEEE Transactions on Communications*, Vol. 48, Mar. 2000, pp. 502–513.

[23] Abdi A. and M. Kaveh, "A space-time correlation model for multielement antenna systems in mobile fading channels," *IEEE Journal on Selected Areas in Communications*, Vol. 20, Apr. 2002, pp. 550–560.

[24] Kang H., G. L. Stüber, T. G. Pratt, and M. A. Ingram, "Studies on the capacity of MIMO systems in mobile-to-mobile environment," *IEEE Wireless Communications and Networking Conference, WCNC 2004*, Vol. 1, Atlanta, GA, Mar. 2004, pp. 363–368.

[25] Maurer J., T. Fügen, and W. Wiesbeck, "Narrow-band measurement and analysis of the inter-vehicle transmission channel at 5.2 GHz," *IEEE 55th Vehicular Technology Conference, VTC Spring 2002*, Vol. 3, Birmingham, AL, May 2002, pp. 1274–1278.

[26] Acosta G., K. Tokuda, and M. A. Ingram, "Measured joint Doppler-delay power profiles for vehicle-to-vehicle communications at 2.4 GHz," *IEEE Global Telecommunications Conference, GLOBECOM '04*, Vol. 6, Dallas, TX, Nov. 2004, pp. 3813–3817.

[27] Pätzold M., U. Killat, and F. Laue, "An extended Suzuki model for land mobile satellite channels and its statistical properties,"*IEEE Transactions on Vehicular Technology*, Vol. 47, May 1998, pp. 617–630.

[28] Akki A. S., "Statistical properties of mobile-to-mobile land communication channels," *IEEE Transactions on Vehicular Technology*, Vol. 43, November 1994, pp. 826–831.

[29] Zhang J., Q. Zhang, B. Li, X. Luo, and W. Zhu, "Energy-efficient routing in mobile and ad-hoc networks: mobility-assisted case," *IEEE Transactions on Vehicular Technology*, Vol. 55, Jan. 2006, pp. 369–379.

[30] Akyildiz I. F. and W. Wang, "The predictive user mobility profile framework for wireless multimedia networks," *IEEE Transactions on Networking*, Vol. 12, Dec. 2004, pp. 1021–1035.

[31] Holtzman J. and A. Sampath, "Adaptive averaging methodology for handoffs in cellular systems," *IEEE Transactions on Vehicular Technology*, Vol. 44, Feb. 1995, pp. 59–66.

[32] Chu M. J. and W. E. Stark, "Effect of mobile velocity on communications in fading channels," *IEEE Transactions on Vehicular Technology*, Vol. 49, Jan. 2000, pp. 202–210.

[33] Krasny L., H. Arslan, D. Koilpillai, and S. Chennakeshu, "Doppler spread estimation in mobile radio systems," *IEEE Communication Letters*, Vol. 5, May 2001, pp. 197–199.

[34] Abdi A., H. Zhang, and C. Tepedelenlioglu, "A unified approach to the performance analysis of speed estimation techniques in mobile communication," *IEEE Transactions on Communications*, Vol. 56, Jan. 2008, pp. 126–135.

[35] Anim-Appiah K. D., "On generalized covariance-based velocity estimation," *IEEE Transactions on Vehicular Technology*, Vol. 48, Sept. 1999, pp. 1546–1557.

[36] Xiao C., K. D. Mann, and J. C. Olivier, "Mobile speed estimation for TDMA-based hierarchical cellular systems," *IEEE Transactions on Vehicular Technology*, Vol. 50, July 2001, pp. 981–991.

[37] Tepedelenlioglu C. and G. B. Giannakis, "On velocity estimation and correlation properties of narrow-band mobile communication channels," *IEEE Transactions on Vehicular Technology*, Vol. 50, July 2001, pp. 1039–1052.

[38] Zheng Y. R. and C. Xiao, "Mobile speed estimation for broadband wireless communications over rician fading channels," *IEEE Transactions on Wireless Communications*, Vol. 8, Jan. 2009, pp. 1–5.

[39] Baddour K. and N. C. Beaulieu, "Nonparametric Doppler spread estimation for flat fading channels," *IEEE Wireless Communications and Networking, WCNC 2003*, Vol. 2 New Orleans, LA, Mar. 2003, pp. 953–958.

[40] Dogandzic A. and B. Zhang, "Estimating Jakes' Doppler power spectrum parameters using the Whittle approximation," *IEEE Transactions on Signal Processing*, Vol. 53, March 2005, pp. 987–1005.

[41] Zajić A. G., "Estimation of mobile velocities and direction of movement in mobile-to-mobile wireless fading channels," *IEEE Transactions on Vehicular Technology*, Vol.61, Jan. 2012, pp. 130–139.

[42] Jakes, W. C., *Microwave Mobile Communications*, 2nd ed., Piscataway, NJ: Wiley-IEEE Press, 1994.

[43] Zajić A. G. and G. L. Stüber, "Three-dimensional modeling, simulation, and capacity analysis of space-time correlated mobile-to-mobile channels," *IEEE Transactions on Vehicular Technology*, Vol. 57, July 2008, pp. 2042–2054.

[44] Zajić A. G. and G. L. Stüber, "Three-dimensional modeling and simulation of wideband MIMO mobile-to-mobile channels," *IEEE Transactions on Wireless Communications*, Vol. 8, Mar. 2009, pp. 1260–275.

[45] Parsons J. D. and A.M.D. Turkmani, "Characterisation of mobile radio signals: model description," *IEE Proceedings I, Communications, Speech, and Vision*, Vol. 138, Dec. 1991, pp. 549–556.

[46] Kuchar A., J.-P. Rossi, and E. Bonek, "Directional macro-cell channel characterization from urban measurements," *IEEE Transactions on Antennas and Propagation*, Vol. 48, Feb. 2000, pp. 137–146.

[47] Hoeher P., "A statistical discrete-time model for the WSSUS multipath channel," *IEEE Transactions on Vehicular Technology*, Vol. 41, Nov. 1992, pp. 461–468.

[48] Chiavaccini E. and G.M. Vitetta, "GQR models for multipath Rayleigh fading channels," *IEEE Journal on Selected Areas in Communications*, Vol. 19, June 2001, pp. 1009–1018.

[49] Kennedy R. S., *Fading dispersive communication channels*, New York: Wiley-Interscience, 1969.

[50] Latinovic Z., A. Abdi, and Y. Bar-Ness, "On the utility of the circular ring model for wideband MIMO channels," *IEEE 60th Vehicular Technology Conference, VTC2004-Fall*,vol. 1, Los Angeles, CA, Sept. 2004, pp. 96–100.

[51] Acosta G. and M. A. Ingram, "Model development for the wideband expressway vehicle-to-vehicle 2.4 GHz channel," *IEEE Wireless Communications and Networking Conference, WCNC 2006.*,Vol. 3, Las Vegas, NV NE, USA, April 2006, pp. 1283–1288.

[52] Balanis C. A., *Antenna Theory: Analysis and Design*, Wiley, 1997.

[53] Zajić A. G., "Impact of mutual coupling on MIMO vehicle-to-vehicle systems," *IEEE Global Telecommunications Conference, GLOBECOM 2009*, Honolulu, HI, December 2009, pp. 1–6.

[54] Zajić A. G., "Effects of mutual coupling on capacity of MIMO vehicle-to-vehicle systems," *IEEE Military Communications Conference, MILCOM 2009*, Boson, MA, October 2009, pp. 1–5.

[55] Wallace J. W. and M. A. Jensen, "Mutual coupling in MIMO wireless systems: A rigorous network theory analysis," *IEEE Transactions on Wireless Communications*, Vol. 3, July 2004, pp. 1317–1325.

[56] Lau B. K., J. B. Andersen, G. Kristensson, and A. F. Molisch, "Impact of matching network on bandwidth of compact antenna arrays," *IEEE Transactions on Antennas and Propagation*, Vol. 54, Nov. 2006, pp. 3225–3238.

[57] Djordjević A. R., M. B. Bazdar, T. K. Sarkar, and R. F. Harrington, *AWAS for Windows: Analysis of Wire Antennas and Scatterers, Software and User's Manual*, Boston: Artech House, 2002.

[58] Rappaport, T., *Wireless Communications: Principles and Practice*, Upper Saddle River, NJ: Pretince Hall PTR, 2001.

[59] Wang R. and D. Cox, "Channel modeling for ad hoc mobile wireless networks," *IEEE 55th Vehicular Technology Conference, VTC Spring 2002*, Vol. 1, Birmingham, AL, May 2002, pp. 21–25.

[60] Patel C. S. et al., "Simulation of Rayleigh-faded mobile-to-mobile communication channels," *IEEE Transaction on Communication*, Vol. 53, November 2005, pp. 1876–1884.

[61] Zajić A. G. and G. L. Stüber, "A new simulation model for mobile-to-mobile Rayleigh fading channels," *IEEE Wireless Communications and Networking Conference, WCNC 2006*, Vol. 3, Las Vegas, NE, April 2006, pp. 1266–1270.

[62] Xiao C., Y. R. Zheng and N. Beaulieu, "Statistical simulation models for Rayleigh and Ricean fading," *IEEE International Conference on Communications, ICC '03.*, Vol. 65, Anchorage, AK, May 2003, pp. 3524–3529.

[63] Mardia K. V. and E. P. Jupp, *Directional Statistics*, New York: John Wiley and Sons, 1999.

[64] Dohler M. and Y. Li, *Cooperative Communications: Hardware, Channel & PHY*.New York: John Wiley and Sons, 2010.

[65] Harley, P., "Short distance attenuation measurements at 900 MHz and 1.8 GHz using low antenna heights for microcells," *IEEE Journal on Selected Areas in Communications*, Vol. 7, Jan. 1989, pp. 5–11.

[66] Davis, J. S. and J. P. M. G. Linnartz, "Vehicle to vehicle RF propagation measurements," *Conference Record of the Twenty-Eighth Asilomar Conference on Signals, Systems and Computers, 1994*, Vol. 1, Pacific Grove, CA, Nov. 1994, pp. 470–74.

[67] Kovács I. Z., P. C. F. Eggers, K. Olesen, and L. G. Patersen, "Investigations of outdoor-to-indoor mobile-to-mobile radio communication channels," *IEEE 56th Vehicular Technology Conference, VTC 2002-Fall*, Vancouver, Canada, Vol. 1, Sept. 2002, pp. 430–434.

[68] Kovács I. Z., *Radio Channel Characterization for Private Mobile Radio Systems: Mobile-to-Mobile Radio Link Investigations.* PhD Dissertation, Aalborg University, Denmark, 2002.

[69] Maurer J., T. Fugen, T. Schafer, and W. Wiesbeck, "A new inter-vehicle communications (IVC) channel model," *IEEE 60th Vehicular Technology Conference, VTC-2004-Fall*, Los Angeles, CA, Vol. 2, Sept. 2004, pp. 9–13.

[70] Maurer J., T. Fugen, and W. Wiesbeck, "Physical layer simulations of IEEE802.11a for vehicle-to-vehicle communications," *IEEE 62nd Vehicular Technology Conference, VTC-2005-Fall*, Dallas, TX, Vol. 3, Sept. 2005, pp. 1849–1853.

[71] Acosta G. and M. A. Ingram, "Six time- and frequency-selective empirical channel models for vehicular wireless LANs," *IEEE Vehicular Technology Magazine*, Vol. 2, Dec. 2007, pp. 4–11.

[72] Acosta G. and M. A. Ingram, "Six time- and frequency-selective empirical channel models for vehicular wireless LANs," *IEEE 66th Vehicular Technology Conference VTC-2007-Fall*, Baltimore, MD, Sept. 2007, pp. 2134–2138.

[73] Sen I. and D. W. Matolak, "Vehicle-vehicle channel models for the 5-GHz band," *IEEE Transactions on Intelligent Transportation Systems*, Vol. 9, June 2008, pp. 235–245.

[74] Matolak D.W., I. Sen, W. Xiong, and N. T. Yaskoff, "5 GHz wireless channel characterization for vehicle to vehicle communications," *IEEE Military Communications Conference, MILCOM 2005*, Vol. 5, Atlantic City, NJ, Oct. 2005, pp. 3016–3022.

[75] Matolak D.W., "Channel modeling for vehicle-to-vehicle communications," *IEEE Communications Magazine*, Vol. 46, May 2008, pp. 76–83.

[76] Wang B. I. Sen, and D.W. Matolak, "Performance evaluation of 802.16e in vehicle to vehicle channels," *IEEE 66th Vehicular Technology Conference VTC-2007-Fall*, Baltimore, MD, Sept. 2007, pp. 1406–1410.

[77] IST-4-027756 Wireless World Initiative Radio, WINNER, Available: www.ist-winner.org.

[78] D.1.1.2 WINNER II Channels, IST-4-027756 WINER, Available: http://www.ist-winner.org/WINNER2-Deliverables/D1.1.2.zip.

CHAPTER 4

Air-to-Ground Radio Channels

Air-to-ground (A-to-G) communications systems have drawn great attention as a possible alternative to traditional terrestrial or satellite-based communication systems [1–4]. A-to-G systems can have a large footprint on earth, offering the possibility of providing services over large regions at low traffic cost. The major advantages for future services integrating A-to-G communication systems are cost-effective coverage of rural and maritime regions (similar to satellite systems), system flexibility due to the platforms' mobility on demand (e.g., in emergency situations), and the possibility of upgrading the platforms' payload to reduce the risk of technology obsolescence (this is the main problem of satellite systems). An A-to-G communication system consists of sky and ground segments, as shown in Figure 4.1. Ground segments consist of mobile units, such as cellphones or vehicles, while the sky segment consists of a platform (i.e., an airplane or a lighter-than-air airship) positioned in the stratosphere. This platform is often called a high-altitude platform (HAP) or high-altitude very long endurance (HAVE) vehicle. HAPs usually fly at an altitude ranging between 17 and 30 km in the stratosphere and are able to stay aloft for long periods of time. There are several projects addressing the design of telecommunication networks based on the use of these manned and unmanned aerial vehicles [5–8]. A successful design of such A-to-G communication systems requires a detailed understanding of multipath propagation between sky and ground segments.

A-to-G channels have characteristics that are very different from F-to-M cellular and satellite radio channels. In A-to-G channels, the transmitter (T_x) and possibly the receiver (R_x) are in motion as shown in Figure 4.2. Furthermore, in contrast to F-to-M radio channels, A-to-G channels are not significantly affected by shadowing due to high elevation angles. In contrast to satellite channels, A-to-G channels provide much lower propagation loss because HAP stations are located in the stratosphere so the signals do not go through the (high-loss) ionosphere. As shown in Figure 4.2, the T_x communicates with the R_x through either a LoS link or through a NLoS link when natural and manmade objects are situated between the T_x and R_x. This chapter details how A-to-G channels differ from F-to-M cellular radio channels and how to model and simulate A-to-G channels.

4.1 A-to-G Propagation Modeling

Like in F-to-M cellular radio channels, propagation in A-to-G channels is governed by the same four propagation mechanisms described in Section 2.1 (i.e., free-space propagation, reflection, diffraction, and scattering). The differences between F-to-M cellular and A-to-G channels arise due to the differences in propagation environments, and these differences are important for modeling of large-scale and small-scale fading in A-to-G channels.

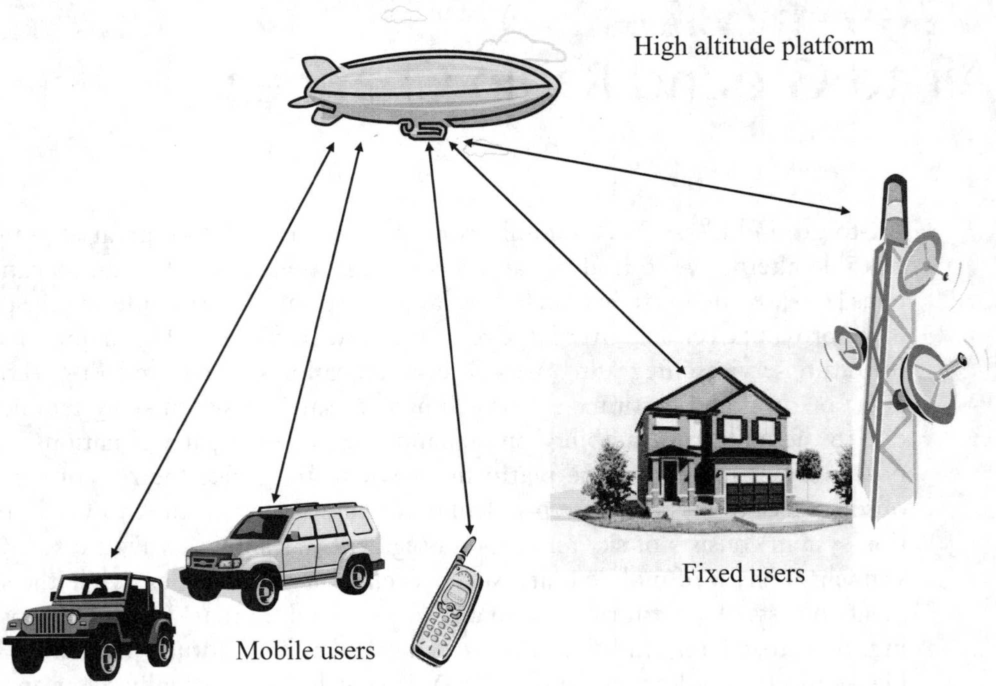

Figure 4.1 Illustration of an A-to-G radio system.

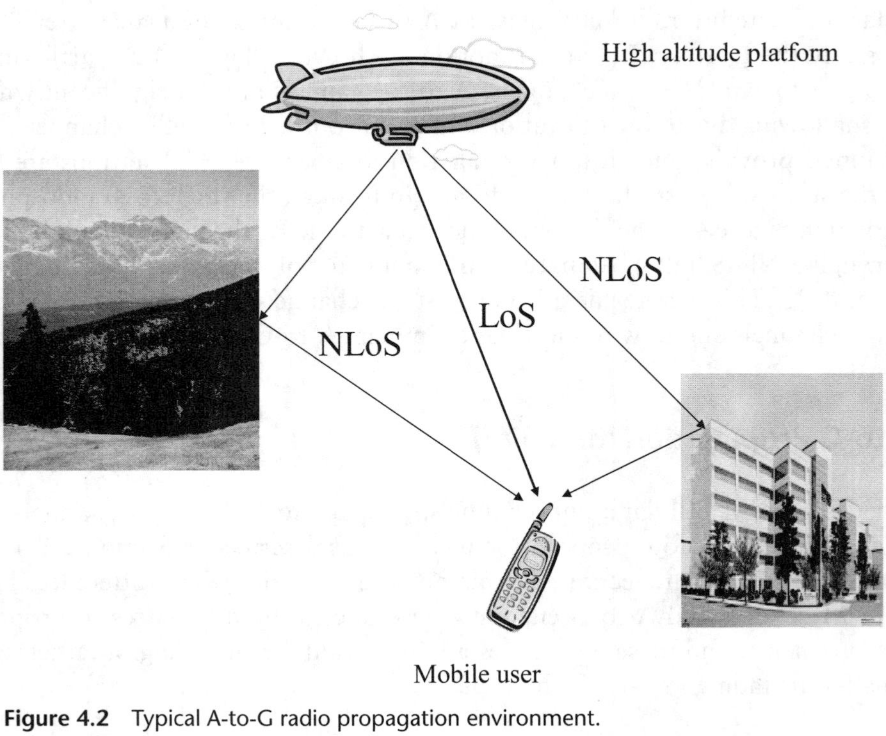

Figure 4.2 Typical A-to-G radio propagation environment.

4.1.1 Large-Scale Fading—Pathloss

The minimum pathloss that A-to-G communication link experiences is equal to the free space loss, which is defined in Section 2.2.1. The A-to-G link experiences additional pathloss due to passing through troposphere.

The troposphere is the nonionized lower portion of the earth's atmosphere covering altitudes from the ground surface up to a height of about 15 km of the atmosphere. It consists of a mixture of particles having a wide range of sizes and characteristics, from the molecules in atmospheric gases to raindrops and hail. The total pathloss resulting from an electromagnetic wave passing through such a medium can be classified into *rain attenuation*, *gaseous absorption*, and *scintillation* [9].

- *Rain attenuation* is the attenuation of the received signal amplitude caused by absorption and scattering of the radio waves from raindrops. The attenuation is measured as a function of rainfall rate and increases with the increase of operating frequency, with increased rainfall rates, and with decrease of elevation angles [10]. Modeling of rain attenuation is discussed in Section 4.1.1.1
- *Gaseous absorbtion* occurs when molecules found in the atmosphere absorb energy from radio waves passing through them, thereby causing attenuation. The principal cause of signal attenuation due to atmospheric gases is molecular absorption. Among different gases found in the atmosphere, only water vapor and oxygen have resonant frequencies in the frequency band of interest (up to 100 GHz), and gaseous attenuation in the atmosphere is normally neglected at frequency bands below 10 GHz. The amount of fading due to gases depends on altitude above sea level, frequency, temperature, pressure, and water vapor concentration. Modeling of gaseous absorbtion is discussed in Section 4.1.1.2.
- *Scintillation* is the result of small random variations in the refraction index of the air caused by the air turbulence and/or rain. This effect generates constructive and destructive interference at the receiver, causing the received signal strength variations. Modeling of scintillation is discussed in Section 4.1.1.3.

4.1.1.1 Modeling of Rain Attenuation

Rain attenuation is a function of the number and size of raindrops and the path length through the rain. It was found experimentally that, if the density and shape of the raindrops in a given region is constant, the received power $P_r(d)$ decays exponentially with distance d; that is,

$$P_r(d) = P_r(0)e^{-\alpha d} \tag{4.1}$$

where l/α is the distance at which the receive power reduces to e^{-1} of its initial value. Then, the pathloss in decibels can be calculated as

$$\begin{aligned} L(d) &= 10\log_{10}\frac{P_t}{P_r} \\ &= 4.343\alpha d \end{aligned} \tag{4.2}$$

The value of α can be calculated as

$$\alpha = \int_{r=0}^{\infty} N(r)C(r)dr \tag{4.3}$$

where $N(r)$ is the drop size distribution (i.e., the number of drops in the area with radius r per meter of path length) and $C(r)$ is the effective attenuation per raindrop cross-section. One way to solve the integral in (4.3) is to take a particular drop size distribution and integrate the result. Large raindrops were found to be less probable than small ones, and exponential distribution was found to be a good approximation for this finding [11]. Hence, the drop size distribution is modeled as

$$N(r) = N_0 \exp\left(-\frac{r}{r_m}\right) \tag{4.4}$$

where $N_0 = 8 \times 10^3 \text{m}^{-2}\text{mm}^{-1}$ and $r_m = 0.122R^{0.21}\text{mm}$ are empirical parameters and R denotes raindrop rate. The cross-section attenuation $C(r)$ can be found theoretically by calculating the attenuation due to each spherical raindrop illuminated by a plane wave and then adding those attenuations together. For low frequencies, where the drop is small compared with the incident wavelength, most of the attenuation is due to absorption in rain. In this region, the cross-section attenuation increases with frequency according to the Rayleigh approximation; that is,

$$C(r) \propto \frac{r^3}{\lambda} \tag{4.5}$$

where λ is the wavelength. At higher frequencies, when the raindrops are similar in size with wavelength, the cross-section attenuation increases more slowly, converging toward a constant value known as the optical limit. At these frequencies, scattering forms a significant part of the attenuation, so the cross-section attenuation is calculated using scattering theory.

In practice, pathloss is often modeled using empirical models that implicitly combine all of these effects. These empirical models can have quite sophisticated expressions and can be found, for example, in [12, 13].

4.1.1.2 Modeling of Gaseous Absorbtion

Gaseous molecules found in the atmosphere may absorb energy from radio waves passing through them, thereby causing attenuation. This attenuation is greatest for polar molecules (such as water), where the molecules align with an applied electric field and thus absorb some of the field's energy. Since the electric field in the radio wave changes its direction twice per period, this realignment of molecules occurs continuously and can cause a significant loss. Also, this realignment occurs faster at higher frequencies, so the absorption loss has an overall tendency to increase with frequency. Nonpolar molecules, such as oxygen, may also absorb electromagnetic energy due to the existence of magnetic moments that interact with the magnetic field in the radio wave. Each type of absorbing molecule tends to contribute both a general background level of absorption, which rises with frequency, and several resonance peaks, each corresponding to different modes of vibration specific to the molecule. In normal atmospheric conditions, only oxygen and water contribute significantly to absorption, although other atmospheric gases may be significant in very dry air at frequencies above 70 GHz.

The total attenuation due to gaseous absorbtion L_a (in decibels) for a particular path can be calculated by integrating the total specific attenuation over the total path length r; that is,

$$L_a(r) = \int_0^r \gamma_a(l)dl$$
$$= \int_0^r (\gamma_w(l) + \gamma_o(l))dl \tag{4.6}$$

where $\gamma_w(l)$ and $\gamma_o(l)$ are attenuation in dB/km for water vapor and oxygen, respectively. The attenuations for water vapor and oxygen are functions of frequency and the empirical values for them can be found in [9].

4.1.1.3 Modeling of Scintillation

As pointed out in Section 4.1.1, scintillation is the result of small random variations in the refraction index. These variations can be caused by air turbulence (wind) and rain. When the wind blows, the layers of the troposphere are mixed, leading to rapid variations of the refractive index over small distances and over short intervals of time. On the other hand, rain leads to slower variations of refractive index over small distances and over short intervals of time.

Scintillation is strongly frequency dependent. Higher frequencies are more affected then lower frequencies. The magnitude of scintillation is measured by its standard deviation or intensity (in decibels). The distribution of the fluctuation in decibels is approximately a Gaussian distribution with the intensity of scintillation as a standard deviation. The scintillation intensity σ_s can be calculated from the ITU-R model [13] as follows:

$$\sigma_s = \frac{\sigma_{ref} f^{7/12} g(D)}{\sin(\theta)^{1.2}} \tag{4.7}$$

where f is the carrier frequency, θ is the elevation angle, σ_{ref} is a typical standard deviation for a given region and depends on temperature, atmospheric pressure, and water vapor pressure, and $g(D)$ accounts for averaging of the scintillation intensity for large aperture diameter D.

Scintillation is most noticeable in warm, humid climates and is greatest during summer days. One way to reduce the effect of scintillation is to use an antenna with a wide aperture, since this produces averaging of the scintillation across the slightly different paths taken to each point across the aperture. Another approach is to use spatial diversity, where the signals from two antennas are combined to reduce the overall fade depth [9]. The best results for a given antenna separation are produced using vertically separated antennas because of the strong relationship between scintillation and the horizontal stratification of the troposphere.

4.1.2 Large-Scale Fading—Shadowing

In addition to signal power loss due to passage through the troposphere, partial or total signal obstruction can be caused by large objects, such as buildings and hills, as well as vegetation and foliage. As a result, the power of the received signal varies around the mean power predicted by the pathloss. This type of variation in the received signal power is called *shadowing* and is usually formulated as log-normally distributed over the ensemble of typical locals. Shadowing creates holes in coverage areas and results in poor coverage and poor carrier-to-interference ratio (CIR) in different places. The probability density function of the received signal envelope affected by shadowing follows lognormal distribution given in (2.2) of Section 2.2.2. Shadowing is particularly dominant in the Ka-band (26–40 GHz), where signals are easily dispersed.

4.1.3 Small-Scale Fading—Multipath Propagation

In A-to-G communications, the motion of the transmitter and/or the receiver causes rapid fluctuations in the amplitude of the received signal, which is called small-scale fading. This small-scale fading is caused by the interference between two or more versions of the transmitted signal that reach the receiver at slightly different times because they travel along different paths.

The multipath that affects A-to-G systems is caused by reflections and scattering from buildings, trees, parts of a HAPs, and so on. Reflection occurs when an electromagnetic wave impinges on a surface that is smooth, where roughness dimensions are relatively large compared to the wavelength [14]. In contrast, scattering occurs when an electromagnetic wave impinges on a surface that is rough. To determine the surface roughness, the Rayleigh criterion (defined in Section 2.1.4) is often used, which determines when a surface can be considered smooth. For frequencies in the Ka-band (26–40 GHz), the critical height that determines roughness is relatively small (0.2 cm), implying that most of the surfaces can be considered rough. On the other hand, for a 2.1 GHz IMT-2000 system, the critical roughness height can be as much as 3.5 cm [15], implying that a surface that appears rough for the Ka-band is smooth for 2.1 GHz and will produce specular reflections and hence produce multipath. Many building components, such as windows, walls, and metal beams (including parts of the HAPs payload), are smooth enough to give rise to specular reflections. A study by Andreyev and Bugaev [16] provides a full-wave model of the reflected fields. Their results, both measured and modeled, show that a typical wall may have a reflection coefficient of up to 98 percent if the whole of the main beam is incident on the wall, hence giving rise to interference. Hence, specular reflections should be accounted for when designing an A-to-G communication system.

Similarly to F-to-M channels, the A-to-G multipath channel can be modeled as a linear time-variant filter whose complex baseband impulse response is [17]

$$h(t,\tau) = \sum_{l=1}^{L} C_l e^{-j\phi_l(t)} \delta(\tau - \tau_l) \tag{4.8}$$

where $h(t, \tau)$ is the channel response at time t due to an impulse applied at time $t - \tau$, and $\delta(\cdot)$ is the Dirac delta function. The channel amplitude, phase, and time delay strongly depend on time and location and are modeled statistically, which is the topic of Section 4.2.

4.2 Statistical Characterization of Multipath Propagation in A-to-G Radio Channels

Performance of A-to-G communication links depends on the operating frequency, geographical location, climate, elevation angle to the HAPs, and so on. The link reliability decreases with the increase in operating frequency and with the decrease of elevation angles. In addition, the random and unpredictable nature of the propagation environment increases the modeling complexity of A-to-G channels. Therefore, it is suitable to describe propagation phenomena in stochastic manner to account for random fluctuations in the A-to-G channel. This section details statistical characterization of multipath propagation in A-to-G radio channels.

4.2.1 Received Envelope Distribution

Similarly to F-to-M channels, the received envelope in A-to-G channels can be modeled as Rayleigh-faded under NLoS conditions and as Ricean-faded under LoS conditions. However, in contrast to F-to-M channels where urban areas are characterized with Rayleigh fading (because the LoS is often obstructed by buildings), the A-to-G channel experiences Ricean fading in urban areas (because the LoS is not obstructed and the reflected signals are significantly attenuated). Similarly, while the F-to-M channel experiences Ricean fading in suburban areas (the LoS is not obstructed), the A-to-G channel experiences Rayleigh fading (because there are one or two reflected signals that can be stronger than LoS). Finally, in rural areas, shadowing is the predominant factor in A-to-G channels and a lognormal distribution is used to characterize fading in this environment.

For Ricean fading, several values of parameter K are reported in the literature for the 1–2 GHz (L-band) frequency range [18–21]. In contrast to F-to-M channels, in A-to-G channels the Ricean parameter K depends on the elevation angles. In [18], the Ricean parameter K takes values 8.5 dB and 6 dB for elevation angles of 45° and 55°, respectively. In [19], the parameter K takes values 12 dB and 20 dB for elevation angles 60° and 80°, respectively, both under clear-sky conditions. In [20], the Ricean parameter K is 14 dB for open areas. Finally, in [21], the Ricean parameter K is 11 dB for 43° elevation angle.

4.2.2 Statistical Modeling of A-to-G Multipath Channels

Since A-to-G systems share some similarities with both F-to-M cellular and satellite systems, channel models already developed for these systems have been applied to A-to-G systems. However, satellite channel models [21–26] do not characterize properly A-to-G propagation because of different propagation loss and multipath propagation characteristics. Similarly, 2-D and 3-D F-to-M channel models [27–30] are not applicable to A-to-G systems due to very different elevation angles. Hence, in the following sections we detail channel models derived for A-to-G communications.

4.2.3 Two-Ray Model for A-to-G Multipath Fading Channels

The two-ray model for A-to-G channels is proposed in [31]. This model assumes that the HAP is quasi-stationary with an altitude of 20–25 km. The geometrical characteristics of the channel are shown in Figure 4.3. The model assumes two-ray propagation (i.e., the LoS ray and one reflected ray). The parameters that describe the geometry of the system are

- α: the elevation angle in degrees;
- h: the HAP height in meters;

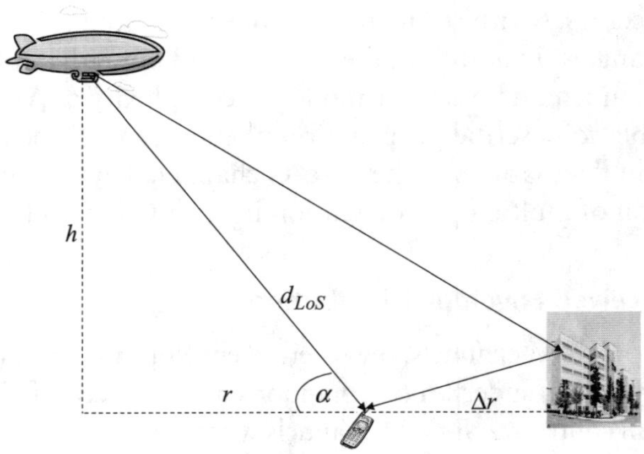

Figure 4.3 A geometry model for two-ray A-to-G channel.

- r: the horizontal distance between the transmitter and the receiver in meters;
- Δr: the horizontal distance between the receiver and the reflector in meters.

Propagation Delay

The horizontal distance between the transmitter and the receiver (i.e., the radius of the coverage area of the HAP) can be computed as

$$r = 4R\left(\cos^{-1}\left(\frac{R}{R+h}\cos\alpha\right) - \alpha\right) \tag{4.9}$$

where $R = 6378$ km is the radius of the earth. From Figure 4.3, for a given elevation angle α, coverage can be approximated as

$$r \approx \frac{h}{\tan\alpha} \tag{4.10}$$

In this approximation, the curvature of the earth is neglected. Once the radius of the coverage area is obtained, the distance between the HAP and the user can be computed. The LoS distance d_{LoS} is given by

$$\begin{aligned} d_{LoS} &= \sqrt{r^2 + h^2} \\ &= \frac{h}{\sin\alpha} \end{aligned} \tag{4.11}$$

Similarly, the horizontal distance between the receiver and the reflector can be calculated as

$$d_{\text{ref}} = \sqrt{(r+\Delta r)^2 + h^2} + \Delta r \tag{4.12}$$

Since electromagnetic waves travel at the speed of light, using these calculations, we can estimate the time needed for each ray to reach the receiver. Hence, we can calculate the relative delay of each reflected ray with respect to the LoS. This method can also be used to estimate the power delay profile of the system, for wideband channel modeling. The total delay $\Delta\tau$ is given by

$$\Delta\tau = \frac{1}{c}\left(\sqrt{(r+\Delta r)^2 + h^2} + |\Delta r| - \frac{h}{\sin\alpha}\right) \tag{4.13}$$

where c is the speed of light. The best delay is achieved when the HAP height approaches infinity (i.e., $h \to \infty$) and is equal to $\Delta\tau = (1 + \cos\alpha)|\Delta r|/c$. On the other hand, the worst delay occurs when the HAP height approaches zero. (i.e., $h \to \infty$) and is equal to $\Delta\tau = 2|\Delta r|/c$.

Received Envelope Distribution

The two-ray model assumes that the received envelope has a Ricean distribution. The Ricean factor (i.e., the ratio of the LoS path and the reflected path) is a function of the elevation angle α and can be calculated in dB as

$$K(\alpha) = 20\log_{10}\left[\frac{h/\sin\alpha}{\sqrt{(r+\Delta r)^2 + h^2} + |\Delta r|}\right] \tag{4.14}$$

Here we can observe that, when the elevation angle approaches $\alpha = 90°$, the Ricean parameter K approaches infinity and the received amplitude follows Gaussian distribution. If the elevation angle is between $12° < \alpha < 90°$, the received amplitude follows Ricean distribution. Finally, for low elevation angles (i.e., $\alpha < 12°$) the Ricean factor approaches zero and the received amplitude follows Rayleigh distribution.

4.2.4 A Scattering Model for A-to-G Multipath Fading Channels

The two-ray model described in the previous section assumes that there is only one reflection from the objects on the ground. However, that is often not a realistic assumption, especially in urban areas where users are usually surrounded by tall buildings. To address this issue, the model in [32] focuses on theoretical derivation of small-scale fading for A-to-G channels while taking into account the presence of scatterers in the terrain.

The model in [32] is based on the theoretical model proposed by Rappaport and Liberty [27] but assumes that the receiver and the transmitter are no longer both in the horizontal plane and so the height of the transmitter must be taken into account. The geometry of the model in [32] is shown in Figure 4.4. The receiver R_x is located on the ground, in the point with coordinates $(x_0, 0, 0)$. The transmitter T_x is elevated above the ground in the point with coordinates $(0, 0, z_0)$. It is assumed that all scatterers that cause excess delays less than τ are inside the ellipsoid shown in Figure 4.4.

The LoS propagation path has length r_0 that can be calculated as

$$r_0 = \sqrt{x_0^2 + z_0^2} \tag{4.15}$$

Consequently, the LoS propagation delay is obtained as

$$\tau_0 = \frac{r_0}{c} \tag{4.16}$$

where c is the speed of light. Reflected waves with excess delay τ will travel a path length given by

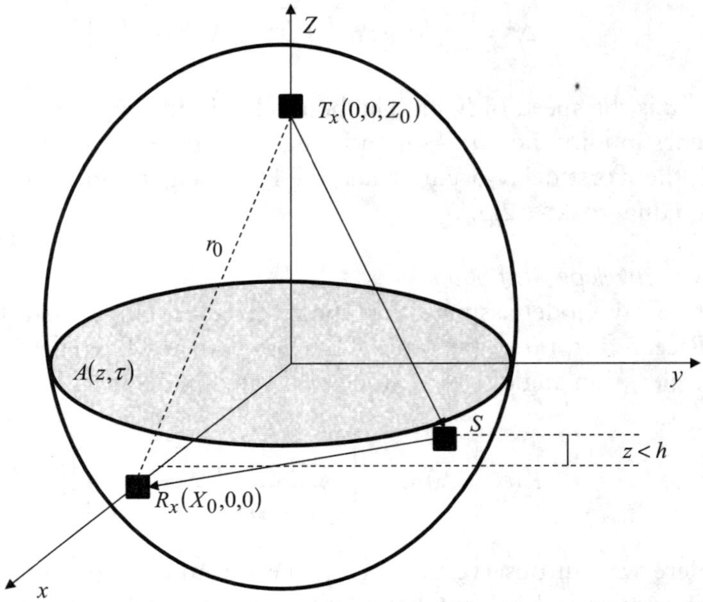

Figure 4.4 Multipath channel model for A-to-G channels.

$$k(\tau) = r_0 + c\tau \tag{4.17}$$

Hence, the ellipsoid shown in Figure 4.4, with T_x and R_x as foci, is defined by the set of points for which the sum of the distance from the T_x and the distance from the R_x equals $k(\tau)$. All the scatterers located inside the ellipsoid generate a total path length smaller than $k(\tau)$. For this reason, a reflected ray with an excess delay smaller than τ must be associated with a scatter S located inside the ellipsoid. Since the scatterers are mainly tall buildings, trees, hills, and so on, they cannot be too high (above a certain height h, there are no scatterers). Thus scatterers can be assumed to be uniformly distributed in a thin layer close to the ground. As a result, the volume $V(\tau)$ containing obstacles that cause scattered rays with the excess delay smaller than τ is the intersection of the ellipsoid shown in Figure 4.4 and the space between the ground and the $z = h$ plane. More formally, the volume $V(\tau)$ is defined as the set of points (x, y, z) that satisfy

$$\sqrt{(x-x_0)^2 + y^2 + z^2} + \sqrt{x^2 + y^2 + (z-z_0)^2} < k(\tau), \quad 0 < z < h \tag{4.18}$$

The volume $V(\tau)$ is obtained by integrating the area $A(z,\tau)$ of the ellipse for z varying from 0 to h, and the result is

$$V(\tau) = \frac{-\pi w(\tau)(r_0 + c\tau)}{4\sqrt{(z_0^2 + w(\tau))^3}}\left[\frac{4}{3}h^3 - 2z_0h^2 - w(\tau)h\right] \tag{4.19}$$

where $w(\tau) = c^2\tau^2 + 2r_0c\tau$.

The cumulative distribution function (cdf) of the excess delay $F_{PS}(\tau)$ can be calculated as the ratio between the volume $V(\tau)$ and the volume that corresponds to the maximum excess delay of the system, $V(\tau_m)$,

$$F_{PS}(\tau) = \frac{V(\tau)}{V(\tau_m)} \tag{4.20}$$

The knowledge of the cdf of the excess delay allows us to evaluate the average number of echoes $n(\tau)$ having excess delay τ. The power of a single echo having excess delay τ, $P_s(\tau)$ can be evaluated using a free-space model with modified path lengths. The power delay profile of the system can then be obtained as the product of $n(\tau)$ and the power $P_s(\tau)$; that is,

$$P(\tau) = n(\tau)P_s(\tau) \tag{4.21}$$

4.2.5 Three-Dimensional Model for MIMO A-to-G Multipath Fading Channels

The design of high-speed A-to-G communication systems often requires the use of multiple antennas as well as multiple-input multiple-output (MIMO) diversity techniques. To address this problem, a 3-D model for MIMO A-to-G frequency-flat multipath fading channels is proposed in [33]. This model considers outdoor radio propagation where direct path and single-bounced reflections are present. The MIMO channel is described by an $L_R \times L_T$ matrix $\mathbf{H}(t) = [h_{ij}(t)]L_R \times T_R$ of complex low-pass faded envelopes.

Figure 4.5 shows the geometry of the 3-D cylinder model for a MIMO A-to-G channel with $L_T = L_R = 2$ antenna elements. This elementary 2 × 2 antenna configuration is used later to construct linear multielement antenna arrays. The spacing between two adjacent antenna elements at the transmitter T_x and the receiver R_x is denoted by δ_T and δ_R, respectively. The T_x and R_x antenna array heights are denoted H_T and H_R, respectively; angles θ_T and θ_R in Figure 4.5 describe the orientation of the T_x and R_x antenna array in the x-y plane, respectively; and angle ψ_R describes the elevation of the R_x's antenna array relative to the x-y plane, respectively. It is assumed that the elevation angle of the T_x's antenna array is small compared to the elevation angle of the T_x and can be neglected. Note also that $H_T >> H_R$ and the elevation angle of the transmitter relative to the receiver can be approximated as

$$\beta_T \approx \arctan\left(\frac{H_T}{D}\right) \tag{4.22}$$

where D is the horizontal distance between the T_x and the R_x.

The cylinder model assumes that the T_x is highly elevated (i.e., a stratospheric platform) and free of local scatterers. On the other hand, the R_x is on the ground and surrounded by local scatterers. Around the receiver, $N \to \infty$ fixed omnidirectional scatterers lie inside of a cylinder of radius R_{Smax}. The nth scatterer is denoted $S^{(n)}$, the distance between the scatterer's projection to the x-y plane and O_R is denoted $d(\hat{S}^{(n)}, O_R) = R_S^{(n)} \in (0, R_{Smax}]$, and its height is denoted $d(\hat{S}^{(n)}, S^{(n)}) = H_S^{(n)} \in (0, H_{Smax}]$, where R_{Smax} and H_{Smax} represent the radius and the height of the cylinder, respectively. Note that $H_{Smax} << H_T$ and that the elevation angle of the nth scatterers relative to O_R is given by

$$\beta_S^{(n)} \approx \arctan\left(\frac{H_S^{(n)}}{R_S^{(n)}}\right) \tag{4.23}$$

Note that the cylinder model assumes almost the same geometry as the model in Section 4.2.4. For $H_{Smax} << H_T$, the volume $V(\tau)$ can be approximated by a cylinder because $V(\tau)$ is a small horizontal slice of the ellipsoid, and the vertical curvature of the surface at the circumference of the slice can be neglected. The T_x and R_x are moving with speeds v_T and v_R in directions described by angles γ_T and γ_R, respectively.

Since the aircraft flies in a circle, v_T is related to the angular velocity of the transmitter. Then, we can observe that

$$\begin{aligned} v_T &= \omega R_T \\ &= 2\pi R_T / T \end{aligned} \tag{4.24}$$

where ω is the angular frequency, T is the period, and R_T is the radius of the rotation. Therefore, γ_T is always perpendicular to R_T.

In Figure 4.5, angle $\alpha_T^{(n)}$ is the azimuth angle of departure (AAoD) of the waves that impinge on scatterer $S^{(n)}$ and angle $\alpha_R^{(n)}$ is the azimuth angle of arrival (AAoA) of the waves that bounce off the scatterer $S^{(n)}$. It is assumed that $\alpha_T^{(n)}$, $\alpha_R^{(n)}$, $R_S^{(n)}$, $H_S^{(n)}$, and $\beta_S^{(n)}$ are random variables. From Figure 4.5, we can observe that $\alpha_T^{(n)}$ is dependent on $\alpha_R^{(n)}$ and $R_S^{(n)}$, while $\beta_S^{(n)}$ is dependent on $R_S^{(n)}$ and $H_S^{(n)}$. However, $\alpha_R^{(n)}$, $R_S^{(n)}$, and $H_S^{(n)}$ are independent random variables. Finally, α_{Rq}^{LoS} denotes the AAoA of the LoS path (not shown in Figure 4.5 to reduce clutter).

This model assumes that the radius of the cylinder $R_{S\max}$ is much smaller than the distance D (i.e., $R_{S\max} \ll D$). Furthermore, it is assumed that the spacings between antennas (δ_T and δ_R) are much smaller than the distance $R_S^{(n)}$ (i.e., δ_T, $\delta_R \ll \min\{R_S^{(n)}\}$), which allows us to approximate the AAoD of the LoS as $\alpha_{Tp}^{LoS} \approx \pi$. For ease of reference, the parameters of the proposed model are summarized in Table 4.1.

The complex faded envelope under frequency-flat fading can therefore be written as a superposition of LoS and NLoS rays. For the channel between the pth transmit and qth receive antenna, the channel impulse response is [33]

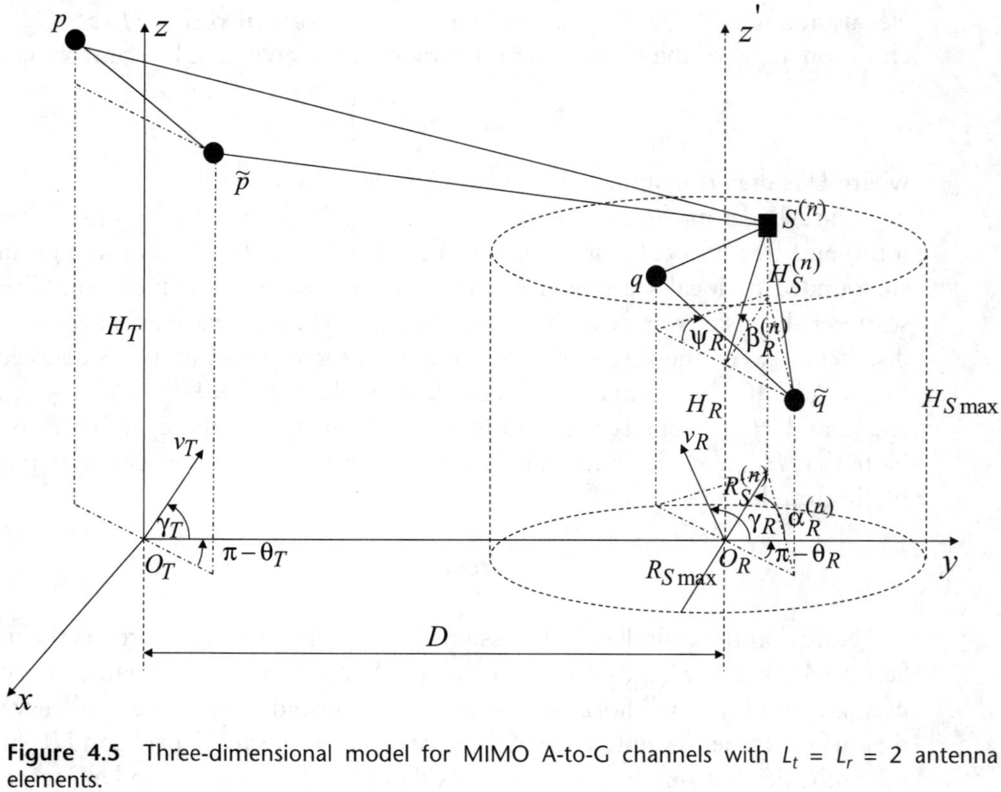

Figure 4.5 Three-dimensional model for MIMO A-to-G channels with $L_t = L_r = 2$ antenna elements.

Table 4.1 Definition of the Parameters Used in the Cylinder Geometrical Model

D	The distance between the horizontal projections of the centers of the T_x and R_x antenna arrays.
R_{Smax}, H_{Smax}	The radius and the height of the cylinder with scatterers, respectively.
δ_T, δ_R	The spacing between two adjacent antenna elements at the T_x and R_x, respectively.
θ_T, θ_R	The azimuth angles of the transmit and the receive antenna elements (relative to the x-axis), respectively.
ψ_R	The elevation angle of receive antenna elements (relative to the x-y plane).
v_T, v_R	The velocities of the transmitter and receiver, respectively.
γ_T, γ_R	Moving directions of the transmitter and receiver, respectively.
$\alpha_T^{(n)}, \alpha_R^{(n)}$	The azimuth angle of departure (AAoD) and the azimuth angle of arrival (AAoA), respectively.
β_T	The elevation angle of the transmitter (relative to the x-y plane).
H_T, H_R, $H_{S^{(n)}}$	The heights of the transmit antenna array, the receive antenna array, and the nth scatterer, respectively.

$$h_{pq}(t) = h_{pq}^{LoS}(t) + h_{pq}^{NLoS}(t) \tag{4.25}$$

where the respective components are

$$h_{pq}^{LoS}(t) = \sqrt{\Omega_{pq}}\, a_{pq}^{LoS} e^{\left(j2\pi f_{pq}^{LoS} t - j\frac{2\pi}{\lambda} d_{pq}^{LoS}\right)} \tag{4.26}$$

$$h_{pq}^{NLoS}(t) = \sqrt{\Omega_{pq}} \lim_{N\to\infty} \sum_{n=1}^{N} a_{pq}^{(n)} e^{j2\pi f^{(n)} t - j\frac{2\pi}{\lambda} d_{pq}^{(n)} + j\phi_{pq}^{(n)}} \tag{4.27}$$

where Ω_{pq} is the power associated with the pqth planar wave. For $\max\{\delta_T, \delta_R\} << \min\{R_S^{(n)}\}$, we can assume that the total power is equally split among all antenna elements (i.e., $\Omega_{pq} = \Omega$ for all $p \in \{1,..., L_t\}$ and $q \in \{1,..., L_r\}$). The channel gains, Doppler shift frequencies, phases, and path lengths in this model can be calculated as follows:

- *Channel gains*. The channel gains in this model are normalized such that the ensemble averages of the $a_{pq}^{(n)}$'s are given by

$$\sum_{n=1}^{N} E[a_{pq}^{(n)2}] = 1 \tag{4.28}$$

In addition, K_{pq} describes how the total signal power between the pth transmit and the qth receive antenna is split between the LoS and NLoS components: the power allocated to the LoS component between the pth transmit and qth receive antenna is $K_{pq}/(1 + K_{pq})$, while the total power allocated to the NLoS components between the pth transmit and qth receive antenna is $1/(1 + K_{pq})$. For $\max\{\delta_T, \delta_R\} << \min\{R_t, R_r\}$, we can assume that the Ricean fading factor is the same for all antenna elements (i.e., $K_{pq} = K$ for all $p \in \{1,..., L_t\}$ and $q \in \{1, ..., L_r\}$). Having all this in mind, the channel gains can be written as follows:

$$a_{pq}^{LoS} = \sqrt{\frac{K}{K+1}} \tag{4.29}$$

$$a_{pq}^{(n)} = \sqrt{\frac{1}{K+1}}\sqrt{\frac{1}{N}} \tag{4.30}$$

- *Doppler shifts*. Doppler shifts depend on the geometrical relation between directions of movement of T_x and R_x (γ_T and γ_R) and the directions determined by the angles of departure and angles of arrival. For this model, the Doppler shifts can be written as

$$f_{pq}^{LoS} = f_{T\max}\cos(\pi - \alpha_{Rq}^{LoS} - \gamma_T) + f_{R\max}\cos(\alpha_{Rq}^{LoS} - \gamma_R) \tag{4.31}$$

$$\begin{aligned} f^{(n)} &= f_{T\max}\cos(\alpha_T^{(n)} - \gamma_T) + f_{R\max}\cos(\alpha_R^{(n)} - \gamma_R)\cos\beta_S^{(n)} \\ &= f_{T\max}\cos(R_S^{(n)}\sin\gamma_T\sin\alpha_R^{(n)}/D + \gamma_T) \\ &+ f_{R\max}\cos(\alpha_R^{(n)} - \gamma_R)(\cos(\arctan(H_S^{(n)}/R_S^{(n)})) \end{aligned} \tag{4.32}$$

where frequencies $f_{T\max} = v_T/\lambda$ and $f_{R\max} = v_R/\lambda$ are the maximum Doppler frequencies associated with the T_x and R_x, respectively, and λ is the carrier wavelength. For $\max\{\delta_T, \delta_R\} << \min\{R_S^{(n)}\}$, the LoS AAoAs are approximately equal to π (i.e., $\alpha_{Rq}^{LoS} = \pi$).

- *Phases*. This model assumes that the phase shifts $\phi_{pq}^{(n)}$ are random variables uniformly distributed on the interval $[0, 2\pi)$ and independent from any other random variable in this model.
- *Path lengths*. From Figure 4.5, the path lengths of the respective waves can be written as

$$d_{pq}^{LoS} = d(p,q) \tag{4.33}$$

$$d_{pq}^{(n)} = d(p,n) + d(n,q) \tag{4.34}$$

where $d(a, b)$ denotes the distance between two points a and b. As pointed out earlier, this model assumes that $R_{S\max} << D$ (local scattering condition). Under this assumption, the distances $d(p,q)$, $d(p,n)$, and $d(n,q)$ can be approximated as follows [33]:

$$d(p,q) \approx \frac{D - D_{T1} + D_{R1}}{\cos\beta_T} \tag{4.35}$$

$$d(p,n) \approx \frac{D - D_{T1} - D_{T2}R_S^{(n)}\alpha_R^{(n)}}{\cos\beta_T} \tag{4.36}$$

$$\begin{aligned} d(n,q) &\approx R_S^{(n)}/\cos\beta_S^{(n)} - D_{R1}\cos\alpha_R^{(n)}\cos\beta_S^{(n)} \\ &- D_{R2}\sin\beta_S^{(n)} - D_{R3}\sin\alpha_R^{(n)}\cos\beta_S^{(n)} \\ &= R_S^{(n)}/\cos(\arctan(H_S^{(n)}/R_S^{(n)})) \\ &- D_{R1}\cos\alpha_R^{(n)}\cos(\arctan(H_S^{(n)}/R_S^{(n)})) \\ &- D_{R2}\sin(\arctan(H_S^{(n)}/R_S^{(n)})) \\ &- D_{R3}\sin\alpha_R^{(n)}\cos(\arctan(H_S^{(n)}/R_S^{(n)})) \end{aligned} \tag{4.37}$$

where parameters p and q take values from the sets $p \in \{1,\ldots, L_t\}$ and $q \in \{1,\ldots, L_r\}$, respectively, and

$$D_{T1} = (0.5L_t + 0.5 - p)\delta_T \cos\theta_T \tag{4.38}$$

$$D_{T2} = (0.5L_t + 0.5 - p)\delta_T \sin\theta_T / D \tag{4.39}$$

$$D_{R1} = (0.5L_r + 0.5 - q)\delta_R \cos\theta_R \cos\psi_R \tag{4.40}$$

$$D_{R2} = (0.5L_r + 0.5 - q)\delta_R \sin\psi_R \tag{4.41}$$

$$D_{R3} = (0.5L_r + 0.5 - q)\delta_R \sin\theta_R \cos\psi_R \tag{4.42}$$

4.2.6 Spatial-Temporal Fading Characteristics of A-to-G Multipath Fading Channels

Using the model described in Section 4.2.5, we can now derive the key temporal and spatial characteristics of A-to-G frequency-flat multipath fading channels.

4.2.6.1 Space-Time Correlation Function

Assuming a 3-D nonisotropic scattering environment, we derive the space-time correlation function (ST-CF) of the complex faded envelope in (4.25). The normalized space-time correlation function between two complex faded envelopes $h_{pq}(t)$ and $h_{\tilde{p}\tilde{q}}(t)$ is defined as

$$R_{pq,\tilde{p}\tilde{q}}(\delta_T, \delta_R, \tau) = \frac{\mathrm{E}[h_{pq}(t) h_{\tilde{p}\tilde{q}}(t+\tau)^*]}{\sqrt{\mathrm{Var}[h_{pq}(t)]\mathrm{Var}[h_{\tilde{p}\tilde{q}}(t+\tau)^*]}} \tag{4.43}$$

where $(\cdot)^*$ denotes complex conjugate operation, $\mathrm{E}[\cdot]$ is the statistical expectation operator, $\mathrm{Var}[\cdot]$ is the statistical variance operator. Since $h_{pq}^{LoS}(t)$ and $h_{pq}^{NLoS}(t)$ are independent complex Gaussian random processes with zero means, (4.43) can be simplified to

$$R_{pq,\tilde{p}\tilde{q}}(\delta_T,\delta_R,\tau) = R^{LoS}_{pq,\tilde{p}\tilde{q}}(\delta_T,\delta_R,\tau) + R^{NLoS}_{pq,\tilde{p}\tilde{q}}(\delta_T,\delta_R,\tau) \tag{4.44}$$

where $R^{LoS}_{pq,\tilde{p}\tilde{q}}(\delta_T,\delta_R,\tau)$ and $R^{NLoS}_{pq,\tilde{p}\tilde{q}}(\delta_T,\delta_R,\tau)$ denote the normalized ST-CFs of the LoS and NLoS components, respectively.

To calculate the normalized spatial-temporal autocorrelation function, we start from (4.44) and follow the derivations in [33]. The LoS component can be obtained as follows

$$\begin{aligned}R^{LoS}_{pq,\tilde{p}\tilde{q}}(\delta_T,\delta_R,\tau) &= \frac{\mathrm{E}[h^{LoS}_{pq}(t)h^{LoS}_{\tilde{p}\tilde{q}}(t+\tau)^*]}{\sqrt{\mathrm{Var}[h^{LoS}_{pq}(t)]\mathrm{Var}[h^{LoS}_{\tilde{p}\tilde{q}}(t+\tau)^*]}} \\ &= \frac{K}{K+1} e^{j\frac{2\pi}{\lambda}(\tilde{p}-p)\delta_T\cos\theta_T - j\frac{2\pi}{\lambda}(\tilde{q}-q)\delta_R\cos\theta_R\cos\psi_R} \\ &\quad\times e^{j2\pi\tau f_{T\max}\cos\gamma_T - j2\pi\tau f_{R\max}\cos\gamma_R}\end{aligned} \tag{4.45}$$

Similarly, the NLoS component can be obtained as follows

$$\begin{aligned}R^{NLoS}_{pq,\tilde{p}\tilde{q}}(\delta_T,\delta_R,\tau) &= \frac{\mathrm{E}[h^{NLoS}_{pq}(t)h^{NLoS}_{\tilde{p}\tilde{q}}(t+\tau)^*]}{\sqrt{\mathrm{Var}[h^{NLoS}_{pq}(t)]\mathrm{Var}[h^{NLoS}_{\tilde{p}\tilde{q}}(t+\tau)^*]}} \\ &= \frac{1}{K+1}\lim_{N\to\infty}\frac{1}{N}\sum_{n=1}^{N} \\ &\quad \mathrm{E}_{\alpha_R^{(n)},R_S^{(n)},H_S^{(n)}}\Bigg[e^{-j2\pi\tau f_{T\max}\left(\frac{R_S^{(n)}\sin\gamma_T\sin\alpha_R^{(n)}}{D}+\cos\gamma_T\right)} \\ &\quad\times e^{-j2\pi\tau f_{R\max}\cos(\alpha_R^{(n)}-\gamma_R)\cos\left[\arctan\left(H_S^{(n)}/R_S^{(n)}\right)\right]} \\ &\quad\times e^{j\frac{2\pi(\tilde{p}-p)\delta_T\cos\theta_T}{\lambda\cos\beta_T}} e^{j\frac{2\pi(\tilde{p}-p)\delta_T\sin\theta_T R_S^{(n)}\sin\alpha_R^{(n)}}{\lambda D\cos\beta_T}} \\ &\quad\times e^{j\frac{2\pi(\tilde{q}-q)\delta_R\sin\psi_R\sin\left[\arctan\left(H_S^{(n)}/R_S^{(n)}\right)\right]}{\lambda}} \\ &\quad\times e^{j\frac{2\pi(\tilde{q}-q)\delta_R\sin\theta_R\cos\psi_R\sin\alpha_R^{(n)}\cos\left[\arctan\left(H_S^{(n)}/R_S^{(n)}\right)\right]}{\lambda}} \\ &\quad\times e^{j\frac{2\pi(\tilde{q}-q)\delta_R\cos\theta_R\cos\psi_R\cos\alpha_R^{(n)}\cos\left[\arctan\left(H_S^{(n)}/R_S^{(n)}\right)\right]}{\lambda}}\Bigg]\end{aligned} \tag{4.46}$$

The model in [33] assumes that the number of scatterers is infinite. Thus, the discrete variables $\alpha_R^{(n)}$, $R_S^{(n)}$, and $H_S^{(n)}$ can be replaced with continuous random variables α_R, R_S, and H_S, respectively. Since these parameters are random variables, they can be characterized with a joint probability density function (pdf) $p(\alpha_R,R_S,H_S)$. However, since α_R, R_S, and H_S are independent, the joint pdf $p(\alpha_R,R_S,H_S)$ can be decomposed into the product of three pdfs, (i.e., $p(\alpha_R)p(R_S)p(H_S)$). Hence, the space-time correlation function of the NLoS component becomes

$$\begin{aligned}
R_{pq,\tilde{p}\tilde{q}}^{NLoS}(\delta_T,\delta_R,\tau) = \\
&= \frac{1}{K+1}\int_0^{H_{S\max}}\int_0^{R_{S\max}}\int_{-\pi}^{\pi} p(\alpha_R)p(R_S)p(H_S) \\
&\times e^{-j2\pi\tau f_{T\max}\left(\frac{R_S\sin\gamma_T\sin\alpha_R}{D}+\cos\gamma_T\right)} \\
&\times e^{-j2\pi\tau f_{R\max}\cos(\alpha_R-\gamma_R)\cos[\arctan(H_S/R_S)]}e^{j\frac{2\pi(\tilde{p}-p)\delta_T\cos\theta_T}{\lambda\cos\beta_T}} \\
&\times e^{j\frac{2\pi(\tilde{p}-p)\delta_T\sin\theta_T R_S\sin\alpha_R}{\lambda D\cos\beta_T}}e^{j\frac{2\pi(\tilde{q}-q)\delta_R\sin\psi_R\sin[\arctan(H_S/R_S)]}{\lambda}} \\
&\times e^{j\frac{2\pi(\tilde{q}-q)\delta_R\sin\theta_R\cos\psi_R\sin\alpha_R\cos[\arctan(H_S/R_S)]}{\lambda}} \\
&\times e^{j\frac{2\pi(\tilde{q}-q)\delta_R\cos\theta_R\cos\psi_R\cos\alpha_R\cos[\arctan(H_S/R_S)]}{\lambda}} d\alpha_R dR_S dH_S
\end{aligned} \tag{4.47}$$

Several different distributions, such as uniform [34], von Mises [35], and Laplacian [36], were used in prior work to characterize the azimuth angle α_R. In [33], the von Mises pdf is used because it approximates many of the aforementioned distributions and is empirically justified in urban and suburban areas [28]. The von Mises pdf can be written as [37]

$$p(\alpha_R) = \frac{\exp[k_R\cos(\alpha_R-\mu_R)]}{(2\pi I_0(k_R))} \tag{4.48}$$

where $\alpha_R \in [-\pi, \pi)$ represents AAoAs, $I_0(\cdot)$ is the zeroth-order modified Bessel function of the first kind, $\mu_R \in [-\pi, \pi)$ is the mean angle at which the scatterers are distributed, and k_R controls the spread of scatterers around the mean. The distance R_S is characterized using the hyperbolic pdf [38], which is empirically justified in [39]. This approach is more realistic than other commonly used distributions such as the uniform distribution [27, 32], because it allows scatterers to assume higher concentration in a specific (determined by a parameter) area in the vicinity of the R_x. The hyperbolic pdf is defined as [38]

$$p(R_S) = \frac{a}{\tanh(aR_{S\max})\cosh^2(aR_S)}, \quad 0 < R_S \le R_{S\max} \tag{4.49}$$

In this hyperbolic pdf, parameter a controls the spread (standard deviation) of the scatterers around the R_x, and its applicable value is in the interval (0,1). Decreasing the value of a increases the spread of the pdf of R_S. As the elevation angle of the platform decreases, the effective scatterers are expected to be more widely distributed, which corresponds to decrease in value a.

Depending on the environment (urban or suburban) and the geographical location, several different distributions, such as Gaussian [40], log-normal [41, 42], and Rayleigh distribution [43], are used to characterize the hight of scatterers H_S. In [33], the lognormal pdf is used, which is defined as

$$p(H_S) = \frac{e^{-\frac{[\ln(H_S)-\ln(H_{S\text{mean}})]^2}{2\sigma^2}}}{H_S\sigma\sqrt{2\pi}}, \quad 0 < H_S \le H_{S\max} \tag{4.50}$$

where parameters $H_{S\text{mean}}$ and σ are the mean and the standard deviation of H_S, respectively.

Finally, using (4.48)–(4.50) and the equality $\int_{-\pi}^{\pi} e^{a\sin(c)+b\cos(c)}dc = 2\pi I_0\left(\sqrt{a^2+b^2}\right)$ (equation 3.338-4 in [44]), the space-time correlation function of the NLoS component becomes [33]

$$R_{pq,\tilde{p}\tilde{q}}^{NLoS}(\delta_T,\delta_R,\tau) = r_1\int_0^{H_{S\max}}\int_0^{R_{S\max}} r_2 I_0\left(\sqrt{r_3^2+r_4^2}\right)dR_S dH_S \tag{4.51}$$

where parameters r_1, r_2, r_3, and r_4 are, respectively,

$$r_1 = \frac{a\cdot e^{j2\pi\left[\frac{(\tilde{p}-p)\delta_T\cos\theta_T}{\lambda\cos\beta_T} - \tau f_{T\max}\cos\gamma_T\right]}}{\sigma\sqrt{2\pi}\tanh(aR_{S\max})I_0(k_R)(K+1)} \tag{4.52}$$

$$r_2 = \frac{e^{-\frac{[\ln(H_S)-\ln(H_{S\text{mean}})]^2}{2\sigma^2}}\, e^{j\frac{2\pi(\tilde{q}-q)\delta_R\sin\psi_R\sin[\arctan(H_S/R_S)]}{\lambda}}}{H_S\cosh^2(aR_S)} \tag{4.53}$$

$$\begin{aligned} r_3 = {} & j\frac{2\pi(\tilde{p}-p)\delta_T\sin\theta_T R_S}{\lambda D\cos\beta_T} - j2\pi\tau f_{T\max}\sin\gamma_T R_S/D \\ & + j2\pi(\tilde{q}-q)\delta_R\sin\theta_R\cos\psi_R\cos[\arctan(H_S/R_S)]/\lambda \\ & - j2\pi\tau f_{R\max}\sin\gamma_R\cos[\arctan(H_S/R_S)] + k_R\sin\mu_R \end{aligned} \tag{4.54}$$

$$\begin{aligned} r_4 = {} & j2\pi(\tilde{q}-q)\delta_R\cos\theta_R\cos\psi_R\cos[\arctan(H_S/R_S)]/\lambda \\ & - j2\pi\tau f_{R\max}\cos\gamma_R\cos[\arctan(H_S/R_S)] + k_R\cos\mu_R \end{aligned} \tag{4.55}$$

To obtain the space-time correlation function of the NLoS component, the integral in (4.51) has to be numerically evaluated, since there is no closed-form solution. Although this was not described in [33], a closed-form expression for the space-time correlation function of the NLoS component can be obtained if the elevation angles of arrival β_S are directly modeled as the random variables instead of using two random variables R_S and H_S to define the elevation angles of arrival as in (4.23). Prior work uses several different scatterer distributions, such as uniform, cosine, and Gaussian, to characterize the random elevation angles of arrival $\beta_S^{(n)}$. Here, we use the cosine pdf [45]

$$p(\beta_S) = \begin{cases} \frac{\pi}{4|\beta_{Sm}|}\cos\left(\frac{\pi}{2}\frac{\beta_S}{\beta_{Sm}}\right), & |\beta_S| \le |\beta_{Sm}| \le \frac{\pi}{2} \\ 0 & , \text{otherwise} \end{cases} \tag{4.56}$$

where β_{Sm} is the maximum elevation angle.

Then, the space-time correlation function of the NLoS component can be obtained as

$$R_{pq,\tilde{p}\tilde{q}}^{NLoS}(\delta_T,\delta_R,\tau)=\frac{\cos\left(\frac{2\pi}{\lambda}\beta_{Sm}(q-\tilde{q})d_{Rz}\right)}{1-\left(\frac{4\beta_{Sm}(q-\tilde{q})d_{Rz}}{\lambda}\right)^2}e^{j\frac{2\pi}{\lambda}(p-\tilde{p})d_{Tx}+j2\pi\tau f_{T\max}\cos\gamma_T}$$
$$\times\frac{1}{I_0(k_R)(K+1)}I_0\left(\sqrt{x^2+y^2}\right) \quad (4.57)$$

where the parameters x and y are

$$x=j(2\pi/\lambda)(q-\tilde{q})d_{Rx}+j2\pi\tau f_{R\max}\cos\gamma_R+k_R\cos\mu_R \quad (4.58)$$

$$y=j2\pi\left[(q-\tilde{q})d_{Ry}/\lambda+(p-\tilde{p})d_{Ty}(R_S/D)/\lambda\right.$$
$$\left.+\tau\left(f_{R\max}\sin\gamma_R+f_{T\max}(R_S/D)\sin\gamma_T\right)\right]+k_R\sin\mu_R \quad (4.59)$$

and $d_{Tx}=\delta_T\cos\theta_T$, $d_{Ty}=\delta_T\sin\theta_T$, $d_{Rx}=\delta_R\cos\psi_R\cos\theta_R$, $d_{Ry}=\delta_R\cos\psi_R\sin\theta_R$, $d_{Rz}=\delta_R\sin\psi_R$.

As an example, Figure 4.6 depicts the A-to-G spatial correlation as a function of the antenna separation at the T_x for different degrees of local scattering in the azimuth domain around the R_x. The curves in Figure 4.6 are obtained with the following

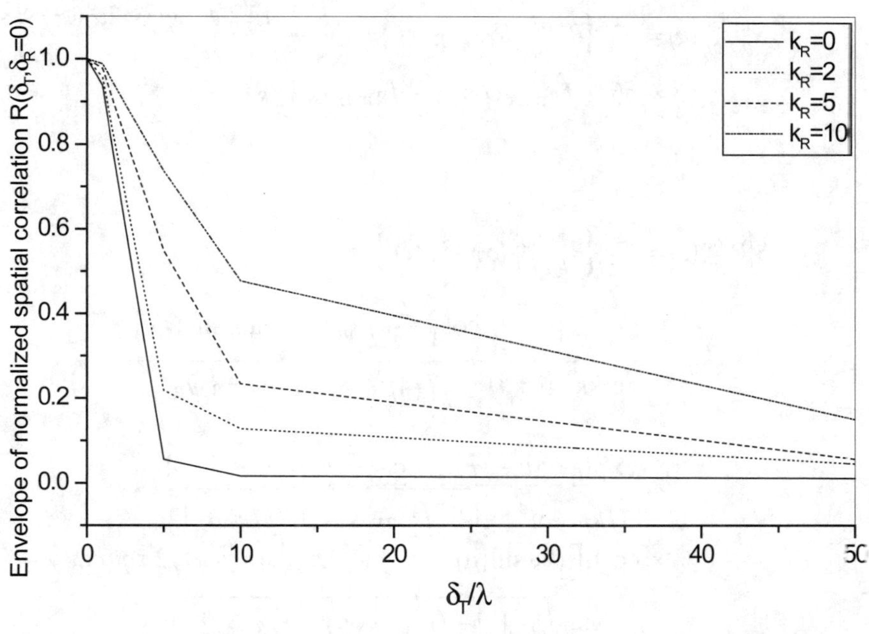

Figure 4.6 Envelope of the spatial correlation function as a function of the antenna separation at the T_x for different degrees of local scattering in the azimuth domain around the R_x.

parameters: $\lambda = 0.3$m, $K = 0$, $\theta_T = \theta_R = 90°$, $\psi_R = 0°$, $\delta_R = 0$, $\delta_T \in \{0,1,5,10,50\}\cdot\lambda$, $\gamma_T = \gamma_R = 90°$, $v_T \approx 0$ m/s, $v_R = 30$ m/s, $k_R \in \{0,1,5,10\}$, $\mu_T = \mu_R = 0$, $\Delta_S = 0.01$, $\beta_{Sm} = 60°$, and $L_t = L_r = 2$. From Figure 4.6 we can observe that, as k_R increases, (the scattering becomes increasingly nonisotropic), the separation at T_x needed for uncorrelated subchannels also increases. For highly concentrated scatterers around the mean angle (i.e., for $k_R = 10$) the required separation at the T_x for uncorrelated subchannels is $\delta_T > 50\lambda$. The effect of nonisotropic scattering on the correlation at the receiver is the same as that described in [35] for fixed-to-mobile cellular systems. In particular, as k_R increases, the correlation significantly increases.

4.2.6.2 Space-Doppler Power Spectral Density (sD-psd)

The space-Doppler power spectral density (sD-psd) of the complex faded envelope $h_{pq}(t)$ is the Fourier transform of the space-time correlation function $R_{pq,\tilde{p}\tilde{q}}(\delta_T, \delta_R, \tau)$. From (4.44) follows that the sD-psd is a sum of the sD-psd functions of the LoS and NLoS components; that is,

$$S_{pq,\tilde{p}\tilde{q}}(v) = S^{LoS}_{pq,\tilde{p}\tilde{q}}(v) + S^{NLoS}_{pq,\tilde{p}\tilde{q}}(v) \tag{4.60}$$

After computing the Fourier transform of the space-time correlations of the LoS and NLoS components in (4.45) and (4.57), respectively, the sD-psds of the LoS and NLoS components can be written as

$$\begin{aligned} S^{LoS}_{pq,\tilde{p}\tilde{q}}(v) &= \mathcal{F}_\tau\left\{R^{LoS}_{pq,\tilde{p}\tilde{q}}(\delta_T, \delta_R, \tau)\right\} = \frac{K}{K+1} e^{j\frac{2\pi}{\lambda}[(p-\tilde{p})\delta_T \cos\theta_T - (q-\tilde{q})\delta_R \cos\psi_R \cos\theta_R]} \\ &\times \delta(v + f_{T\max}\cos\gamma_T - f_{R\max}\cos\gamma_R) \end{aligned} \tag{4.61}$$

$$\begin{aligned} S^{NLoS}_{pq,\tilde{p}\tilde{q}}(v) &= \mathcal{F}_\tau\left\{R^{NLoS}_{pq,\tilde{p}\tilde{q}}(\delta_T, \delta_R, \tau)\right\} \\ &= \frac{1}{I_0(k_R)(K+1)} \frac{\cos\left(\dfrac{2\pi}{\lambda}\beta_{Sm}(q-\tilde{q})\delta_R \sin\psi_R\right)}{1 - \left(\dfrac{4\beta_{Sm}(q-\tilde{q})\delta_R \sin\psi_R}{\lambda}\right)^2} \\ &\times \frac{\exp\{j2\pi(v - f_{T\max}\cos\gamma_T)A_{NLoS} + j2\pi p\}}{\pi f_{R\max}\sqrt{1 - [(v - f_{T\max}\cos\gamma_T)/f_{R\max}]^2}} \\ &\times \cosh\Big[(k_R \sin(\mu_R - \gamma_R) + j2\pi q_x \sin\gamma_R - j2\pi q_y \cos\gamma_R) \\ &\quad \times \sqrt{1 - [(v - f_{T\max}\cos\gamma_T)/f_{R\max}]^2}\,\Big] \end{aligned} \tag{4.62}$$

where $\mathcal{F}\{\cdot\}$ is the Fourier transform, $\cosh(\cdot)$ is the hyperbolic cosine, $\delta(\cdot)$ is the Dirac delta function, $|v + f_{\text{Rmax}}\cos\gamma_R| \leq f_{\text{Tmax}}$, $|v - f_{\text{Tmax}}\cos\gamma_T| \leq f_{\text{Rmax}}$, and

$$p = (p - \tilde{p})\delta_T \cos\theta_T/\lambda \tag{4.63}$$

$$A_{NLoS} = (2\pi q_x \cos\gamma_R + 2\pi q_y \sin\gamma_R - jk_R \cos(\gamma_R - \mu_R))/2\pi f_{\text{Rmax}} \tag{4.64}$$

$$q_x = (q - \tilde{q})\delta_R \cos\psi_R \cos\theta_R/\lambda \tag{4.65}$$

$$q_y = \left[(q - \tilde{q})\delta_R \cos\psi_R \sin\theta_R + (p - \tilde{p})\delta_T \sin\theta_T R_{S\max}/D\right]/\lambda \tag{4.66}$$

As an example, Figure 4.7 plots a SISO sD-psd, assuming 3-D nonisotropic scattering and LoS conditions between the T_x and R_x. This sD-psd is typical for A-to-G propagation environments. Here, we assume that the wavelength is $\lambda = 0.3$m, the Ricean parameter is $K = 2$, the transmitter and the receiver directions are $\gamma_T = \gamma_R = 90°$, the transmitter velocity is $v_T \approx 0$ m/s, the receiver velocity is $v_R = 30$ m/s, the spread around scatterers around the receiver is $k_R = 2$, and the elevation angle is $\beta_{Sm} = 60°$. The results show this spectrum is similar to the U-shaped spectrum of F-to-M cellular channels.

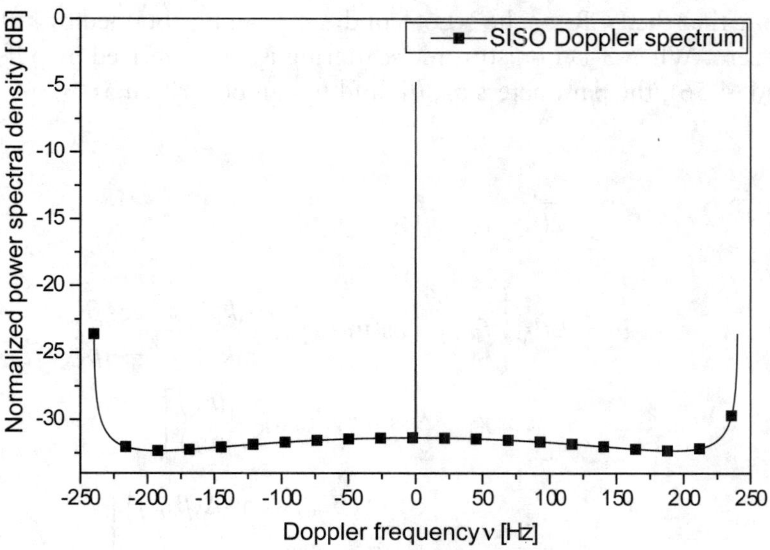

Figure 4.7 The normalized sD-psd in an A-to-G environment.

4.2.6.3 Level Crossing Rate (LCR)

The level crossing rate of the channel envelope can be calculated using similar reasoning as in Section 3.1.8.4. The LCR for LoS conditions (Ricean fading, correlated in-phase and quadrature components) can be written as [46]

$$L(R) = \frac{R\sqrt{2\chi}}{\pi^{3/2} b_0} e^{-(R^2+\delta^2)/(2b_0)} \int_0^{\pi/2} \cosh\left(\frac{R\delta}{b_0}\cos\theta\right) \times \left[e^{-(\xi\delta\sin\theta)^2} + \sqrt{\pi}\xi\delta\sin\theta\,\mathrm{erf}(\xi\delta\sin\theta)\right] d\theta \tag{4.67}$$

where cosh(·) is the hyperbolic cosine function and erf(·) is the error function. The rest of the parameters are $\delta = \sqrt{K/(K+1)}$, $\chi = b_2 - b_1^2/b_0$, $\xi = (2\pi f_{pq}^{LoS} - b_1/b_0)/\sqrt{2\chi}$, and

$$b_0 \triangleq \mathrm{E}\left[h_i(t)^2\right] = \mathrm{E}\left[h_q(t)^2\right] \tag{4.68}$$

$$b_1 \triangleq \mathrm{E}\left[h_i(t)\dot{h}_q(t)\right] = \mathrm{E}\left[h_q(t)\dot{h}_i(t)\right] \tag{4.69}$$

$$b_2 \triangleq \mathrm{E}\left[\dot{h}_i(t)^2\right] = \mathrm{E}\left[\dot{h}_q(t)^2\right] \tag{4.70}$$

where $\dot{h}_i(t)$ and $\dot{h}_q(t)$ are first derivatives of $h_i(t)$ and $h_q(t)$ with respect to time t. Note that the LCR for the NLoS conditions can be obtained by setting parameter K to zero. When 3-D nonisotropic scattering is characterized by pdfs given in (4.48), and (4.56), the parameters b_0, b_1, and b_2 can be written as

$$b_0 = \frac{1}{2(K+1)} \tag{4.71}$$

$$b_1 = 2\pi b_0 \left\{ f_{R\max}\cos(\mu_R - \gamma_R)\frac{I_1(k_R)}{I_0(k_R)}\frac{\pi^2\cos\beta_{Sm}}{\pi^2 - 4\beta_{Sm}^2} + \left[\cos\gamma_T - \Delta_S\sin\gamma_T\sin\mu_R\frac{I_1(k_R)}{I_0(k_R)}\right] \times f_{T\max}\frac{\pi^2\left[\cos(\Delta_S\beta_{Sm}) + \cos(\Delta_S\beta_{Sm})\right]}{2\left[\pi^2 - 4(\Delta_S)^2\beta_{Sm}^2\right]} \right\} \tag{4.72}$$

$$
\begin{aligned}
b_2 = (2\pi)^2 b_0 \Bigg\{ & f_{R\max}^2 \frac{\pi^2 \cos(2\beta_{Sm}) + \pi^2 - 16\beta_{Sm}^2}{2\left(\pi^2 - 16\beta_{Sm}^2\right)} \frac{I_0(k_R) + \cos(2(\mu_R - \gamma_R)) I_2(k_R)}{2I_0(k_R)} \\
& + \frac{4\beta_{Sm}^2 + 8\Delta_S \beta_{Sm}^2 - \pi^2 + 4(\Delta_S)^2 \beta_{Sm}^2}{4\left(8\beta_{Sm}^2 \pi^2 - 16\beta_{Sm}^4 - \pi^4 - 16(\Delta_S)^4 \beta_{Sm}^4 + 8\pi^2 (\Delta_S)^2 \beta_{Sm}^2 + 32(\Delta_S)^2 \beta_{Sm}^4\right)} \\
& \times \left[\cos((\Delta_S - 1)\beta_{Sm}) + \cos((\Delta_S + 1)\beta_{Sm}) \right. \\
& \left. + \cos((\Delta_S - 1)\beta_{Sm}) + \cos((\Delta_S + 1)\beta_{Sm})\right] \\
& \times \left[f_{R\max} f_{T\max} \Delta_S \sin\gamma_T \left(\frac{I_2(k_R)}{I_0(k_R)} \sin(2\mu_R - \gamma_R) + \sin\gamma_R \right) \right. \\
& \left. - 2 f_{R\max} f_{T\max} \cos\gamma_T \cos(\mu_R - \gamma_R) \frac{I_1(k_R)}{I_0(k_R)} \right] \\
& + \left[\frac{\cos[2(\Delta_S \beta_{Sm})]\pi^2 + 2\pi^2 - 32(\Delta_S)^2 \beta_{Sm}^2}{4\left[\pi^2 - 16(\Delta_S)^2 \beta_{Sm}\right]} \right. \\
& \left. + \frac{\cos[2(\Delta_S \beta_{Sm})]\pi^2}{4\left[\pi^2 - 16(\Delta_S)^2 \beta_{Sm}\right]} \right] \\
& \times \left[f_{T\max}^2 \cos\gamma_T^2 + f_{T\max}^2 (\Delta_S)^2 \sin^2\gamma_T \frac{I_0(k_R) - \cos(2\mu_R) I_2(k_R)}{2I_0(k_R)} \right. \\
& \left. \left. - f_{T\max}^2 \Delta_S \sin(2\gamma_T) \sin\mu_R \frac{I_1(k_R)}{I_0(k_R)} \right] \right\}
\end{aligned}
\tag{4.73}
$$

where $\Delta_S = R_{S\max}/D$.

As an example, Figure 4.8 plots a LCR, assuming 3-D nonfisotropic scattering and LoS conditions between the T_x and R_x. Here, we use the same model parameters as in Figure 4.7.

4.3 Simulators for Air-to-Ground Radio Channels

Simulation of A-to-G channels is a crucial part of characterization of A-to-G systems. The channel models detailed in Section 4.2.2 are very useful mathematical tools for understanding propagation mechanisms and for statistical characterization of channels. However, these models are not practical for simulation. Some of them have long simulation times, while the others assume an infinite number of scatterers, making these models impractical for simulators. We now describe models that can be used for simulation of A-to-G channels.

The following approaches are available when simulating the A-to-G fading channels:

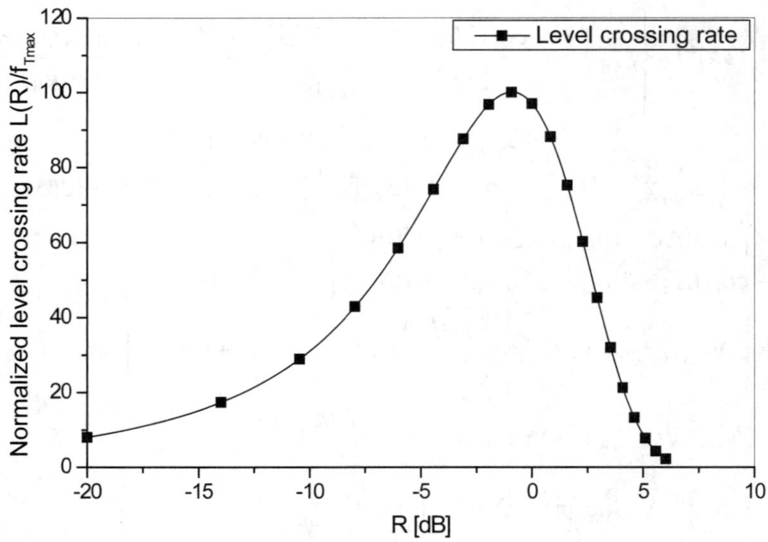

Figure 4.8 The level crossing rate in an A-to-G environment.

- *Markov chains.* Markov chains are often used to model A-to-G channels [31]. As the transmitter and the receiver move from one location to another, the environment properties vary resulting in a nonstationary statistical character of the received signal. Experimental data, however, indicate that A-to-G channels can be viewed as quasi-stationary The quasi-stationary nature of the model is based on the assumption of slowly varying environment characteristics such as elevation angle, obstacle type, and surface roughness. The environment attributes are assumed to be constant within a small area. Consequently, channels can be viewed as operating in one of a finite number of possible channel states and can be modeled as a finite state Markov model, as proposed in [20, 21]. These models can accurately generate amplitude and phase channel distortions. However, Markov chains are only suitable for modeling very slowly fading channels or in applications that require analysis over only a short duration of time.
- *Sum-of-sinusoids Method.* SoS models generate the channel waveform by superimposing a finite number of properly selected sinusoids. These models have low complexity and produce channel waveforms that can model both fast and slow variations of the channel with high accuracy. Due to these advantages, this chapter focuses on SoS models.

4.3.1 Sum-of-Sinusoids Simulators for A-to-G Channels

The propagation model for MIMO A-to-G channel described in Section 4.2.5 assumes an infinite number of scatterers, which prevents practical implementation. It is desirable to design simulation model with a finite (preferably small) number of scatterers, while still matching the statistical properties of the reference model. Assuming 3-D nonisotropic scattering and using the reference model described in Section 4.2.5, the following function can be used as a channel impulse response: $h_{pq}(t) = h_{pq}^{(I)}(t) + jh_{pq}^{(Q)}(t)$ where

$$
\begin{aligned}
h_{pq}^{(I)}(t) = {} & \frac{\rho_{NLoS}}{\sqrt{N_A N_E}} \sum_{n,i=1}^{N_A,N_E} \cos\Bigg[\frac{\pi(L_t+1-2p)}{\lambda} D_T + \frac{\pi(L_r+1-2q)}{\lambda} D_R \\
& + 2\pi t\, f_{T\max}\left(\Delta_S \sin\gamma_T \sin\alpha_R^{(n)} + \cos\gamma_T\right)\cos\left(\Delta_S \beta_S^{(i)}\right) \\
& + 2\pi t\, f_{R\max} \cos\left(\alpha_R^{(n)} - \gamma_R\right)\cos\beta_S^{(i)} + \phi_{n,i} \Bigg] \\
& + \rho_{LoS} \cos\Bigg[2\pi t (f_T^{LoS} + f_R^{LoS}) \\
& + \frac{\pi(L_t+1-2p)}{\lambda} d_{Tx} - \frac{\pi(L_r+1-2q)}{\lambda} d_{Rx} \Bigg]
\end{aligned}
\tag{4.74}
$$

$$
\begin{aligned}
h_{pq}^{(Q)}(t) = {} & \frac{\rho_{NLoS}}{\sqrt{N_A N_E}} \sum_{n,i=1}^{N_A,N_E} \sin\Bigg[\frac{\pi(L_t+1-2p)}{\lambda} D_T + \frac{\pi(L_r+1-2q)}{\lambda} D_R \\
& + 2\pi t\, f_{T\max}\left(\Delta_S \sin\gamma_T \sin\alpha_R^{(n)} + \cos\gamma_T\right)\cos\left(\Delta_S \beta_S^{(i)}\right) \\
& + 2\pi t\, f_{R\max} \cos\left(\alpha_R^{(n)} - \gamma_R\right)\cos\beta_S^{(i)} + \phi_{n,i} \Bigg] \\
& + \rho_{LoS} \sin\Bigg[2\pi t (f_T^{LoS} + f_R^{LoS}) \\
& + \frac{\pi(L_t+1-2p)}{\lambda} d_{Tx} - \frac{\pi(L_r+1-2q)}{\lambda} d_{Rx} \Bigg]
\end{aligned}
\tag{4.75}
$$

are the in-phase (I) and quadrature (Q) components of the channel impulse response. The rest of the parameters are

$$\rho_{NLoS} = \sqrt{\frac{1}{K+1}} \tag{4.76}$$

$$\rho_{LoS} = \sqrt{\frac{K}{1+K}} \tag{4.77}$$

$$D_T = d_{Ty} \Delta_S \sin\alpha_R^{(n)} + d_{Tx} \tag{4.78}$$

$$D_R = d_{Rx} \cos\alpha_R^{(n)} + d_{Ry} \sin\alpha_R^{(n)} + d_{Rz} \sin\beta_S^{(i)} \tag{4.79}$$

$$f_T^{LoS} = f_{T\max} \cos(\pi - \alpha_{Rq}^{LoS} - \gamma_T) \tag{4.80}$$

$$f_R^{LoS} = f_{R\max} \cos(\alpha_{Rq}^{LoS} - \gamma_R) \tag{4.81}$$

$$d_{Tx} = \delta_T \cos\theta_T \tag{4.82}$$

$$d_{Ty} = \delta_T \sin\theta_T \tag{4.83}$$

$$d_{Rx} = \delta_R \cos\psi_R \cos\theta_R \tag{4.84}$$

$$d_{Ry} = \delta_R \cos\psi_R \sin\theta_R \tag{4.85}$$

$$d_{Rz} = \delta_R \sin\psi_R \tag{4.86}$$

The following two modeling methods that trade simulation precision for simulation speed and complexity are considered:

- *Deterministic MIMO 3-D SoS model.* The deterministic SoS model needs only a single simulation trial to obtain a channel impulse response with desired statistical properties. The AAoAs ${\alpha_R}^{(n)}$ are modeled using the von Mises pdf and are generated as follows:

$$\alpha_R^{(n)} = F^{-1}\left(\frac{n-0.5}{N_A}\right) \tag{4.87}$$

  for $n = 1,..., N_A$, where $F(\cdot)^{-1}$ denotes the inverse cumulative von Mises distribution function and is evaluated using the method from [47]. The EAoAs $\beta_R^{(i)}$, are modeled using pdf in (4.56) and are generated as follows:

$$\beta_S^{(i)} = \frac{2\beta_{Sm}}{\pi}\arcsin\left(\frac{2i-1}{N_E} - 1\right) \tag{4.88}$$

  for $i = 1,..., N_E$. Phases $\phi_{n,i}$ are generated as independent uniform random variables on the interval $[-\pi,\pi)$.

- *Statistical MIMO 3-D SoS model.* The statistical SoS model generates channel realizations with statistical properties that vary with every simulation run but converge to the desired properties when averaged over a sufficient channel ensemble. The AAoAs and the EAoAs are generated as follows:

$$\alpha_R^{(n)} = F^{-1}\left(\frac{n+\psi_A-1}{N_A}\right) \tag{4.89}$$

$$\beta_S^{(i)} = \frac{2\beta_{Sm}}{\pi}\arcsin\left(\frac{2(i+\psi_E-1)}{N_E} - 1\right) \tag{4.90}$$

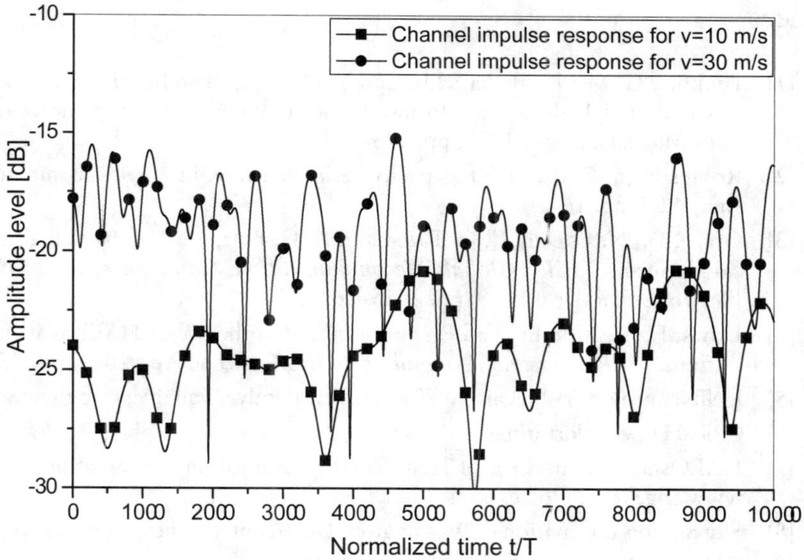

Figure 4.9 Channel impulse response for two different receiver velocities.

for $n = 1,\ldots,N_A$ and $i = 1,\ldots,N_E$. The parameters ψ_A and ψ_E are independent random variables uniformly distributed on the interval [0,1). The function $F^{-1}(\cdot)$ denotes the inverse von Mises cumulative distribution function.

To illustrate how the statistical simulator works, we plot two different channel impulse responses that illustrate how changes in the receiver's velocity impact the channel impulse response. We use the statistical simulator and select the following channel parameters: carrier frequency is 2 GHz, sampling frequency is 100 kHz, number of samples is 10,000, and distance between the transmitter and the receiver is 20,000 km. The isotropic and non-line-of-sight type of propagation is selected. The remaining simulator parameters are $\theta_T = \theta_R = 90°$, $\psi_R = 0°$, $\delta_R = \delta_T = 0$, $\gamma_T = \gamma_R = 90°$, $v_T \approx 0$ m/s, $k_R = 0$, $\mu_T = \mu_R = 0$, $\Delta_S = 0.01$, $\beta_{Sm} = 60°$, $L_t = L_r = 1$, and $v_R \in$ {10 m/s, 30 m/s}. The results obtained for the channel impulse response are shown in Figure 4.9. The results show that, as the velocity of the receiver increases, the variability of the channel impulse response increases.

4.4 Summary

This chapter first briefly described the propagation principles that govern signal propagation in A-to-G channels, and then proceeded to describe important concepts of propagation modeling such as large-scale and small-scale fading effects. Then, statistical characterization of multipath propagation in A-to-G channels was presented, and several models and their statistics such as correlation function, Doppler spectrum, and level crossing rate were detailed. Finally, practical simulators, for A-to-G multipath fading channels were presented, with a small number of propagation rays that still match the statistical properties of theoretical A-to-G channel models. Note that modeling of frequency-selective A-to-G channels and verification of A-to-G models against measurements is still an open area for research.

References

[1] Djuknic, G. M., J. Freidenfelds, and Y. Okunev, "Establishing wireless communications services via high altitude platforms: A concept whose time has come?," *IEEE Communications Magazine*, Sept. 1997, pp. 128–135.

[2] Robinson Jr., C. A., "High capacity aerial vehicles aid wireless communications," *Signal*, Apr. 1997, pp. 16–20.

[3] Grace, D., N. E. Daly, T. C. Tozer, and A. G. Burr, "LMDS from high altitude aeronautical platforms," *IEEE Global Telecommunications Conference, GLOBECOM '99*, Rio de Janeiro, Brazil, Dec. 1999, pp. 2625–2629.

[4] Dovis, F., M. Mondin, and P. Mulassano, "On the use of HALE platforms as GSM base-stations," *IEEE Personal Communications Magazine*, Apr. 2001, pp. 37–44.

[5] HeliNet project. Politecnico di Torino, Turin, Italy, Available: http://www.helinet.polito.it.

[6] Global Hawk. Northrop Grumman, Los Angeles, CA, Available: http://www.northgrup.com.

[7] HALOstar Network. Angel Technologies Corporation, Washington, DC, Available: http://www.angeltechnologies.com.

[8] Sky Station International, Washington, DC, Available: http://www.skystation.com.

[9] Saunders, R., and A. Aragón-Zavala, *Antennas and Propagation for Wireless communication systems*, 2nd ed., John Wiley and Sons Ltd, London: 2007.

[10] Ippolito, J. L., Jr., *Satellite Communications Systems Engineering: Atmospheric Effects, Satellite link Design and System Performance*, London: John Wiley and Sons Ltd., 2008.

[11] Marshall, J. S., and W. M. K. Palmer, "The distribution of raindrops with size," *Journal of Metrology*, 1948, pp. 1965–1966.

[12] International Telecommunication Union, *ITU-R Propagation Recommendation P. 838: Specific Attenuation Model for Rain for Use in Prediction Methods*, Geneva, 1992.

[13] International Telecommunication Union, *ITU-R Propagation Recommendation P. 618-5: Propagation Data and Prediction Methods Required for the Design of Earthspace Telecommunication Systems*, Geneva, 1997.

[14] Cuevas-Ruíz, J. L., and J. A. Delgado-Penín, "Channel model based on semi-Markovian processes, an approach for HAPS systems," *XIV International Conference on Electronics, Communications, and Computers, CONIELECOMP'04*, Vevacruz, Mexico, February 2004, pp. 52–56.

[15] Cuevas-Ruíz, J. L., and J. A. Delgado-Penín, "Channel model for millimetre bands for HAPS based systems," *XVIII Simposium Nacional Unión Cientiífica International de Radio URSI 2003*, La Corufia, Spain, September 2003.

[16] Andreyev, G. A., and A. S. Bugaev, "mm-Wave propagation in urban and rural environment," *COST-280 Joint Workshop with COST-272*, Noordwijk, Netherlands, 2003.

[17] Stüber, G. L., *Principles of Mobile Communication*, 2nd ed., Norwell, MA: Kluwer, 2001.

[18] Jahn, A., "Propagation considerations and fading countermeasures for mobile multimedia service," *International Journal of Satellite Communications*, Vol. 19, 2001, pp. 223–250.

[19] Butt, G., M. A. N. Parks, and B. G. Evans, "Measurements and modeling of wireless channel for satellite mobile and personal communication systems" *10th International Conference on Digital Satellite Communications*, May 1995, pp. 106–114.

[20] Vucetic, B., and J. Du, "Channel modeling and simulation in satellite mobile communication systems," *IEEE Journal on Selected Areas on Communications*, Vol. 10, October 1992, pp. 1209–1218.

[21] Lutz, E., D. Cygan, M. Dippold, F. Dolainsky, and W. Papke, "The land mobile satellite communication channel-recording, statistics and channel model," *IEEE Transactions on Vehicular Technology*, Vol. 40, May 1991, pp. 375–386.

[22] Xie, Y., and Y. Fang, "A general statistical channel model for mobile satellite system," *IEEE Transactions on Vehicular Technology*, Vol. 49, May 2000, pp. 744–752.

[23] Li, W., C. L. Law, J. T. Ong, and V. Dubey, "Ka band land mobile satellite channel model with rain attenuation and other impairments in the equatorial zone," *IEEE 51st Vehicular Technology Conference Proceedings, VTC'00*, Tokyo, Japan, May 2000, pp. 2468–2472.

[24] Loo, C., "A statistical model for a land mobile satellite link," *IEEE Transactions on Vehicular Technology*, Vol. 34, August 1985, pp. 122–127.

[25] Loo, C., "Land mobile satellite channel measurement at Ka band using olympus," *IEEE Vehicular Technology Conference, VTC'94*, Stockholm, Sweden, June 1994, pp. 919–923.

[26] Loo, C., and J. Butterworth, "Land mobile satellite channel measurements and modeling", *Proceedings of the IEEE*, Vol. 86, July 1998, pp. 1442–1446.

[27] Rappaport, T. S., and J. C. Liberti, "A geometrical-based model for line-of sight multipath radio channel," *Proceedings of IEEE 46th Vehicular Technology Conference VTC'96*, Atlanta, GA, Apr./May 1996, pp. 844–48.

[28] Abdi, A., and M. Kaveh, "A space-time correlation model for multielement antenna systems in mobile fading channels," *IEEE Journal of Selected Areas in Communications*, Vol. 20, April 2002, pp. 550–560.

[29] Abdi, A., J. A. Barger, and M. Kaveh, "A parametric model for the distribution of the angle of arrival and the associated correlation function and power spectrum at the mobile station," *IEEE Transactions on Vehicular Technology*, Vol. 51, May 2002, pp. 425–434.

[30] Leong, S.-Y., Y. R. Zheng, and C. Xiao, "Space-time fading correlation functions of a 3-D MIMO channel model," *Proceeding of IEEE Wireless Communications and Networking Conference WCNC'04*, Atlanta, GA, 2004, pp. 112–1132.

[31] Aragn-Zavala, A., J. L. Cuevas-Ruz, and J. A. Delgado-Penn, *High-altitude Platforms for Wireless Communications*, John Wiley and Sons Ltd, London: 2008.

[32] Dovis, F., R. Fantini, M. Mondin, and P. Savi, "Small-scale fading for high-altitude platform (HAP) propagation channels," *IEEE Journal of Selected Areas in Communication*, Vol. 20, Apr. 2002, pp. 641–647.

[33] Michailidis, E. T., and A. G. Kanatas, "Three-dimensional HAP-MIMO channels: modeling and analysis of space-time correlation," *IEEE Transactions on Vehicular Technology*, Vol. 59, Jun 2010, pp. 2232–2242.

[34] Salz, J., and J. H. Winters, "Effect of fading correlation on adaptive arrays in digital mobile radio," *IEEE Transactions on Vehicular Technology*, Vol. 43, Nov. 1994, pp. 1049–1057.

[35] Abdi, A., and M. Kaveh, "A space-time correlation model for multielement antenna systems in mobile fading channels," *IEEE Journal on Selected Areas in Communications*, Vol. 20, Apr. 2002, pp. 550–560.

[36] King, P. R., B. G. Evans, and S. Stavrou, "Physical-statistical model for the land mobile-satellite channel applied to satellite/HAP MIMO," *Proceeding of 11th European Wireless Conference*, Nicosia, Cyprus, Apr. 2005, pp. 198–204.

[37] Von Mises, R. "Über die 'Ganzzahligkeit' der Atomgewicht und verwandte Fragen," *Physikal. Z.*, Vol. 19, 1918, pp. 490–500.

[38] Mahmoud, S. S., Z. M. Hussain, and P. O'Shea, "Space-time model for mobile radio channel with hyperbolically distributed scatterers," *IEEE Antennas Wireless Propagation Letters*, Vol. 1, May 2002, pp. 211–214.

[39] Mahmoud, S. S., F. S. Al-Qahtani, Z. M. Hussain, and A. Gopalakrishnan, "Spatial and temporal statistics for the geometrical-based hyperbolic macrocell channel model," *Elsevier Digital Signal Processing*, Vol. 18, March 2008, pp. 151–167.

[40] Vzquez-Castro, M. A., F. Perez-Fontan, and S. R. Saunders, "Shadowing correlation assessment and modeling for satellite diversity in urban environments," *Wiley International Journal of Satellite Communications and Networking*, Vol. 20, Mar./Apr. 2002, pp. 151–166.

[41] Vzquez-Castro, M. A., D. Belay-Zeleke, and A. Curieses-Guerrero, "Availability of systems based on satellites with spatial diversity and HAPS," *IEEE Electronics Letters*, Vol. 38, March 2002, pp. 286–288.

[42] Tzaras, C., B. G. Evans, and S. R. Saunders, "Physical-statistical analysis of land mobile-satellite channel," *IEEE Electronics Letters*, Vol. 34, June 1998, pp. 1355–357.

[43] Saunders, S. R., and B. G. Evans, "Physical model of shadowing probability for land mobile satellite propagation," *IEEE Electronics Letters*, Vol. 32, Aug. 1996, pp. 1548–1549.

[44] Gradshteyn, I. S., and I. M. Ryzhik, *Table of Integrals, Series, and Products*, A. Jeffrey, Ed., 5th ed., New York: Academic, 1994.

[45] Parsons, J. D., and A.M.D. Turkmani, "Characterisation of mobile radio signals: model description," *IEEE Proceedings I, Communications, Speech, and Vision*, Vol. 138, Dec. 1991, pp. 549–556.

[46] Pätzold, M., U. Killat, and F. Laue, "An extended Suzuki model for land mobile satellite channels and its statistical properties," *IEEE Transactions on Vehicular Technology*, Vol. 47, May 1998, pp. 617–630.

[47] Mardia, K. V., and E. P. Jupp, *Directional Statistics*, New York: John Wiley and Sons, 1999.

CHAPTER 5

Underwater Vehicle-to-Underwater Vehicle Acoustic Channels

Many scientific, defense, and safety endeavors require communications between submerged devices and/or vehicles. An example of communications between fixed submerged devices are sensor networks deployed on the ocean bottom for seismic monitoring (e.g., to predict tsunamis), monitoring of deep-water oil wells, analysis of resource deposits, oceanographic studies, tactical surveillance, and pollution monitoring. Examples of communications between underwater vehicles (UV) are unmanned or autonomous underwater vehicles (UUV or AUV) for deep-water construction, repairs of deep-water oil wells, scientific or resource exploration, defense applications (e.g., between submarines), and so on. From the communications perspective, the ultimate goal is to create an underwater network that consists of distributed underwater sensors and underwater vehicles that perform collaborative monitoring and communication as illustrated in Figure 5.1.

Unfortunately, underwater communications are extremely challenging—use of wires/cables either severely limits or entirely precludes most of the important underwater communication scenarios, whereas existing wireless communications still provide extremely low data rates, work over very limited distances, and/or have low energy efficiency. The main reason for such poor performance is that the underwater environment is very harsh for traditional wireless communication approaches: radio signal range is only several meters because of severe losses, especially in sea water (which is a relatively good conductor), while optical communication has a very short range because of high dispersion due to murkiness and particles suspended in water. As a result, underwater wireless communications mostly use acoustics (sound), which can travel long distances in water. Unfortunately, acoustic communication suffers from its own set of problems—the available spectrum is measured in kilohertz instead of gigahertz, propagation is orders of magnitude slower than with radio or optics, and there are many sources of signal distortion that further limit data rates that can be achieved using simple modulations and coding.

More advanced modulations and coding, especially those that exploit multipath propagation and available bandwidth can dramatically improve data rates in acoustic underwater communication. However, the acoustic underwater propagation environment differs significantly from the air- and space-based radio propagation environment, so this environment must be accounted for in order to exploit advanced modulations and coding. This chapter describes the principles that govern signal propagation in UV-to-UV acoustic channels, discusses how UV-to-UV acoustic channels differ from terrestrial RF radio channels, and then details how to model, simulate, and measure UV-to-UV acoustic channels.

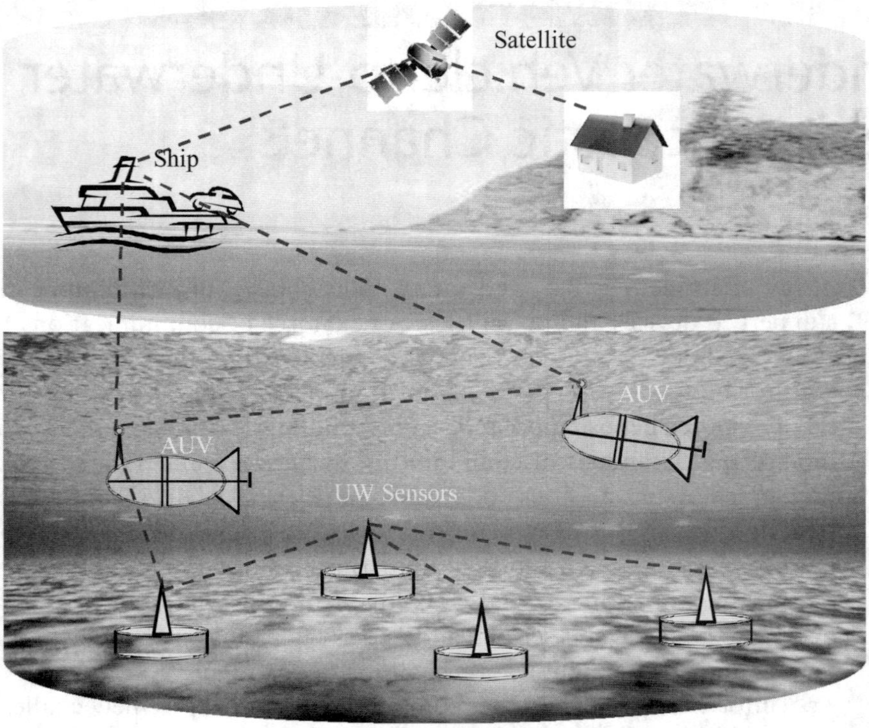

Figure 5.1 An acoustic network of underwater sensors and autonomous underwater vehicles.

5.1 Underwater Propagation Principles

The underwater environment is very challenging for electromagnetic (radio) and optical communications. Electromagnetic waves propagate to long distances only at extra low frequencies (30–300 Hz through conductive sea water), which requires large antennas and high transmission power. Optical waves do not suffer from such high attenuation but are affected by scattering. As a result, underwater communications primarily use acoustic (pressure) waves [1].

The underwater acoustic (UWA) channel is still one of the most challenging wireless communication environments for several reasons:

- Acoustic signals propagate through water at a very low speed (1500 m/s), five orders of magnitude slower than radio waves. This low propagation speed significantly increases the latency in the communication system.
- Acoustic propagation through the sea water is characterized by significant frequency-dependent attenuation. Three primary mechanisms that attenuate underwater acoustic signals are spreading loss, absorption loss, and scattering loss, which will be discussed in Section 5.2.1.
- The available bandwidth of an UWA channel depends critically on transmission loss, which increases with both range and frequency, and severely limits the available bandwidth [2, 3]. For example, long-range systems operating over several tens of *kilometers* may have only a kilohertz of bandwidth, while short-range systems operating over several tens of *meters* may have one or a few hundred kilohertz of bandwidth [4].

- Acoustic signals are greatly affected by variations in the speed of sound at different water depths, by reflections from the ocean surface and ocean floor, and by the location of the transmitter and receiver. As a result, a received signal is impacted by time-varying multipath, which may result in intersymbol interference and Doppler spreads that are much more severe than in radio channels, especially in shallow water channels [2].

Together, all these factors produce a communication medium that combines the worst of traditional radio propagation environments: the poor link quality of a terrestrial system and the high latency of a satellite system. To make the problem even more complex, there is no "typical" underwater acoustic environment, so there is no "typical" underwater acoustic communication channel. In different environments, different physical processes pose significant problems to reliable communications, resulting in different challenges to a system. Thus, a system that works effectively in one environment (e.g., a shallow water environment) may fail completely in another environment (e.g., a deep water environment). The design of reliable general purpose systems that work effectively across a broad spectrum of ocean environments is an open problem.

5.1.1 An Underwater Communication System

A simple schematic of an underwater communication scheme, involving a projector (transmitter) and a hydrophone (receiver), is presented in Figure 5.2. The projector takes the data and creates modulated signals. The modulated signals are then amplified to a level sufficient for successful reception and converted from the electrical to acoustical signals. On the receiver side, the sensitivity of the hydrophone converts

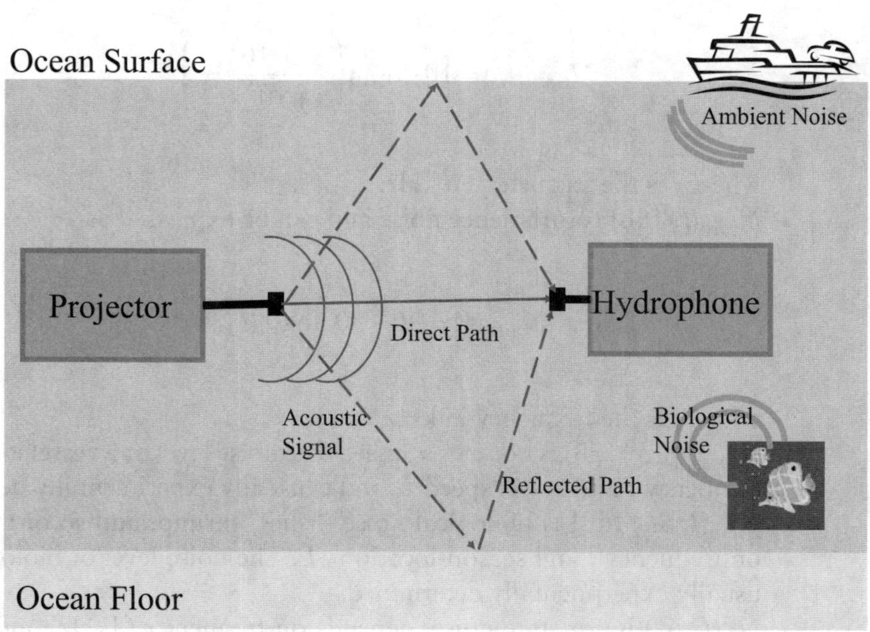

Figure 5.2 A simple schematic of an underwater communication scheme.

the sound pressure that reaches the hydrophone to electrical energy. This signal is then detected, amplified, and demodulated at the receiver.

5.1.2 Underwater Noise

Underwater communication is impacted by frequency-dependent noise, which consists of *ambient noise* and *site-specific noise*.

Ambient noise is always present in the background of the quiet deep sea and comes from the sources such as turbulence, breaking waves, rain, and distant shipping. On the other hand, site-specific noise is present only in certain places. For example, ice cracking in polar regions or snapping shrimp in warmer waters create site-specific acoustic noise.

The noise level contains a lot of information concerning the state of the ocean surface, the atmosphere over the ocean, tectonic processes in the earth's crust under the ocean, the behavior of marine animals, and so on. The total noise level in decibels can be written as [5]

$$N(f,\upsilon_s,SL,\upsilon_w) = N_{\text{traffic}}(f) + N_{\text{turb}}(f) + N_{\text{vessel}}(f,\upsilon_s) + N_{\text{bio}}(f,SL) + N_{ss}(f,\upsilon_w) + N_{\text{term}}(f) \quad (5.1)$$

where

- $N_{\text{traffic}}(f)$ [dB] is shipping noise (traffic) at 10–300 Hz and can be expressed as

$$N_{\text{traffic}}(f) = 10 \cdot \log_{10}\left(\frac{3 \cdot 10^8}{1+10^4 \cdot f^4}\right) \quad (5.2)$$

  where f is the frequency in kHz.
- $N_{\text{turb}}(f)$ [dB] is turbulence noise and can be expressed as

$$N_{\text{turb}}(f) = 30 - 30 \cdot \log_{10}(f) \quad (5.3)$$

  where f is the frequency in kHz.
- $N_{\text{vessel}}(f, \upsilon_s)$ [dB] is noise of a vessel. The noise level of a vessel is a function of frequency f and vessel speed υ_s and is usually experimentally determined.
- $N_{\text{bio}}(f, SL)$ [dB] is biological noise (fishes, shrimp, and so on) and depends on frequency f and season-location SL. The noise level of biological noise is usually experimentally determined.
- $N_{\text{ss}}(f, \upsilon_w)$ is sea state noise, depends on frequency f [kHz] and wind speed υ_w [kn], and can be expressed as

$$N_{ss}(f, v_w) = 40 + 10 \cdot \log_{10}\left(\frac{v_w^2}{1+f^{5/3}}\right) \tag{5.4}$$

- $N_{term}(f)$ is thermal noise and occurs due to molecular agitation (Brownian motion). Thermal noise is a function of frequency f [kHz] and can be expressed as

$$N_{term}(f) = -15 + 20 \cdot \log_{10}(f) \tag{5.5}$$

At low frequencies (0.1–10 Hz), the main noise sources are earthquakes, underwater volcanic eruptions, distant storms, and turbulence in the ocean and atmosphere. In the 50–300 Hz frequency band, underwater noise is mainly due to distant ship traffic. In the 0.5–50 kHz frequency band, underwater noise is directly associated with the state of the ocean surface and the wind in the immediate area. At frequencies above 100 kHz, thermal noise is dominant.

In underwater communication system simulations, ambient noise is often approximated as Gaussian, but not white. Unlike ambient noise, site-specific noise often contains significant non-Gaussian components, so it cannot be modeled as Gaussian noise.

5.1.3 Propagation Principles

Acoustic waves propagate toward the receiver (hydrophone) via four main propagation mechanisms: *direct propagation*, *refraction*, *reflection*, and *scattering*. Direct propagation occurs when sound energy from the transmitter makes its way directly to the receiver along a LoS path. Refraction occurs due to spatial variability of the speed of sound. Finally, reflection and scattering are created by sound interaction with the surface, bottom, or other objects in the channel (such as fish). The received acoustic waves are a superposition of the wave coming from the LoS path and all waves coming from other directions due to refraction, reflection, and scattering. This effect is known as *multipath propagation* and is similar in its effects to (but much more pronounced than) multipath propagation in radio channels.

5.1.4 Refraction

The refraction of sound in the ocean occurs due to spatial variability of the speed of sound. When a sound wave propagating in a medium with speed of sound c_1 impinges upon another medium having different speed of sound c_2, the sound wave changes its direction, as shown in Figure 5.3. Variations of the sound velocity c in the ocean are relatively small (i.e., c lies between 1450 m/s and 1540 m/s). However, such small changes of c still result in significant refraction effects.

Numerous laboratory and field measurements have shown that the sound speed increases with temperature T, hydrostatic pressure P (which is a function of depth z), and the salinity S. An empirical formula for the speed of sound is [5]

$$\begin{aligned} c = {} & 1449.2 + 4.6 \cdot T - 0.055 \cdot T^2 + 0.00029 \cdot T^3 \\ & + (1.34 - 0.01 \cdot T)(S - 35) + 0.016 \cdot z \end{aligned} \tag{5.6}$$

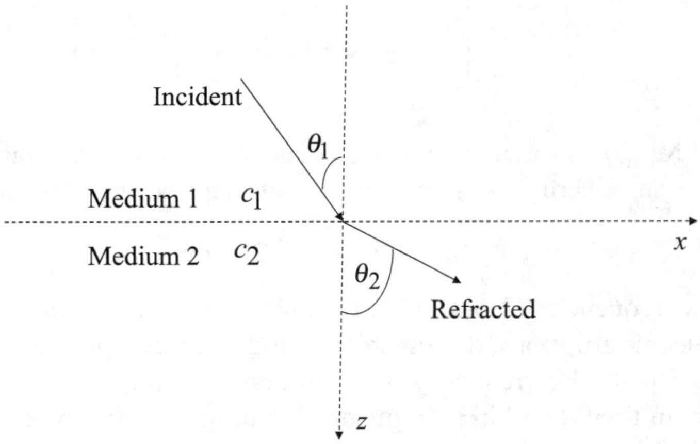

Figure 5.3 Refraction at two different interfaces.

where temperature T is expressed in °C, salinity S in percentiles, depth z in meters, and sound velocity c in m/s. Equation (5.6) is valid for $0°C \leq T \leq 35°C$, $0 \leq S \leq 45‰$, and $0 \leq z \leq 1000$m.

5.1.4.1 Typical Vertical Profiles of Sound Velocity and Corresponding Conditions of Sound Propagation

As pointed out in Section 5.1.4, sound velocity varies with temperature, salinity, and depth. The sound velocity profile $c(z,T(z), S(z))$ and the distribution of the sound velocity gradient with depth are the most important parameters for determining the type of sound propagation in the ocean. For one type of profile $c(z,T(z), S(z))$, sound can propagate hundreds of kilometers, while for another type, sound at the same frequency can propagate only a few tens of kilometers (or even less).

The sound velocity profiles are different in various regions of the ocean, and they also vary with time. The largest fluctuations are in the upper ocean, mainly due to seasonal variations of temperature and salinity. For ocean depths smaller

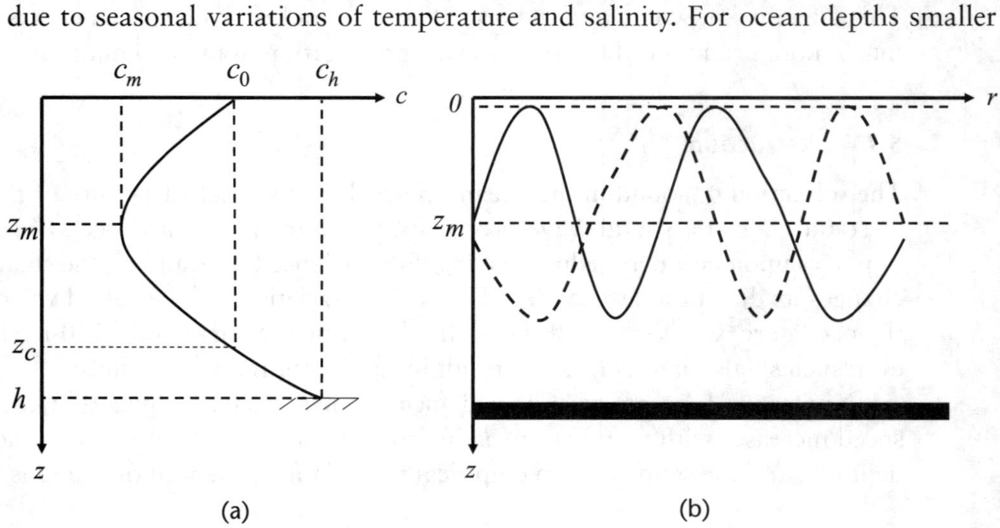

Figure 5.4 (a) Profile of an underwater sound channel of the first kind ($c_0 < c_h$), and (b) ray diagram of the underwater sound channel of the first kind.

than 200m, temperature is the dominant factor that impacts the velocity of sound. Between 200m and 400m, both temperature and depth have a significant impact on the velocity of sound. Finally, at ocean depths bigger than 400m, the sound velocity almost purely depends on the depth. The velocity of sound also depends on salinity S. As S increases, c also increases, but the shape of the speed-of-sound profile remains unchanged.

Next, we describe four main forms of the sound-velocity profile and the corresponding acoustic channel types: *underwater sound channel*, *surface sound channel*, *antiwaveguide sound channel*, and *shallow water sound channel*.

5.1.4.2 Underwater Sound Channel

In deep-water regions, the typical sound velocity profile is characterized by the existence of a minimum sound velocity at a certain depth z_m, as shown in Figure 5.4(a). The depth z_m is an axis of the underwater sound channel. Above this axis, the sound velocity increases mainly due to temperature increase. Below this axis, the sound velocity increases mainly due to hydrostatic pressure. If a sound source is located on the axis of the underwater sound channel or near it, some part of the sound energy is trapped in the underwater sound channel because of refraction (i.e., refraction bends the sound waves back toward the axis). Hence, the sound propagates within the channel, not reaching the bottom or surface, and therefore, not undergoing scattering and absorption at these boundaries [5]. Depending on the ratio of the sound velocity at the surface c_0 and the sound velocity at the bottom c_b, there are two types of underwater sound channels:

- When c_0 is smaller than c_b, the rays leave the source at small or moderate grazing angles and return to the channel axis repeatedly, causing a waveguide-type propagation. This type of channel is called *underwater sound channel of the first kind*.

  In this type of underwater sound channel, waveguide propagation can be observed in the interval depths of $0 \le z \le z_c$. The depths $z = 0$ and $z = z_c$ are the boundaries of the underwater sound channel. The channel traps all sound rays that leave the source (which is located on the underwater sound channel axis) at grazing angles $\chi < \chi_{\max}$, where

$$\chi_{\max} = \sqrt{\frac{2(c_0 - c_m)}{c_m}} \tag{5.7}$$

  and c_m and c_0 are the sound velocities at the axis and boundaries of the channel, respectively. Hence, the greater the difference $c_0 - c_m$, the larger the interval of angles in which the rays are trapped is (i.e., the waveguide is more effective).
- When c_0 is greater than c_b, the sound bounces between the bottom and the depth z_c, as shown in Figure 5.5. This type of channel is called *underwater sound channel of the second kind*. As shown in Figure 5.5(b), trapped rays do not extend above the depth z_c.

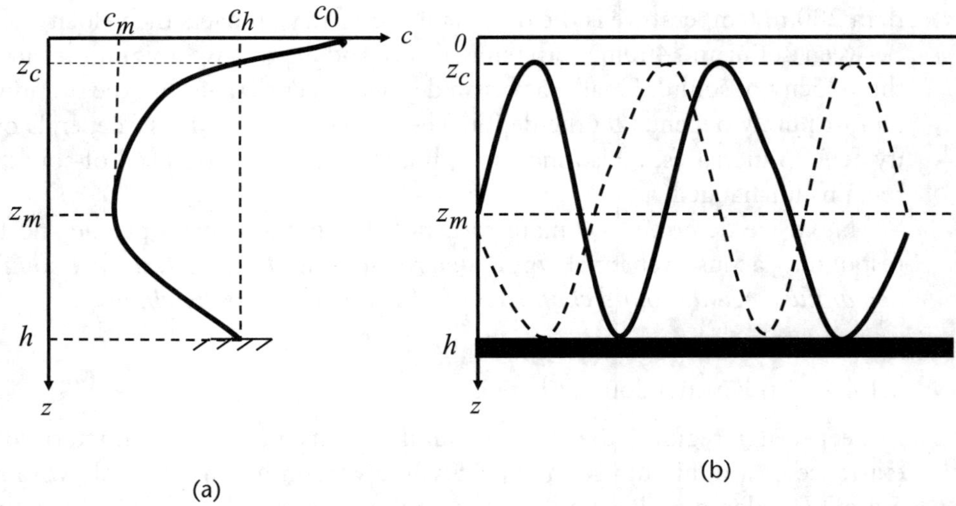

Figure 5.5 (a) Profile of an underwater sound channel of the second kind ($c_0 > c_h$), and (b) ray diagram of the underwater sound channel of the second kind.

5.1.4.3 Surface Sound Channel

This type of sound channel occurs when the axis is at the surface (i.e., when $c_m = c_0$). A typical profile for this channel is shown in Figure 5.6. The sound velocity increases with the depth until $z = h$ and then begins to decrease, as shown in Figure 5.6(a). Rays leaving the source at grazing angles $\chi < \chi_b$ (where χ_b is the grazing angle of the boundary ray tangent to the lower boundary of the channel as shown in Figure 5.6(b)) propagate with multiple reflections from the surface [5]. If the ocean surface is smooth, these rays remain in the sound channel regardless of distance from the source and give rise to waveguide propagation. If the ocean surface is rough, part of the sound energy is scattered into angles $\chi > \chi_b$ at each contact with the surface and, therefore, leaves the sound channel. As a result, the sound field decays in the sound

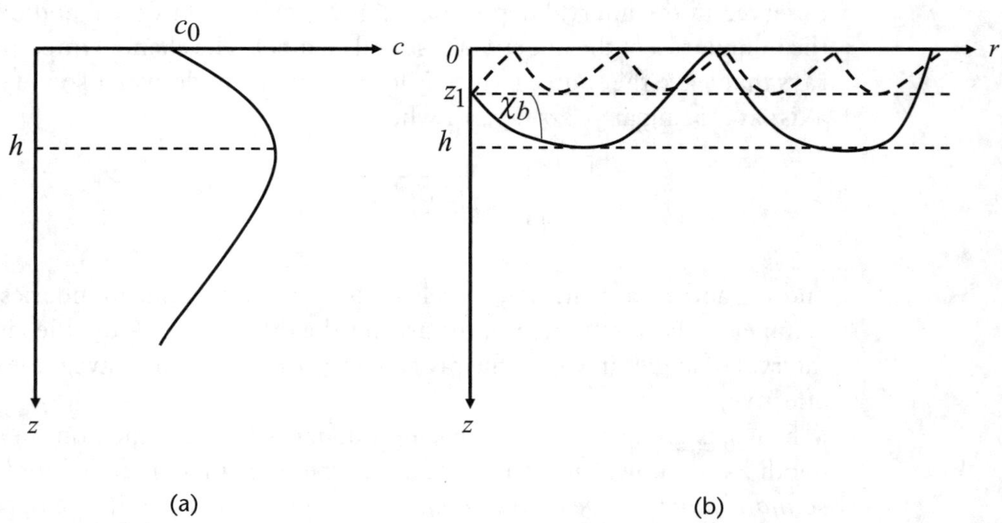

Figure 5.6 (a) Profile of a surface sound channel, and (b) ray diagram of the surface sound channel.

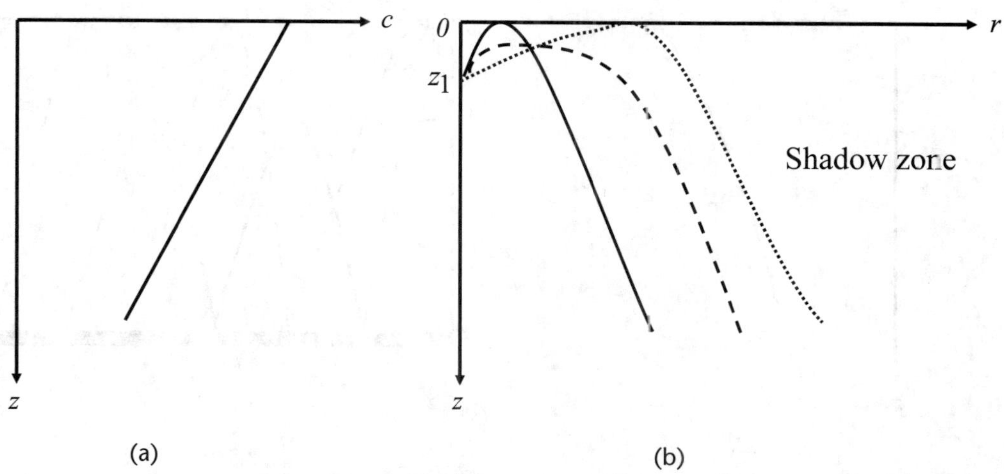

Figure 5.7 (a) Profile of an antiwaveguide sound channel, and (b) ray diagram of the antiwaveguide sound channel.

channel, but in the area below the channel its level increases. This type of channel occurs in the tropical and moderate zones of the ocean, where the temperature and salinity in the upper ocean layer are almost constant.

5.1.4.4 Antiwaveguide Sound Channel

The antiwaveguide sound channel occurs when the sound velocity monotonically decreases with depth as shown in Figure 5.7(a). Such sound velocity profiles are often a result of intensive heating by solar radiation of the upper ocean layer. From Figure 5.7(b) we can observe that all rays refract downward. The ray that propagates farthest in this type of channel is one that is tangent to the surface. For typical conditions in the ocean, the distance from the source to the shadow zone is only a few kilometers.

The shadow zone shown in Figure 5.7(b) is not a region of zero sound intensity. In the ideal case, the sound energy penetrates into the shadow zone due to diffraction. In practice, sound waves reflected from the bottom and scattered by random inhomogeneities of the medium also produce some sound in this zone.

5.1.4.5 Shallow Water Sound Channel

This type of sound propagation corresponds to the case when each ray emitted from the source, when continued long enough, is reflected at the bottom. A typical sound velocity profile is shown in Figure 5.8(a), and the corresponding ray diagram is shown in Figure 5.8(b). This type of propagation is characteristic for shallow seas and the ocean shelves, especially during the summer-autumn period when the upper water layers are well heated.

5.1.5 Reflection

Reflection arises when a sound wave propagating in one medium (e.g., water, impinges upon another medium having different properties such as air or ground). In

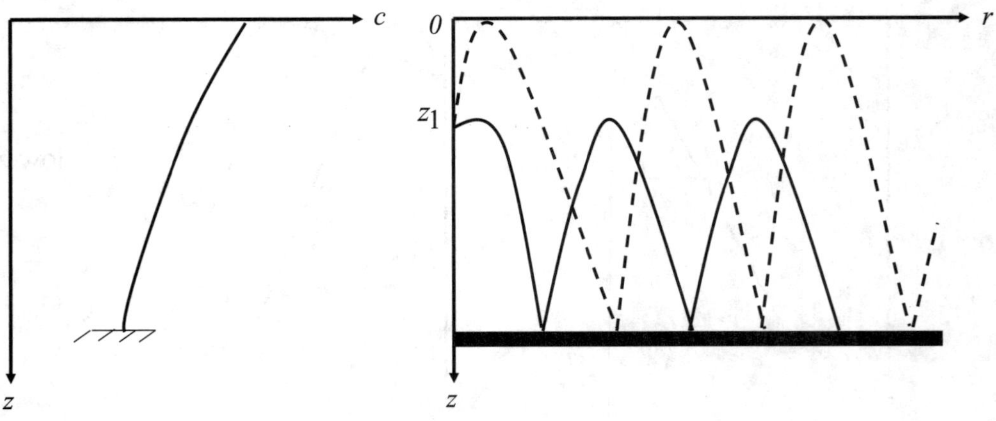

Figure 5.8 (a) Profile of a shallow water sound channel, and (b) ray diagram of the shallow water sound channel.

addition of being refracted (continuing component of the wave), the sound wave is also reflected, as shown in Figure 5.9.

The surface and bottom of the ocean are highly complex boundaries. They are usually rough, and underwater ground is essentially an inhomogeneous medium. However, to analytically obtain the reflection coefficient, we can roughly approximate the surface and bottom as flat surfaces. This allows us to model them as reflective surfaces with specular reflection. Furthermore, we will consider here the simplest case of plane waves.

Figure 5.9 shows reflection at an interface separating two homogeneous fluid media with densities ρ_i and sound speeds c_i, for i = 1, 2. The pertinent incident angles in the x–z plane are denoted by θ_i, for i = 1, 2. Assuming that the incident wave has the unit amplitude, and denoting the amplitudes of the reflected and the refracted wave as R and T, respectively, we can write the acoustic pressures as [5]

$$p_i = \exp(jk_1(x\sin\theta_1 + z\cos\theta_1) - j2\pi ft) \tag{5.8}$$

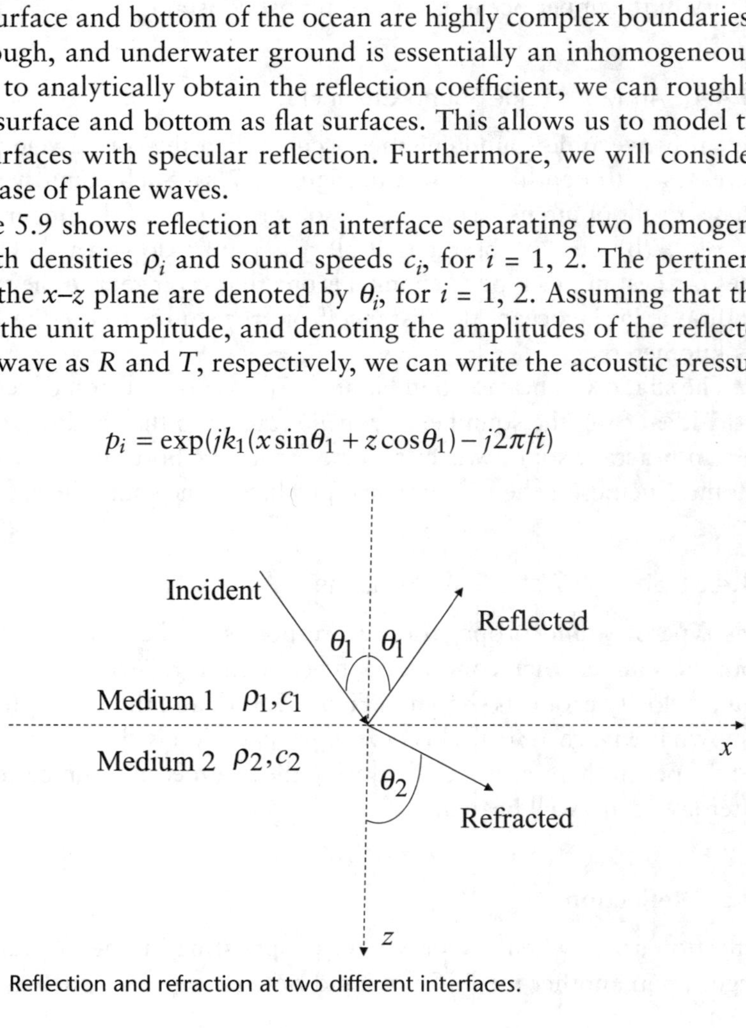

Figure 5.9 Reflection and refraction at two different interfaces.

$$p_r = R\exp(jk_1(x\sin\theta_1 - z\cos\theta_1) - j2\pi ft) \tag{5.9}$$

$$p_t = T\exp(jk_2(x\sin\theta_2 + z\cos\theta_2) - j2\pi ft) \tag{5.10}$$

where $k_1 = 2\pi f/c_1$ and $k_2 = 2\pi f/c_2$. In the subsequent equations, the common time factor $\exp(-j2\pi ft)$ is omitted for brevity.

The unknown quantities R, T, and θ_2 are determined from boundary conditions that require continuity of pressure and vertical particle velocity at the interface $z = 0$. With the total pressure in medium 1 given by $p_1 = p_i + p_r$ and the pressure in medium 2 by $p_2 = p_t$, the boundary conditions can be written as

$$p_1 = p_2 \tag{5.11}$$

$$\frac{1}{j2\pi f\,\rho_1}\frac{dp_1}{dz} = \frac{1}{j2\pi f\,\rho_2}\frac{dp_2}{dz} \tag{5.12}$$

where d/dz denotes the first derivative with respect to z. It can be shown that the requirement of continuity of pressure and vertical particle velocity at $z = 0$ leads to

$$1 + R = T\exp(j(k_2\cos\theta_2 - k_1\cos\theta_1)x) \tag{5.13}$$

$$\frac{k_1}{\rho_1}(1 - R)\cos\theta_1 = \frac{k_2}{\rho_2}T\cos\theta_2 \tag{5.14}$$

Since the left side is independent of x, this yields to *Snell's law of refraction*:

$$k_1\cos\theta_1 = k_2\cos\theta_2 \tag{5.15}$$

$$\frac{\sin\theta_1}{\sin\theta_2} = \frac{k_1}{k_2} = \frac{c_1}{c_2} = n \tag{5.16}$$

This law simply states that the horizontal component of the wave vector does not vary across the interface. From (5.13) and (5.15), we can write

$$1 + R = T \tag{5.17}$$

From (5.16), we can write

$$m(1 - R)\cos\theta_1 = nT\cos\theta_2 \tag{5.18}$$

where $m = \rho_2/\rho_1$. From (5.17) and (5.18), the reflection and transmission coefficients R and T can be written, respectively, as

$$R = \frac{m\cos\theta_1 - n\cos\theta_2}{m\cos\theta_1 + n\cos\theta_2} = \frac{m\cos\theta_1 - \sqrt{n^2 - \sin^2\theta_1}}{m\cos\theta_1 + \sqrt{n^2 - \sin^2\theta_1}} \tag{5.19}$$

$$T = \frac{2m\cos\theta_1}{m\cos\theta_1 + n\cos\theta_2} = \frac{2m\cos\theta_1}{m\cos\theta_1 + \sqrt{n^2 - \sin^2\theta_1}} \tag{5.20}$$

5.1.6 Scattering

Scattered waves arise when a sound wave in water hits a rough surface. There are coherent and incoherent components in the scattered sound field. The coherent component is a wave propagating in the direction of specular reflection and can be calculated through statistical averaging of the scattered field. Often, the effect of scattering from a rough surface is considered as additional loss to the specularly reflected (coherent) component resulting from the scattering of energy away from the specular direction. If the ocean bottom or surface are modeled as randomly rough surfaces, and if the roughness is small with respect to the acoustic wavelength, the reflection loss can be considered to be modified in a simple fashion by the scattering process. A formula often used to describe reflectivity from a rough boundary is

$$\hat{R}(\theta) = R(\theta)e^{-0.5\Gamma^2} \tag{5.21}$$

where $\hat{R}(\theta)$ is the new reflection coefficient reduced because of scattering at the randomly rough interface. Γ is the Rayleigh roughness parameter defined as

$$\Gamma = 2k\sigma \sin\theta \tag{5.22}$$

where $k = 2\pi/\lambda$, σ is the root mean square displacement of the rough surface and θ is the grazing angle of the sound wave relative to the horizontal plane. For large values of the Rayleigh parameter ($\Gamma \gg 1$), the coherent component is close to zero, and the scattered field is almost completely incoherent. Note that this is the generalized definition of Rayleigh roughness parameter, first defined in Section 2.1.4.

5.2 UV-to-UV Acoustic Propagation Modeling

Similarly to fading in F-to-M, V-to-V, and A-to-G channels, fading in UV-to-UV acoustic channels can be classified into large-scale and small-scale fading. We now describe modeling of large-scale and small-scale fading in UV-to-UV acoustic channels.

5.2.1 Large-Scale Fading—Pathloss

A distinguishing property of acoustic channels is the fact that pathloss depends on the signal frequency. *Spreading loss*, *absorption loss*, and *scattering loss* are the three primary mechanisms that attenuate underwater acoustic signals.

5.2.1.1 Spreading Loss

Spreading loss is a weakening of the signal due to the geometrical spreading of wave propagating away from the source. This loss occurs because the fixed amount of transmitted energy spreads over a larger surface area as the signal propagates away from the its source. Two types of spreading are important in underwater propagation: *spherical spreading* and *cylindrical spreading*. The loss due to spherical spreading is dominant at shorter ranges, while the loss due to cylindrical spreading is dominant at medium and longer ranges.

Spherical Spreading

In a homogeneous and infinitely extended medium, the power generated by a point source is radiated in all directions on the surface of a sphere—thus the name *spherical spreading*. Since the sound intensity is proportional to the sound power per area, the intensities at ranges r_0 and r can be written as

$$I_0 = \frac{P}{4\pi r_0^2} \tag{5.23}$$

$$I = \frac{P}{4\pi r^2} \tag{5.24}$$

respectively, where r_0 is the reference distance, P is the acoustic power of the source, I_0 is the acoustic intensity of the source at the reference distance r_0, and I is the acoustic intensity of the source at distance r. Then, the loss due to spherical spreading can be defined as

$$L_s(r) = \frac{I}{I_0} = \left[\frac{r}{r_0}\right]^2 \tag{5.25}$$

At some range from the source, the acoustic signal can no longer spread vertically, and the nature of spreading changes from spherical to cylindrical spreading. This transition typically occurs at ranges greater than the water depth.

Cylindrical Spreading

Cylindrical spreading exists when the medium is confined by two reflecting planes separated by at least $h > 10\lambda$, where λ is the wavelength of the sound wave. Since the sound intensity is proportional to the sound power per area, the intensities at ranges r_0 and r can be written as

$$I_0 = \frac{P}{2\pi h r_0} \tag{5.26}$$

$$I = \frac{P}{2\pi h r} \tag{5.27}$$

respectively, where r_0 is the reference distance, P is the acoustic power of the source, I_0 is the acoustic intensity of the source at the reference distance r_0, I is the acoustic intensity of the source at distance r, and h is the water depth. Then, the loss due to cylindrical spreading is

$$L_c(r) = \frac{I}{I_0} = \frac{r}{r_0} \tag{5.28}$$

5.2.1.2 Absorption and Scattering Loss

When sound propagates in the ocean, part of the acoustic energy is continuously transformed into heat. This absorption is primarily a result of relaxation processes in sea water at frequencies 100 Hz–100 kHz. The *absorption loss* at frequencies of interest for UV-to-UV communications (3 kHz–0.5 MHz) can be written as [5]

$$L_a(r,f) = e^{-r\beta(f)} \tag{5.29}$$

where

$$\beta(f) = 8.68 \cdot 10^3 \left(\frac{Sl \cdot A \cdot f_T \cdot f_c^2}{f_T^2 + f_c^2} + \frac{B \cdot f^2}{f_T} \right)(1 - 6.54 \cdot 10^{-4} \cdot P)[\text{dB/km}] \tag{5.30}$$

and $A = 2.34 \cdot 10^{-6}$, $B = 3.38 \cdot 10^{-6}$, Sl is salinity (in ‰), P is hydrostatic pressure [kg/cm^2], f_c is the carrier frequency [kHz], $f_T = 21.9 \cdot 10^{6-1520/(T+273)}$ is relaxation frequency [kHz], and T is the temperature [°C].

Another reason for the decay of the sound intensity with range in the ocean is the scattering of sound waves by inhomogeneities such as bubbles. Bubbles generated by breaking waves at the sea surface can have a major influence on high-frequency acoustic propagation, both in the open ocean and near shore regions. Layers of bubbles near the surface can result in significant attenuation of surface-scattered signals. In one experiment at 30 kHz, the reported signal attenuation due to scattering at the surface bubble layer was 3 dB per surface bounce [6]. Furthermore, bubble clouds injected down into the water column also significantly attenuate propagating signals with rates as high as 26 dB/m [7]. In shallow water, the injection of bubbles by a breaking wave can thus result in a sudden channel outage.

5.2.1.3 Total Pathloss

The overall pathloss can be written as [8]

$$L(r,f) = \left(\frac{r}{r_0} \right)^k L_a(r,f) \tag{5.31}$$

where k is the spreading factor and ranges between 1 and 2 (cylindrical or spherical spreading) and $L_a(r, f)$ is defined in (5.29).

Two characteristics of spreading and absorption loss are worth noting. First, at short ranges, the spreading loss dominates the absorption loss. Second, even at short ranges (e.g., approximately 400m) the absorption loss at 100 kHz exceeds that at 25 kHz by 15 dB. The practical impact of the frequency dependence of absorption loss is that the communications channel is effectively band-limited, and the available bandwidth is a decreasing function of range. This characteristic can significantly impact the choice of modulation and multiaccess techniques as well as the problem of optimizing network topology.

5.2.2 Large-Scale Fading—Shadowing

Pathloss models described in Section 5.2.1 assume that the pathloss is constant at a given distance and frequency. However, variations in the ocean surface as well as obstacles (such as fish) result in random variations of the received power, which are called *shadow fading*. Very little is known about statistical modeling of shadow fading in underwater communications. Hence, to address this issue, we have recently performed some measurements.

The experimental data was collected during the Naval Research Laboratory (NRL) ACOMMS09 experiment, held near the New Jersey shore in May 2009. The water depth was about 80m. The stationary vertical transmit array was about 45.57m below the surface float, and it had eight hydrophones with the aperture of 0.8m, where we only used the first, fourth, and eighth hydrophone for transmission. In this experiment, we used two stationary receive arrays. The first receive array was 19.66m below the surface float, while the second receive array was 41.96m below the surface float. Both receive arrays had eight hydrophones with the aperture 2.06m, where we only used the first, fourth, and eighth hydrophone for reception. The first receiver was in one direction from the transmitter, 2,000m away, while the second receiver was in the opposite direction from the transmitter, 1,500m away. The center frequency was f_c = 17 kHz, and the signal bandwidth was B = 4 kHz. Finally, the sampling rate was f_s = 93.75 kHz.

Figure 5.10 shows a comparison of the experimentally obtained probability density function and a log-normal probability density function. The results show that the effect of shadowing can be modeled as a log-normal process

$$\alpha(t) = \exp(\mu + \sigma x(t)) \tag{5.32}$$

where $x(t)$ is a real Gaussian random process with zero mean and unit variance, μ is the local mean and represents the effect of the pathloss, and σ is the standard deviation of the log-normal process.

The autocorrelated shadowing quantifies how shadowed power changes over time as the transmitter, receiver, and ocean move (i.e., over a temporal separation Δt). This metric is very important for power control algorithms in general and for

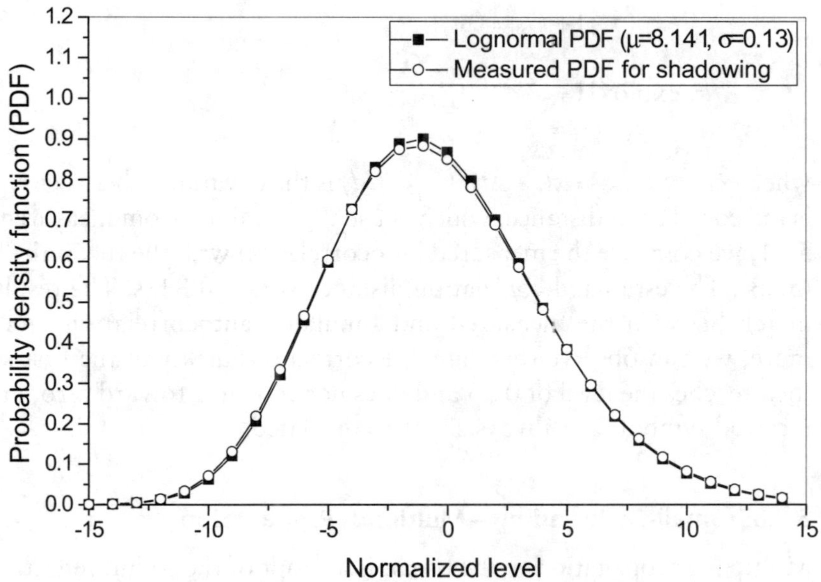

Figure 5.10 Comparison of the measured probability density function of shadowing and the log-normal probability density function.

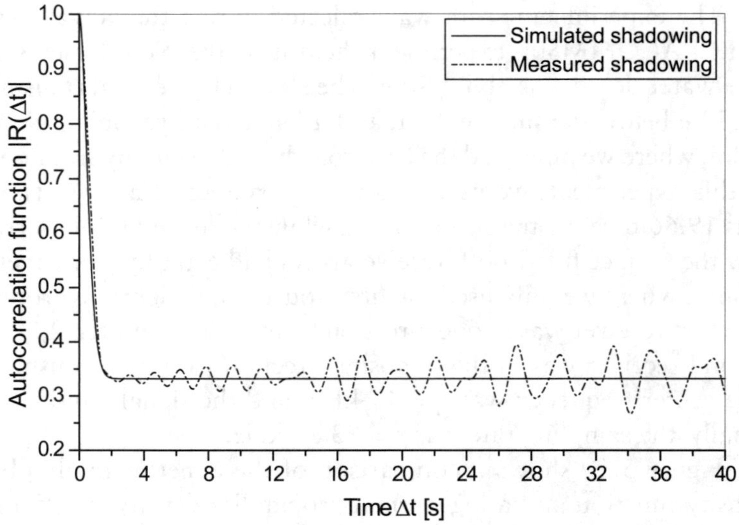

Figure 5.11 Comparison of the measured and simulated shadowing autocorrelation functions.

diversity gains in cooperative systems. The autocorrelation of the shadowing process can be obtained as follows:

$$\begin{aligned} R(\Delta t) &= E[\alpha(t)\alpha(t+\Delta t)^*] \\ &= \mathrm{E}[\exp(\mu+\sigma x(t))\exp(-(\mu+\sigma x(t+\Delta t)))] \\ &= \int_{-\infty}^{\infty}\int_{-\infty}^{\infty} e^{\sigma(x_1-x_2)} \frac{1}{2\pi\sqrt{1-C_{x_1x_2}(\Delta t)^2}} e^{-\frac{x_1^2-2C_{x1x2}(\Delta t)x1x2+x_2^2}{2(1-C_{x1x2}^2(\Delta t))}} dx_1dx_2 \\ &= \exp(\sigma^2[1+C_{x_1x_2}(\Delta t)]) \\ &= \exp(\sigma^2[1+e^{-(\Delta t/Lc)^2}]) \end{aligned} \tag{5.33}$$

where $x_1 = x(t)$, $x_2 = x(t + \Delta t)$, $C_{x_1x_2}(\Delta t)$ is the covariance between x_1 and x_2, and L_c is the correlation distance, which is usually obtained from measurements. In Figure 5.11, we compare the measured autocorrelation with the theoretical one derived in (5.33). The estimated correlation distance is $L_c = 0.828$s. The results show a good match between the measured and simulated autocorrelation functions. Furthermore, we can observe that signal decorrelates quickly in the first several seconds, then reaches the level of 0.33 and does not converge toward zero, implying that the received symbols are always slightly correlated.

5.2.3 Small-Scale Fading—Multipath Propagation

Multipath propagation in the ocean is a result of the sound reflection from the surface, bottom, and/or any other objects in the water and of the sound refraction in the ocean. When a source launches a beam of rays, each ray will follow a slightly different path, and a receiver placed at some distance will observe multiple signal arrivals. In contrast to EM terrestrial propagation where the LoS ray (if it exists) always arrives at the receiver first, in underwater acoustic propagation, a ray traveling over a longer path may do so at a higher speed and thus reach the receiver before the LoS ray.

The impulse response of an acoustic channel is influenced by the geometry of the channel and its reflection and refraction properties, which determine the number of significant propagation paths, and their amplitudes, phases, and delays. Similarly to radio channels, the multipath acoustic channel can be modeled as a linear time-variant filter whose complex baseband impulse response is

$$h(t,\tau) = \sum_{l=1}^{L} C_l e^{-j\phi_l(t)} \delta(\tau - \tau_l) \tag{5.34}$$

where $h(t, \tau)$ is the channel response at time t due to an impulse applied at time $t - \tau$, $\delta(\cdot)$ is the Dirac delta function, and C_l, $\phi_l(t)$, and τ_l are the amplitude, phase, and time delay, respectively, associated with the lth propagation path.

5.2.4 Factors Influencing Small-Scale Fading

Small-scale fading is influenced by many factors, such as multipath propagation, speed of the transmitter and receiver, inherent changes of the ocean, and bandwidth and carrier frequency of the transmitted signal. Here, we will focus on multipath propagation, movement of the transmitter and receiver, and on ocean variations induced by surface waves and how these factors impact small-scale fading.

5.2.4.1 Impact of Multipath Propagation on Small-Scale Fading

As indicated in (5.34), the received signal is a sum of many attenuated, delayed, and phase-shifted replicas of the transmitted signal. A sum of these delayed signals creates *time dispersion* of the original signal. In the frequency domain, this manifests as *frequency selectivity* (i.e., the channel distorts the frequency response characteristic of the transmitted signal).

The total delay spread in the ocean is governed by the longest path delay. A value typically observed in shallow water experiments is on the order of tens of milliseconds. Channel coherence times below 100 ms have been observed [9], but not often. For a general-purpose design, one may consider coherence times on the order of hundreds of milliseconds. Due to long delay spreads and limited available bandwidth, the underwater channel is often classified as frequency-selective. The frequency-selectivity of the channel leads to intrasymbol and intersymbol interference. While the intrasymbol interference is generally irreversible, the intersymbol interference can be reversed using equalizers, spread spectrum techniques, or low-complexity orthogonal frequency division multiplexing (OFDM)-like techniques.

For modern underwater systems, the individual path dispersion is much less than the total delay spread and can be ignored for systems whose maximal frequency lies well below the channel cutoff. However, as transducer technology advances and larger bandwidths become available, this effect may become significant.

5.2.4.2 Impact of the Relative Movement of Transmitter, Receiver, and Ocean Surface on Small-Scale Fading

A transmitted signal is impacted not only by multipath propagation, but also by other factors such as movement of the ocean surface, transmitter, and receiver. The variations in the ocean range from those that occur over very long timescales and do not affect the instantaneous level of a communication signal (e.g., monthly changes in temperature) to those that occur over short timescales and affect the transmitted signal. Prominent among the fast variations of the ocean are those induced by surface waves and by the motion of the transmitter and receiver. These variations effectively cause the displacement of the reflection point, resulting in both scattering of the signal and Doppler spreading due to the changing path length. The magnitude of the Doppler shift is proportional to the ratio $d = v/c$ of the relative transmitter-receiver-surface velocity v to the speed of sound in water c. Because the speed of sound is very low compared to the speed of electromagnetic waves, motion-induced Doppler distortion can be significant.

In UV-to-UV communications, there is always some motion present in the system: autonomous underwater vehicles move at speeds of a few meters per second; submarines move at speeds of tens of meters per second; and even fixed underwater systems experience some motion due to drifting with waves, currents, and tides, and the velocity of such a system can reach several meters per second. In other words, there is always some motion present in the underwater system, and a communication system has to be designed to take this into account. The only comparable situation in radio communications occurs in low Earth orbiting (LEO) satellite systems, where the relative velocity of satellites flying overhead is extremely high (the channel there, however, is not nearly as dispersive).

The way in which motion-induced distortions affect signal detection depends on the actual value of factor d. For comparison, let us look at a highly mobile radio system. At 160 km/h, the Doppler shift is proportional to $d = 1.5/10^7$. This value is low enough that Doppler spreading can be neglected. In contrast to mobile radio systems, a stationary acoustic system may experience unintentional motion of 0.5 m/s, which leads to Doppler shift of $d = 3/10^4$. For an AUV moving at several meters per second, factor d will be on the order of $1/10^3$, a value that cannot be ignored.

The motion-induced distortion mainly impacts the design of synchronization and channel estimation algorithms. In multicarrier systems, the Doppler effect creates a particularly severe distortion. Unlike radio systems, in which time compression/dilation is negligible and the Doppler shift appears equal for all subcarriers, in an acoustic system each subcarrier may experience a significantly different Doppler shift, creating nonuniform Doppler distortion across the signal bandwidth.

As the history of underwater acoustic communications testifies, major advances in signal processing were made when the physical nature of propagation was taken into account through proper channel modeling. Examples are combined modeling of multipath and phase distortion for equalization in single-carrier wideband systems [10], detection of multicarrier signals using explicit modeling of Doppler shift [11], and sparse channel estimation using the fact that the underwater multipath is not contiguous but consists of isolated signal arrivals [12]. Hence, to address the challenges in underwater wireless communications, one must understand, characterize, and model the acoustic wireless communication channel.

5.3 Statistical Characterization of Multipath Propagation in UV-to-UV Acoustic Channels

Sound propagation in the ocean is mathematically described by the wave equation, whose parameters and boundary conditions describe the ocean environment. There are five types of models that describe sound propagation in the sea: fast field program, normal mode, parabolic equation, direct finite-difference (or finite-element), and ray theory. In practice, the ray theory is most often used for sound propagation modeling [5].

In most environments and at the frequencies of interest for communications signals, the ocean can be modeled as a waveguide with reflecting surfaces and a spatially variant sound speed. Recent references commonly use ray tracing for determining the deterministic multipath structure for communication channel modeling [13, 14]. Associated with each of the deterministic propagation paths (macro-multipaths), random signal fluctuations (micro-multipaths), account for the time variability of the channel response. In shallow water, the most dominant contributor to the overall time variability is surface scattering due to waves. In deep water, in addition to surface scattering, internal waves contribute to the time variations of the signal propagation along each of the deterministic paths.

Addressing the challenges in underwater wireless communications requires statistical characterization and modeling of acoustic wireless communication channels. While there exist a large number of both deterministic and statistical models that characterize underwater sound propagation [3, 13–17], a statistical framework for these channels, (i.e., correlation functions, Doppler power spectral density, power delay spectral density, level-crossing rate, and so on) has been only recently proposed. These statistics are necessary tools for a successful design of underwater communication systems. They are commonly used to determine available capacity, investigate spatial diversity, determine propagation delay spread, and so on. Furthermore, the closed-form expressions of these statistics can be used to estimate physical parameters such as angle spread and mean angles of departure and arrival. The following sections detail the most recent results on statistical characterization of multipath propagation in UV-to-UV acoustic channels.

5.3.1 Received Envelope Distribution

A time-varying multipath communication channel is commonly modeled as a tapped delay line, with tap spacing equal to the reciprocal of twice the channel bandwidth and the tap gains modeled as stochastic processes with certain distributions and power spectral densities. Unlike RF radio channels, where a number of models for both the probability distribution (e.g., Rayleigh fading) and the power spectral density of the fading process (e.g., the Jakes model) are well accepted and even standardized, there is no consensus yet on the statistical characterization of underwater acoustic communication channels. For a shallow-water medium-range channel, some experimental results suggest that this channel may be characterized as K-distributed fading [18], while others suggest Ricean or Rayleigh fading [17]. The deep water channel has also been modeled as a Rayleigh fading channel, but the available measurements are scarce.

5.3.2 Statistical Modeling of UV-to-UV Frequency Flat Acoustic Fading Channels

The first attempts to statistically characterize UV-to-UV acoustic channels are reported in [19–23]. Abdi and Guo [19] were the first to propose a statistical model for *time-invariant* stationary shallow water acoustic channels, by taking into account only macro-multipath effects. Later, a *time-varying* shallow water acoustic channel that takes into account both macro- and micro-multipath effects was proposed in [23], which we detail here.

In [23], the UV-to-UV acoustic channel is modeled as a 2-D waveguide bounded from the top and bottom. The surface and bottom boundaries reflect the acoustic signal, which results in multiple *macro-eigenrays* traveling between the transmitter (T_x) and receiver (R_x), as shown in Figure 5.12. In contrast to fixed-to-mobile cellular radio channels and air-to-ground radio channels (where single-bounced eigenrays are dominant [24, 25]), or vehicle-to-vehicle radio channels (where single-bounced and double-bounced eigenrays are dominant [26]), in shallow water acoustic (SWA) channels the energy stays trapped in the waveguide and the number of bounces (i.e., interactions with the surface and bottom) can be large [5]. However, the experimental results for medium- and long-range acoustic channels show that the number of different macro-eigenrays that arrive at the R_x rarely exceeds eight [17, 27, 28]. Hence, we cannot simply assume a very large number of macro-eigenrays. Furthermore, the locations where macro-scatterers interact with the surface/bottom are deterministic and depend only on the waveguide geometry and the number of macro-eigenrays [5]. Hence, the deterministic ray-tracing theory can be used to model these macro-eigenrays.

However, in each of the deterministic macro-eigenrays, random signal fluctuations (micro-eigenrays) account for the time variability of the channel impulse

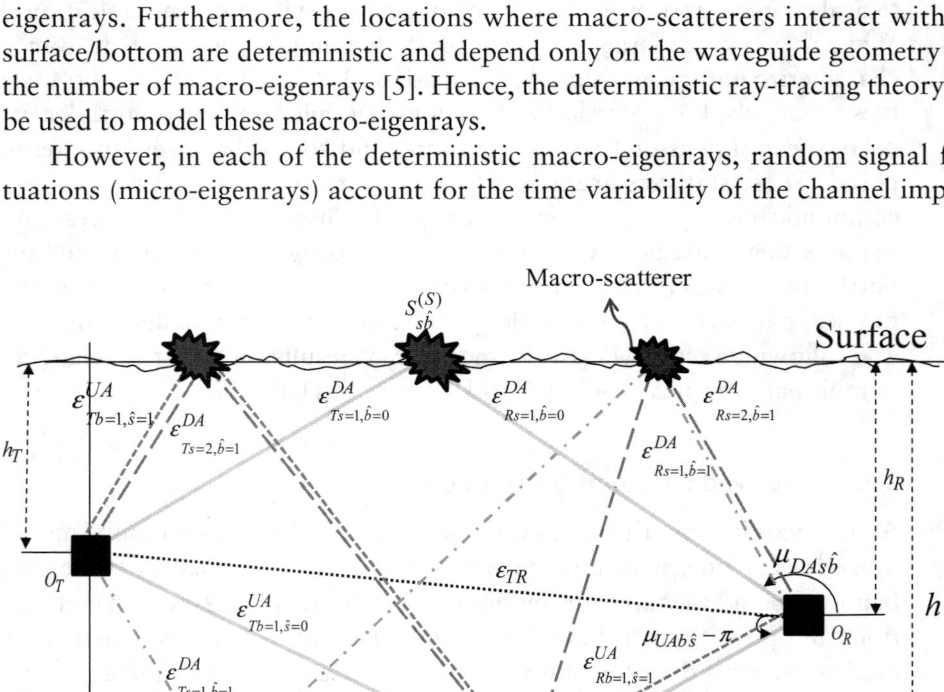

Figure 5.12 Illustration of the line-of-sight path and several macro-eigenrays traveling between the T_x and R_x in a shallow water acoustic channel.

response [5, 15]. For example, the roughness of the sea surface and the sea bottom is characterized by random *micro-scatterers*, as shown in Figure 5.13. Each macro-eigenray is thus modeled as an average of a large number of micro-eigenrays, where the possible positions of micro-scatterers are clustered around the positions of macro-scatterers, as shown in Figure 5.13. Note the contrast to radio channel propagation, where scatterers can be randomly placed anywhere in a 2-D or 3-D plane.

The underwater acoustic channel model in [23] is based on the following assumptions:

- Propagation occurs in a shallow water environment with constant sound speed (i.e., sound energy propagates along the plane waves);
- The T_x is an omnidirectional transducer and therefore produces a spherical wavefront in this isovelocity medium;
- Propagation is characterized with either LoS or NLoS conditions between the T_x and R_x.

In Figure 5.13, the depths of T_x and R_x are denoted by h_T and h_R, respectively, while the orientations of T_x and R_x transducer arrays in the x-z plane are denoted by θ_T and θ_R, respectively, relative to the x-axis. The T_x and R_x are moving with constant speeds v_T and v_R in directions described by angles γ_T and γ_R in the x-z plane (relative to the x-axis), respectively.

At any instant in time t, the R_x receives $2S$ downward arriving (DA) macro-eigenrays (whose last reflection was from the surface), each one having different number of s surface and $\hat{b}$ bottom reflections, where S denotes the maximum number of interactions between any DA macro-eigenray and the surface, $1 \leq s \leq S$, and $s-1 \leq \hat{b} \leq s$. For example, if a DA macro-eigenray has only one interaction

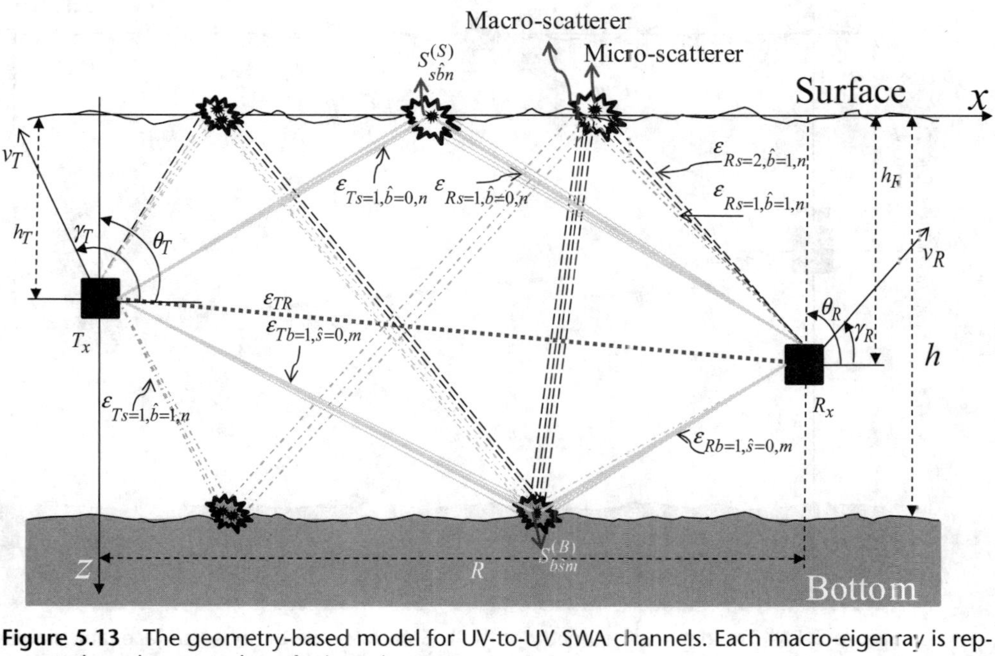

Figure 5.13 The geometry-based model for UV-to-UV SWA channels. Each macro-eigenray is represented as a large number of micro-eigenrays.

with the surface (i.e., S = 1) there are two possible paths that this eigenray traveled. The first path is a single-bounced path, where a DA macro-eigenray started upward, interacted with the surface, and arrived at the R_x (i.e., $s = 1, \hat{b} = 0$). The second path is a double-bounced path, where a DA macro-eigenray started downward, interacted with the bottom, then interacted with the surface, and arrived at the R_x (i.e., $s = 1, \hat{b} = 1$). Note that, for S = 1, both macro-eigenrays will reach the R_x because they have similar energy [5, 17]. Similarly, there are $2B$ upward arriving (UA) macro-eigenrays (i.e., the last reflection is from the bottom) with b bottom and $\hat{s}$ surface reflections, where B denotes the maximum number of interactions between a UA macro-eigenray and the bottom, $1 \le b \le B$ and $b-1 \le \hat{s} \le b$. As noted earlier, each macro-eigenray is modeled as an average of micro-eigenrays. Thus, each UA macro-eigenray is modeled as an average of $N_{s\hat{b}}$ DA micro-eigenrays, and each UA macro-eigenray is modeled as an average of $M_{b\hat{s}}$ UA micro-eigenrays.

The symbols $\alpha_{Ts\hat{b}n}$ and $\alpha_{Tb\hat{s}m}$ are the angles of departure (AoD) of micro-eigenrays that start from T_x and impinge on the scatterers $S^{(S)}_{s\hat{b}n}$ and $S^{(B)}_{b\hat{s}m}$, respectively, whereas $\alpha_{Rs\hat{b}n}$ and $\alpha_{Rb\hat{s}m}$ are the angles of arrival (AoA) of the micro-eigenrays scattered from $S^{(S)}_{s\hat{b}n}$ and $S^{(B)}_{b\hat{s}m}$ and arriving at R_x, respectively. Finally, the symbol α_{TR} denotes the AoA of LoS ray. For ease of reference, Figure 5.14 details the geometry of the LoS path as well as the geometry of single-bounced surface and single-bounced bottom micro-eigenrays (i.e., $s = 1, \hat{b} = 0,\ b = 1, \hat{s} = 0$) scattered from the $S^{(S)}_{s=1,\hat{b}=0,n}$-th and $S^{(B)}_{b=1,\hat{s}=0,m}$-th micro-scatterer, respectively. The geometry of multiple-bounced micro-eigenrays is defined similarly, but omitted from Figure 5.14 for ease of reference.

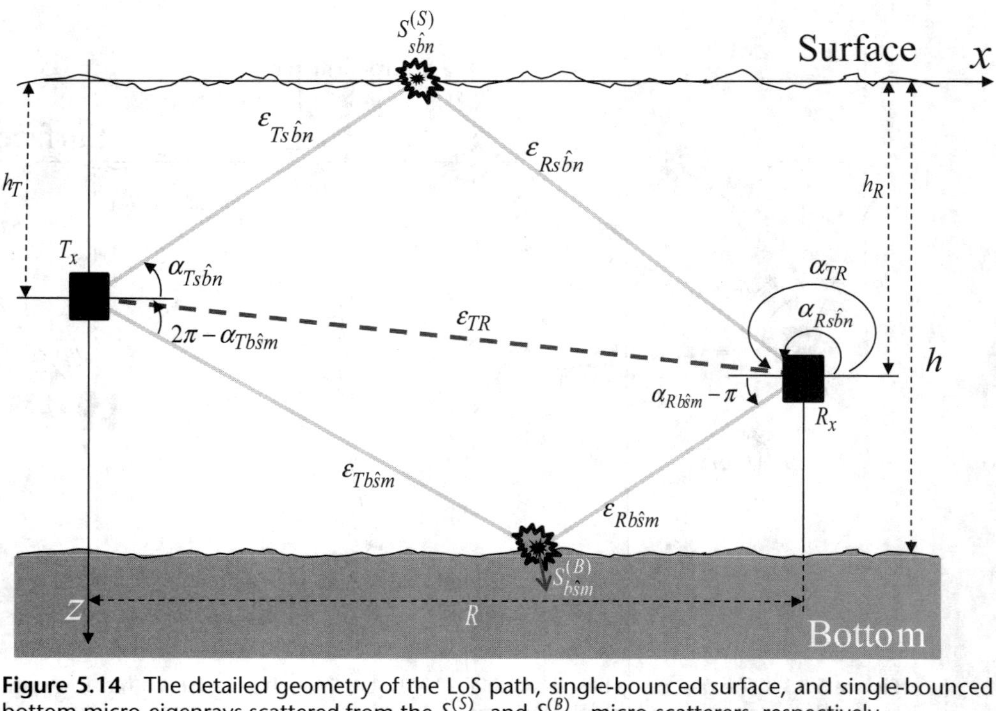

Figure 5.14 The detailed geometry of the LoS path, single-bounced surface, and single-bounced bottom micro-eigenrays scattered from the $S^{(S)}_{s,\hat{b},n}$ and $S^{(B)}_{b,\hat{s},m}$ micro-scatterers, respectively.

The horizontal spacing between the T_x and R_x is denoted by R and the water depth is denoted by h. Here, we are interested in medium- and long-range shallow-water communications. Hence, we assume that the depths h, h_T, and h_R are much smaller than the distance R. The symbols $\varepsilon_{Ts\hat{b}n}$ and $\varepsilon_{Tb\hat{s}m}$ denote distances $T_x - S^{(S)}_{s\hat{b}n}$ and $T_x - S^{(B)}_{b\hat{s}m}$, respectively, where $S^{(S)}_{s\hat{b}n}$ and $S^{(B)}_{b\hat{s}m}$ denote the nth and mth micro-scatterers located around the macro-scatterers $S^{(S)}_{s\hat{b}}$ and $S^{(B)}_{b\hat{s}}$ at the surface and bottom, respectively, for $1 \le n \le N_{s\hat{b}}$ and $1 \le m \le M_{b\hat{s}}$, as shown in Figure 5.13. Finally, ε_{TR} denotes the distance $T_x - R_x$.

The complex faded envelope under frequency-flat fading can therefore be written as a superposition of the LoS, the UA bounced macro-eigenrays (i.e., the last reflection is from the bottom), and the DA bounced macro-eigenrays (i.e., the last reflection is from the surface). Furthermore, each DA macro-eigenray is represented as an average of $N_{s\hat{b}}$ DA micro-eigenrays, while each UA macro-eigenray is represented as an average of $M_{b\hat{s}}$ UA micro-eigenrays. Then, for the channel between the T_x and R_x, the channel impulse response is [23]

$$h(t) = h^{\mathrm{LoS}}(t) + h^{\mathrm{UA}}(t) + h^{\mathrm{DA}}(t) \tag{5.35}$$

where the respective components are [23]

$$h^{\mathrm{LoS}}(t) = \sqrt{\Omega} \cdot a^{\mathrm{LoS}} \cdot e^{j2\pi f^{\mathrm{LoS}} t} \tag{5.36}$$

$$h^{\mathrm{UA}}(t) = \sqrt{\Omega} \sum_{b=1}^{B} \sum_{\hat{s}=b-1}^{b} \lim_{M_{b\hat{s}} \to \infty} \sum_{m=1}^{M_{b\hat{s}}} a^{\mathrm{UA}}_{b\hat{s}m} \cdot e^{j(2\pi f^{\mathrm{UA}}_{b\hat{s}m} t + \phi^{\mathrm{UA}}_{b\hat{s}m})} \tag{5.37}$$

$$h^{\mathrm{DA}}(t) = \sqrt{\Omega} \sum_{s=1}^{S} \sum_{\hat{b}=s-1}^{s} \lim_{N_{s\hat{b}} \to \infty} \sum_{n=1}^{N_{s\hat{b}}} a^{\mathrm{DA}}_{s\hat{b}n} \cdot e^{j(2\pi f^{\mathrm{DA}}_{s\hat{b}n} t + \phi^{\mathrm{DA}}_{s\hat{b}n})} \tag{5.38}$$

and where Ω is the averaged total power ($\sqrt{\Omega}$ is assumed to be constant). The channel gains, Doppler shift frequencies, and phases in this model can be calculated as follows:

- *Channel gains*. The channel gains in this model are normalized such that the ensemble averages of the $a^{\mathrm{UA}}_{b\hat{s}m}$s and $a^{\mathrm{DA}}_{s\hat{b}n}$s are given by

$$\sum_{m=1}^{M_{b\hat{s}}} \mathrm{E}\left[a^{\mathrm{UA}^2}_{b\hat{s}m} \right] = 1, \quad \text{for} \quad \forall b, \hat{s} \tag{5.39}$$

$$\sum_{n=1}^{N_{s\hat{b}}} \mathrm{E}\left[a^{\mathrm{DA}^2}_{s\hat{b}n} \right] = 1, \quad \text{for} \quad \forall s, \hat{b} \tag{5.40}$$

In addition, let's introduce parameter K, which describes how the total signal power is split between the LoS and NLoS components: the power allocated to the LoS component is $K/(1 + K)$, while the total power allocated to the NLoS components is $1/(1 + K)$. Furthermore, it is assumed that the relative power allocated to the UA and DA components is respectively η_B and η_S such that $\eta_B + \eta_S = 1$. These parameters have to be either set during simulations or estimated from measurements. Having all this in mind, the channel gains can be written as follows:

$$a^{LoS} = \sqrt{\frac{K}{K+1}} \tag{5.41}$$

$$a^{UA}_{b\hat{s}m} = \sqrt{\frac{1}{K+1}}\sqrt{\frac{\eta_B}{2B}}\sqrt{\frac{1}{M_{b\hat{s}}}} \tag{5.42}$$

$$a^{DA}_{s\hat{b}n} = \sqrt{\frac{1}{K+1}}\sqrt{\frac{\eta_S}{2S}}\sqrt{\frac{2}{N_{s\hat{b}}}} \tag{5.43}$$

- *Doppler shifts.* Doppler shifts depend on the geometrical relation between directions of movement of T_x and R_x (γ_T and γ_R), the vertical displacements of surface micro-scatterers due to surface motion ($\Delta Z_{s\hat{b}n}(t)$ and $\Delta Z_{b\hat{s}m}(t)$), and the directions of AoDs and AoAs. For this model, the Doppler shifts are

$$f^{LoS} = f_{\mathrm{Tmax}}\cos(\alpha_{TR} + \pi - \gamma_T) + f_{\mathrm{Rmax}}\cos(\alpha_{TR} - \gamma_R) \tag{5.44}$$

$$f^{UA}_{b\hat{s}m} = \frac{1}{\lambda t}\Delta Z_{b\hat{s}m}(t)\sin\alpha_{Rb\hat{s}m} + f_{\mathrm{Tmax}}\cos(\alpha_{Tb\hat{s}m} - \gamma_T) + f_{\mathrm{Rmax}}\cos(\alpha_{Rb\hat{s}m} - \gamma_R) \tag{5.45}$$

$$f^{DA}_{s\hat{b}n} = \frac{1}{\lambda t}\Delta Z_{s\hat{b}n}(t)\sin\alpha_{Rs\hat{b}n} + f_{\mathrm{Tmax}}\cos(\alpha_{Ts\hat{b}n} - \gamma_T) + f_{\mathrm{Rmax}}\cos(\alpha_{Rs\hat{b}n} - \gamma_R) \tag{5.46}$$

where the frequencies $f_{\mathrm{Tmax}} = v_T/\lambda$ and $f_{\mathrm{Rmax}} = v_R/\lambda$ are the maximum Doppler frequencies associated with the T_x and R_x, respectively, and λ is the carrier wavelength. Since the location of the scatterers is not known a priori, the AoDs $\alpha_{Ts\hat{b}n}$ and $\alpha_{Tb\hat{s}m}$, the AoAs $\alpha_{Rs\hat{b}n}$ and $\alpha_{Rb\hat{s}m}$, and the vertical displacements $\Delta Z_{s\hat{b}n}(t)$ and $\Delta Z_{b\hat{s}m}(t)$ are modeled as random variables. However, since the number of micro-eigenrays is approaching infinity, these discrete random variables become continuous with pdfs $p(\alpha_{Ts\hat{b}})$, $p(\alpha_{Tb\hat{s}})$, $p(\alpha_{Rs\hat{b}})$, $p(\alpha_{Rb\hat{s}})$, $p(\Delta Z_{s\hat{b}}(t))$, and $p(\Delta Z_{b\hat{s}}(t))$, respectively. Since the geometry that propagating waves experience while traveling from the transmitter to the receiver dictates the relationship between the AoDs and AoAs, the AoAs ($\alpha_{Rs\hat{b}n}$ and $\alpha_{Rb\hat{s}m}$) are dependent on the AoDs ($\alpha_{Ts\hat{b}n}$ and $\alpha_{Tb\hat{s}m}$, respectively). Assuming that each micro-eigenray, when interacting with the surface and the bottom, has equal incident and reflecting angles, we can observe that $\alpha_{Ts\hat{b}n} = \pi - \alpha_{Rs\hat{b}n}$ and $\alpha_{Tb\hat{s}m} = 3\pi - \alpha_{Rb\hat{s}m}$.
- *Phases.* First, we note that the phase shift introduced by the scatterer located near the T_x is independent of the phase shift introduced by the scatterer located near the R_x. Then, without loss of generality, we can assume that the

phases $\phi_{s\hat{b}n}$ and $\phi_{b\hat{s}m}$ are independent random variables. Since the number of micro-eigenrays approaches infinity, these discrete random variables become continuous with pdfs $p(\phi_{s\hat{b}})$ and $p(\phi_{b\hat{s}})$, respectively. Here, it is assumed that they are uniformly distributed on the interval $[-\pi, \pi)$ and independent from any other random variable.

5.3.3 Temporal Fading Characteristics of UV-to-UV Frequency Flat Acoustic Fading Channels

Using the model described in Section 5.3.2, we can now derive the key temporal characteristics (i.e., temporal autocorrelation function, Doppler spectrum, and level crossing rate) of UV-to-UV frequency-flat acoustic fading channels.

5.3.3.1 Temporal Autocorrelation Function

The temporal autocorrelation function (ACF) of the complex fading channel $h(t)$ is an important quantity since it allows the quantification of necessary communication parameters such as the required number of pilot symbols and/or the system performance degradation.

The normalized ACF is defined as

$$R(\tau) = \frac{E\left[h(t)h(t+\tau)^*\right]}{\sqrt{\mathrm{Var}[h(t)]\mathrm{Var}[h(t+\tau)^*]}} \tag{5.47}$$

where the variance Var[·] and the expectation E[·] are taken with respect to the set of random variables $\alpha_{Rs\hat{b}}$, $\alpha_{Rb\hat{s}}$, $\Delta Z_{s\hat{b}}(t)$, $\Delta Z_{b\hat{s}}(t)$, $\phi_{s\hat{b}}$, and $\phi_{b\hat{s}}$.

The angles of arrival $\alpha_{Rs\hat{b}}$ and $\alpha_{Rb\hat{s}}$ are modeled using the following Gaussian probability density functions [23]

$$p_{\mathrm{top}}(\alpha_{Rs\hat{b}}) = \frac{1}{\sqrt{2\pi\sigma^2_{\mathrm{DA}s\hat{b}}}} \exp\{-(\alpha_{Rs\hat{b}} - \mu_{\mathrm{DA}s\hat{b}})^2/(2\sigma^2_{\mathrm{DA}s\hat{b}})\},\ 0 < \alpha_{Rs\hat{b}} < \pi \tag{5.48}$$

$$p_{\mathrm{bottom}}(\alpha_{Rb\hat{s}}) = \frac{1}{\sqrt{2\pi\sigma^2_{\mathrm{UA}b\hat{s}}}} \exp\{-(\alpha_{Rb\hat{s}} - \mu_{\mathrm{UA}b\hat{s}})^2/(2\sigma^2_{\mathrm{UA}b\hat{s}})\},\ \pi < \alpha_{Rb\hat{s}} < 2\pi \tag{5.49}$$

where $\mu_{\mathrm{DA}s\hat{b}}$ and $\mu_{\mathrm{UA}b\hat{s}}$ are the mean AoAs, and $\sigma_{\mathrm{DA}s\hat{b}}$ and $\sigma_{\mathrm{UA}b\hat{s}}$ are the angle spreads. The vertical displacements $\Delta Z_{s\hat{b}}(t)$ and $\Delta Z_{b\hat{s}}(t)$ are modeled as zero-mean Gaussian random processes with stationary and independent increments and can be written as

$$p(\Delta Z_{s\hat{b}}(t)) = \frac{1}{\sqrt{2\pi t\zeta^2_{\Delta Z_{\mathrm{DA}s\hat{b}}}}} \exp(-(\Delta Z_{s\hat{b}}(t))^2/(2t\zeta^2_{\Delta Z_{\mathrm{DA}s\hat{b}}})) \tag{5.50}$$

$$p(\Delta Z_{b\hat{s}}(t)) = \frac{1}{\sqrt{2\pi t\zeta^2_{\Delta Z_{\mathrm{UA}b\hat{s}}}}} \exp(-(\Delta Z_{b\hat{s}}(t))^2/(2t\zeta^2_{\Delta Z_{\mathrm{UA}b\hat{s}}})) \tag{5.51}$$

where $t\zeta^2_{\Delta Z_{\mathrm{DA}s\hat{b}}}$ and $t\zeta^2_{\Delta Z_{\mathrm{UA}b\hat{s}}}$ denote their respective variances.

Here, we note that the locations of micro-scatterers that contribute to one macro-eigenray are independent from the locations of micro-scatterers that contribute to a different macro-eigenray because different macro-eigenrays have different angles of departure and arrival. Using the assumptions introduced earlier and the Central Limit Theorem [25], we can conclude that each macro-eigenray is an independent complex Gaussian process with a zero mean. A sum of several complex zero-mean Gaussian processes results in a new complex zero-mean Gaussian process, so $h^{\mathrm{UA}}(t)$ and $h^{\mathrm{DA}}(t)$ are also independent complex Gaussian random processes with zero means. Then, the normalized ACF can be simplified to

$$R(\tau) = R^{LoS}(\tau) + R^{UA}(\tau) + R^{DA}(\tau) \tag{5.52}$$

where $R^{LoS}(\tau)$, $R^{UA}(\tau)$, and $R^{DA}(\tau)$ denote the normalized ACFs of the LoS, upward-arriving, and downward-arriving eigenrays, respectively.

The temporal autocorrelation function of the LoS component can be obtained as follows

$$\begin{aligned} R^{\mathrm{LoS}}(\tau) &= \frac{\mathrm{E}\left[h^{\mathrm{LoS}}(t)h^{\mathrm{LoS}}(t+\tau)^*\right]}{\sqrt{\mathrm{Var}[h^{\mathrm{LoS}}(t)]\mathrm{Var}[h^{\mathrm{LoS}}(t+\tau)^*]}} \\ &= \frac{K}{K+1} e^{j2\pi\tau[f_{\mathrm{Tmax}}\cos(\alpha_{TR}-\gamma_T)-f_{\mathrm{Rmax}}\cos(\alpha_{TR}-\gamma_R)]} \end{aligned} \tag{5.53}$$

Noting that the phases $\phi_{b\hat{s}m}$ and $\phi_{s\hat{b}n}$ are independent and uniformly distributed over $[-\pi, \pi)$, the ACFs of the UA and DA components can be written as

$$\begin{aligned} R^{\mathrm{UA}}(\tau) &= \frac{\mathrm{E}\left[h^{\mathrm{UA}}(t)h^{\mathrm{UA}}(t+\tau)^*\right]}{\sqrt{\mathrm{Var}[h^{\mathrm{UA}}(t)]\mathrm{Var}[h^{\mathrm{UA}}(t+\tau)^*]}} \\ &= \frac{\eta_B}{2B(K+1)} \sum_{b=1}^{B} \sum_{\hat{s}=b-1}^{b} \mathrm{E}_{\alpha_{Rb\hat{s}},\Delta Z_{b\hat{s}}(t)} \\ &\left[\exp\left(j2\pi\tau\left[f_{\mathrm{Tmax}}\cos(\alpha_{Rb\hat{s}}+\gamma_T)-f_{\mathrm{Rmax}}\cos(\alpha_{Rb\hat{s}}-\gamma_R)\right]\right)\right. \\ &\left.\exp\left(-j\frac{2\pi}{\lambda}(\Delta Z_{b\hat{s}}(t+\tau)-\Delta Z_{b\hat{s}}(t))\sin\alpha_{Rb\hat{s}}\right)\right] \end{aligned} \tag{5.54}$$

$$R^{\mathrm{DA}}(\tau)=\frac{\mathrm{E}\left[h^{\mathrm{DA}}(t)h^{\mathrm{DA}}(t+\tau)^{*}\right]}{\sqrt{\mathrm{Var}[h^{\mathrm{DA}}(t)]\mathrm{Var}[h^{\mathrm{DA}}(t+\tau)^{*}]}}$$

$$=\frac{\eta_S}{2S(K+1)}\sum_{s=1}^{S}\sum_{\hat{b}=s-1}^{s}\mathrm{E}_{\alpha_{Rs\hat{b}},\Delta Z_{s\hat{b}}(t)}$$

$$\left[\exp\left(j2\pi\tau\left[f_{\mathrm{Tmax}}\cos(\alpha_{Rs\hat{b}}+\gamma_T)-f_{\mathrm{Rmax}}\cos(\alpha_{Rs\hat{b}}-\gamma_R)\right]\right)\right.$$

$$\left.\exp\left(-j\frac{2\pi}{\lambda}(\Delta Z_{s\hat{b}}(t+\tau)-\Delta Z_{s\hat{b}}(t))\sin\alpha_{Rs\hat{b}}\right)\right] \tag{5.55}$$

where $\mathrm{E}_{\alpha_{Rb\hat{s}},\Delta Z_{b\hat{s}}(t)}[\cdot]$ and $\mathrm{E}_{\alpha_{Rs\hat{b}},\Delta Z_{s\hat{b}}(t)}[\cdot]$ denote the statistical expectations. Since random variables $\alpha_{Rb\hat{s}}$, $\alpha_{Rs\hat{b}}$, $\Delta Z_{b\hat{s}}(t)$, and $\Delta Z_{s\hat{b}}(t)$ are independent random variables, these expectations can be evaluated as follows:

$$R^{UA}(\tau)=\frac{\eta_B}{2B(K+1)}\sum_{b=1}^{B}\sum_{\hat{s}=b-1}^{b}$$

$$\int_{\pi}^{2\pi}\exp\left(j2\pi\tau\left[f_{\mathrm{Tmax}}\cos(\alpha_{Rb\hat{s}}+\gamma_T)-f_{\mathrm{Rmax}}\cos(\alpha_{Rb\hat{s}}-\gamma_R)\right]\right)$$

$$\mathrm{E}_{\Delta Z_{b\hat{s}m}(t)}\left[\exp\left(-j\tfrac{2\pi}{\lambda}(\Delta Z_{b\hat{s}m}(t+\tau)-\Delta Z_{b\hat{s}m}(t))\sin\alpha_{Rb\hat{s}}\right]\right)d\alpha_{Rb\hat{s}} \tag{5.56}$$

$$R^{DA}(\tau)=\frac{\eta_s}{2S(K+1)}\sum_{s=1}^{S}\sum_{\hat{b}=s-1}^{s}$$

$$\int_{0}^{\pi}\exp\left(j2\pi\tau\left[f_{\mathrm{Tmax}}\cos(\alpha_{Rs\hat{b}}+\gamma_T)-f_{\mathrm{Rmax}}\cos(\alpha_{Rs\hat{b}}-\gamma_R)\right]\right)$$

$$\mathrm{E}_{\Delta Z_{s\hat{b}n}(t)}\left[\exp\left(-j\tfrac{2\pi}{\lambda}(\Delta Z_{s\hat{b}n}(t+\tau)-\Delta Z_{s\hat{b}n}(t))\sin\alpha_{Rs\hat{b}}\right]\right)d\alpha_{Rs\hat{b}} \tag{5.57}$$

Since $\Delta Z_{b\hat{s}}(t)$ has stationary and independent increments with pdfs defined in (5.51), the expectation $\mathrm{E}_{\Delta Z_{b\hat{s}}(t)}[\cdot]$ in (5.56) can be calculated as follows

$$\mathrm{E}_{\Delta Z_{b\hat{s}m}(t)}\left[\exp\left(-j\frac{2\pi}{\lambda}(\Delta Z_{b\hat{s}m}(t+\tau)-\Delta Z_{b\hat{s}m}(t))\sin\alpha_{Rb\hat{s}m}\right)\right]$$

$$=\int_{-\infty}^{\infty}\exp\left(-j\tfrac{2\pi}{\lambda}(\Delta Z_{b\hat{s}}(t+\tau)-\Delta Z_{b\hat{s}}(t))\sin\alpha_{Rb\hat{s}}\right)$$

$$p(\Delta Z_{b\hat{s}}(t+\tau)-\Delta Z_{b\hat{s}}(t))d(\Delta Z_{b\hat{s}}(t+\tau)-\Delta Z_{b\hat{s}}(t))$$

$$=\exp\left\{-\frac{\tau\zeta^{2}_{\Delta Z_{UAb\hat{s}}}}{2}\left[\tfrac{2\pi}{\lambda}\sin\alpha_{Rb\hat{s}}\right]^{2}\right\} \tag{5.58}$$

Similarly, the expectation $\mathrm{E}_{\Delta Z_{s\hat{b}}(t)}[\cdot]$ in (5.57) becomes

$$\mathrm{E}_{\Delta Z_{s\hat{b}n}(t)}\left[\exp\left(-j\frac{2\pi}{\lambda}(\Delta Z_{s\hat{b}n}(t+\tau)-\Delta Z_{s\hat{b}n}(t))\sin\alpha_{Rs\hat{b}n}\right)\right]$$

$$=\exp\left\{-\frac{\tau\zeta^2_{\Delta Z_{DAs\hat{b}}}}{2}\left[\frac{2\pi}{\lambda}\sin\alpha_{Rs\hat{b}}\right]^2\right\} \tag{5.59}$$

Substituting (5.58) and (5.49) into (5.56), and (5.59) and (5.48) into (5.57), we obtain the ACFs of the UA and DA components, respectively:

$$R^{UA}(\tau)=\frac{\eta_B}{2B(K+1)}\sum_{b=1}^{B}\sum_{\hat{s}=b-1}^{b}\int_{\pi}^{2\pi}\frac{1}{\sqrt{2\pi\sigma^2_{UAb\hat{s}}}}\exp\left(-\frac{(\alpha_{Rb\hat{s}}-\mu_{UAb\hat{s}})^2}{(2\sigma^2_{UAb\hat{s}})}\right)$$
$$\exp\left(j2\pi\tau\left[f_{T\max}\cos(\alpha_{Rb\hat{s}}+\gamma_T)-f_{R\max}\cos(\alpha_{Rb\hat{s}}-\gamma_R)\right]\right)$$
$$\exp\left(-\frac{\tau\zeta^2_{\Delta Z_{UAb\hat{s}}}}{2}\left[\frac{2\pi}{\lambda}\sin\alpha_{Rb\hat{s}}\right]^2\right)d\alpha_{Rb\hat{s}} \tag{5.60}$$

$$R^{DA}(\tau)=\frac{\eta_S}{2S(K+1)}\sum_{s=1}^{S}\sum_{\hat{b}=s-1}^{s}\int_{0}^{\pi}\frac{1}{\sqrt{2\pi\sigma^2_{DAs\hat{b}}}}\exp\left(-\frac{(\alpha_{Rs\hat{b}}-\mu_{DAs\hat{b}})^2}{(2\sigma^2_{DAs\hat{b}})}\right)$$
$$\exp\left(j2\pi\tau\left[f_{T\max}\cos(\alpha_{Rs\hat{b}}+\gamma_T)-f_{R\max}\cos(\alpha_{Rs\hat{b}}-\gamma_R)\right]\right)$$
$$\exp\left(-\frac{\tau\zeta^2_{\Delta Z_{DAs\hat{b}}}}{2}\left[\frac{2\pi}{\lambda}\sin\alpha_{Rs\hat{b}}\right]^2\right)d\alpha_{Rs\hat{b}} \tag{5.61}$$

For small angle spreads, the AoAs $\alpha_{Rs\hat{b}}$ and $\alpha_{Rb\hat{s}}$ are mainly concentrated around the mean AoAs $\mu_{DAs\hat{b}}$ and $\mu_{UAb\hat{s}}$, respectively. Using the first-order Taylor expansion, the AoA angles can be approximated as follows

$$\cos(\alpha_{Rs\hat{b}})\approx\cos(\mu_{DAs\hat{b}})-\sin(\mu_{DAs\hat{b}})(\alpha_{Rs\hat{b}}-\mu_{DAs\hat{b}}) \tag{5.62}$$

$$\sin(\alpha_{Rs\hat{b}})\approx\sin(\mu_{DAs\hat{b}})+\cos(\mu_{DAs\hat{b}})(\alpha_{Rs\hat{b}}-\mu_{DAs\hat{b}}) \tag{5.63}$$

$$\sin(\alpha_{Rs\hat{b}})^2\approx\sin(\mu_{DAs\hat{b}})^2+2\sin(\mu_{DAs\hat{b}})\cos(\mu_{DAs\hat{b}})(\alpha_{Rs\hat{b}}-\mu_{DAs\hat{b}}) \tag{5.64}$$

$$\sin(\alpha_{Rs\hat{b}})^{-1}\approx\sin(\mu_{DAs\hat{b}})^{-1}-\cos(\mu_{DAs\hat{b}})(\alpha_{Rs\hat{b}}-\mu_{DAs\hat{b}})/\sin(\mu_{DAs\hat{b}})^{-2} \tag{5.65}$$

and similar approximations can be applied to the AoAs $\alpha_{Rb\hat{s}}$. Using these trigonometric approximations and the equality $\int e^{j\alpha x}(2\pi\sigma^2)^{-1/2}e^{-x^2/(2\sigma^2)}dx=e^{-\sigma^2\alpha^2/2}$ [19], the ACFs of the UA and DA components can be written as

$$R^{UA}(\tau) = \frac{\eta_B}{2B(K+1)} \sum_{b=1}^{B} \sum_{\hat{s}=b-1}^{b} \exp\left(-\frac{\tau \zeta^2_{\Delta Z_{UAb\hat{s}}}}{2} \left[\frac{2\pi}{\lambda} \sin \mu_{UAb\hat{s}} \right]^2 \right)$$
$$\exp(-j2\pi\tau [f_{\mathrm{Rmax}} \cos(\mu_{UAb\hat{s}} - \gamma_R) - f_{\mathrm{Tmax}} \cos(\mu_{UAb\hat{s}} + \gamma_T)])$$
$$\exp\left\{ -\frac{\sigma^2_{UAb\hat{s}}}{2} \left[2\pi\tau [f_{\mathrm{Rmax}} \sin(\mu_{UAb\hat{s}} - \gamma_R) - f_{\mathrm{Tmax}} \sin(\mu_{UAb\hat{s}} + \gamma_T)] \right. \right.$$
$$\left. \left. + \tau \zeta^2_{\Delta Z_{UAb\hat{s}}} \left[\frac{2\pi}{\lambda} \right]^2 \sin \mu_{UAb\hat{s}} \cos \mu_{UAb\hat{s}} \right]^2 \right\} \quad (5.66)$$

$$R^{DA}(\tau) = \frac{\eta_S}{2S(K+1)} \sum_{s=1}^{S} \sum_{\hat{b}=s-1}^{s} \exp\left(-\frac{\tau \zeta^2_{\Delta Z_{DAs\hat{b}}}}{2} \left[\frac{2\pi}{\lambda} \sin \mu_{DAs\hat{b}} \right]^2 \right)$$
$$\exp\left(j2\pi\tau \left[f_{\mathrm{Tmax}} \cos(\mu_{DAs\hat{b}} + \gamma_T) - f_{\mathrm{Rmax}} \cos(\mu_{DAs\hat{b}} - \gamma_R) \right] \right)$$
$$\exp\left\{ -\frac{\sigma^2_{DAs\hat{b}}}{2} \left[2\pi\tau \left[f_{\mathrm{Rmax}} \sin(\mu_{DAs\hat{b}} - \gamma_R) - f_{\mathrm{Tmax}} \sin(\mu_{DAs\hat{b}} + \gamma_T) \right] \right. \right.$$
$$\left. \left. + \tau \zeta^2_{\Delta Z_{DAs\hat{b}}} \left[\frac{2\pi}{\lambda} \right]^2 \sin \mu_{DAs\hat{b}} \cos \mu_{DAs\hat{b}} \right]^2 \right\} \quad (5.67)$$

From (5.53), (5.66), and (5.67), we can observe that the ACF of a shallow-water acoustic channel follows an *exponential function*. This result is in agreement with the experimentally obtained ACFs [3, 27, 28]. Furthermore, this result differs from the traditional Bessel-shaped ACFs that are typical for radio channels. These differences are due to different propagation mechanisms in SWA and radio channels. To illustrate this exponential behavior, Figure 5.15 plots ACF $R(\tau)$ as a function of time. This curve is obtained assuming that the carrier frequency is f_c = 1.7 kHz and the speed of sound is c = 1500 m/s. The distance between the T_x and R_x is set to

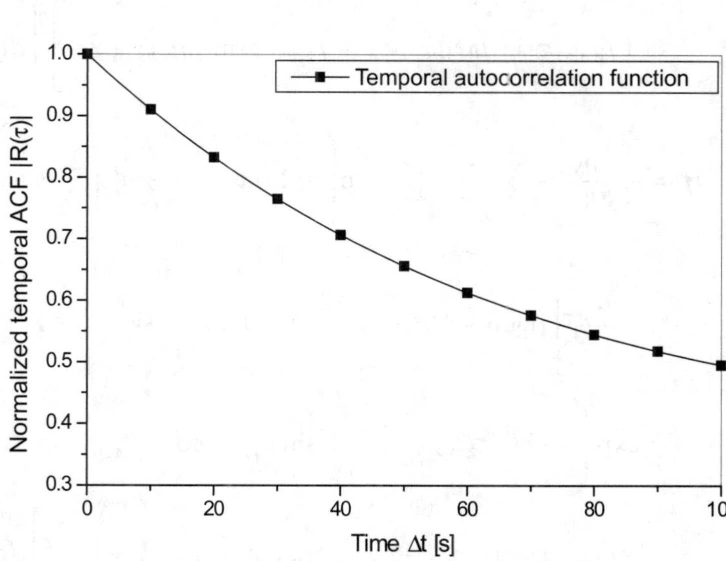

Figure 5.15 The magnitude of temporal autocorrelation function as a function of time.

$R = 2$ km, while the water, T_x, and R_x depths are $h = 100$m, $h = 50$m, and $h_R = 50$m, respectively. It is assumed that the T_x and R_x are stationary (i.e., $v_T = v_R = 0$ m/s). For illustration purposes, the rest of the model parameters are chosen to be: $K = 0.6$, $S = B = 1$, $\zeta_{\Delta Z_{DA10}} = \zeta_{\Delta Z_{DA11}} = \zeta_{\Delta Z_{UA10}} = \zeta_{\Delta Z_{UA11}} = 0.1$, $\eta_S = 0.6$, $\mu_{DA10} = 166°$, $\mu_{DA11} = 163.5°$, $\mu_{UA10} = 191.9°$, $\mu_{UA11} = 192.9°$, $\sigma_{DA10} = 2.5°$, $\sigma_{DA11} = 4°$, $\sigma_{UA10} = 2.4°$, $\sigma_{UA11} = 3°$, $\alpha_{TR} = 180°$.

5.3.3.2 Doppler Power Spectrum

The Doppler power spectrum of the complex channel $h(t)$ is a Fourier ransform of its temporal autocorrelation function $R(\tau)$; that is,

$$S(f) = \int_{-\infty}^{\infty} R(\tau) e^{-j2\pi f \tau} d\tau \tag{5.68}$$

Applying the Fourier transform to (5.52) yields

$$S(f) = S^{LoS}(f) + S^{UA}(f) + S^{DA}(f) \tag{5.69}$$

By calculating the Fourier transform of the ACF in (5.53), we obtain the Doppler power spectrum of the LoS component as follows

$$\begin{aligned} S^{LoS}(f) &= \mathcal{F}_\tau\{R^{LoS}(\tau)\} \\ &= \frac{K}{K+1} \delta\left(f - f_{\text{Rmax}} \cos(\alpha_{TR} - \gamma_R) + f_{\text{Tmax}} \cos(\alpha_{TR} - \gamma_T)\right) \end{aligned} \tag{5.70}$$

The Fourier transforms of the ACFs $R^{UA}(\tau)$ and $R^{DA}(\tau)$ from (5.66) and (5.67) can be written as

$$\begin{aligned} S^{UA}(f) = \frac{\eta_B}{2B(K+1)} \sum_{b=1}^{B} \sum_{\hat{s}=b-1}^{b} \int_{-\infty}^{\infty} \exp\Bigg(-j2\pi f\, \tau - \frac{\tau \zeta^2_{\Delta Z_{UAb\hat{s}}}}{2} \left[\frac{2\pi}{\lambda} \sin \mu_{UAb\hat{s}}\right]^2 \\ - j2\pi\tau \left[f_{\text{Rmax}} \cos(\mu_{UAb\hat{s}} - \gamma_R) - f_{\text{Tmax}} \cos(\mu_{UAb\hat{s}} + \gamma_T) \right] \Bigg) \\ \exp\Bigg\{ -\frac{\sigma^2_{UAb\hat{s}}}{2} \Bigg[\tau \zeta^2_{\Delta Z_{UAb\hat{s}}} \left[\frac{2\pi}{\lambda}\right]^2 \sin \mu_{UAb\hat{s}} \cos \mu_{UAb\hat{s}} \\ + f_{\text{Rmax}} \tau \sin(\mu_{UAb\hat{s}} - \gamma_R) - f_{\text{Tmax}} \tau \sin(\mu_{UAb\hat{s}} + \gamma_T) \Bigg]^2 \Bigg\} d\tau \end{aligned} \tag{5.71}$$

$$\begin{aligned} S^{DA}(f) = \frac{\eta_S}{2S(K+1)} \sum_{s=1}^{S} \sum_{\hat{b}=s-1}^{s} \int_{-\infty}^{\infty} \exp\Bigg(-j2\pi f \tau - \frac{\tau \zeta^2_{\Delta Z_{DAs\hat{b}}}}{2} \left[\frac{2\pi}{\lambda} \sin \mu_{DAs\hat{b}}\right]^2 \\ - j2\pi\tau \left[f_{\text{Rmax}} \cos(\mu_{DAs\hat{b}} - \gamma_R) - f_{\text{Tmax}} \cos(\mu_{DAs\hat{b}} + \gamma_T) \right] \Bigg) \\ \exp\Bigg\{ -\frac{\sigma^2_{DAs\hat{b}}}{2} [\tau \zeta^2_{\Delta Z_{DAs\hat{b}}} \left[\tfrac{2\pi}{\lambda}\right]^2 \sin \mu_{DAs\hat{b}} \cos \mu_{DAs\hat{b}} \\ + f_{\text{Rmax}} \tau \sin(\mu_{DAs\hat{b}} - \gamma_R) - f_{\text{Tmax}} \tau \sin(\mu_{DAs\hat{b}} + \gamma_T)]^2 \Bigg\} d\tau \end{aligned} \tag{5.72}$$

Using the equality $\int_{-\infty}^{\infty} e^{-p^2x^2 \pm qx} dx = \sqrt{\pi} e^{q^2/(4p^2)}/p$ (3.323-2, [29]) to solve the integrals in (5.71) and (5.72), we obtain the Doppler power spectrum of the LoS component as follows:

$$S^{UA}(f) = \frac{\eta_B}{2B(K+1)} \sum_{b=1}^{B} \sum_{\hat{s}=b-1}^{b}$$

$$\exp\left[\frac{(j2\pi f + C_{UAb\hat{s}})^2}{\sigma_{UAb\hat{s}}^4 A_{UAb\hat{s}}^2} \frac{\sqrt{\pi}}{0.5\sigma_{UAb\hat{s}}^2 A_{UAb\hat{s}}} \right] \tag{5.73}$$

$$S^{DA}(f) = \frac{\eta_S}{2S(K+1)} \sum_{s=1}^{S} \sum_{\hat{b}=s-1}^{s}$$

$$\exp\left[\frac{(j2\pi f + C_{DAs\hat{b}})^2}{\sigma_{DAs\hat{b}}^4 A_{DAs\hat{b}}^2} \frac{\sqrt{\pi}}{0.5\sigma_{DAs\hat{b}}^2 A_{DAs\hat{b}}} \right] \tag{5.74}$$

where $A_{UAb\hat{s}}$, $C_{UAb\hat{s}}$, $A_{DAs\hat{b}}$, and $C_{DAs\hat{b}}$ are

$$A_{UAb\hat{s}} = \zeta_{\Delta Z_{UAb\hat{s}}}^2 \left[\frac{2\pi}{\lambda}\right]^2 \sin\mu_{UAb\hat{s}} \cos\mu_{UAb\hat{s}} +$$

$$\frac{2\pi}{\lambda}\left[v_R \sin(\mu_{UAb\hat{s}} - \gamma_R) - v_T \sin(\mu_{UAb\hat{s}} + \gamma_T)\right] \tag{5.75}$$

$$C_{UAb\hat{s}} = \frac{\zeta_{\Delta Z_{UAb\hat{s}}}^2}{2} \left[\frac{2\pi}{\lambda} \sin\mu_{UAb\hat{s}}\right]^2$$

$$+ j\frac{2\pi}{\lambda}\left[v_R \cos(\mu_{UAb\hat{s}} - \gamma_R) - v_T \cos(\mu_{UAb\hat{s}} + \gamma_T)\right] \tag{5.76}$$

$$A_{DAs\hat{b}} = \zeta_{\Delta Z_{DAs\hat{b}}}^2 \left[\frac{2\pi}{\lambda}\right]^2 \sin\mu_{DAs\hat{b}} \cos\mu_{DAs\hat{b}}$$

$$+ \frac{2\pi}{\lambda}\left[v_R \sin(\mu_{DAs\hat{b}} - \gamma_R) - v_T \sin(\mu_{DAs\hat{b}} + \gamma_T)\right] \tag{5.77}$$

$$C_{DAs\hat{b}} = \frac{\zeta_{\Delta Z_{DAs\hat{b}}}^2}{2} \left[\frac{2\pi}{\lambda} \sin\mu_{DAs\hat{b}}\right]^2$$

$$+ j\frac{2\pi}{\lambda}\left[v_R \cos(\mu_{DAs\hat{b}} - \gamma_R) - v_T \cos(\mu_{DAs\hat{b}} + \gamma_T)\right] \tag{5.78}$$

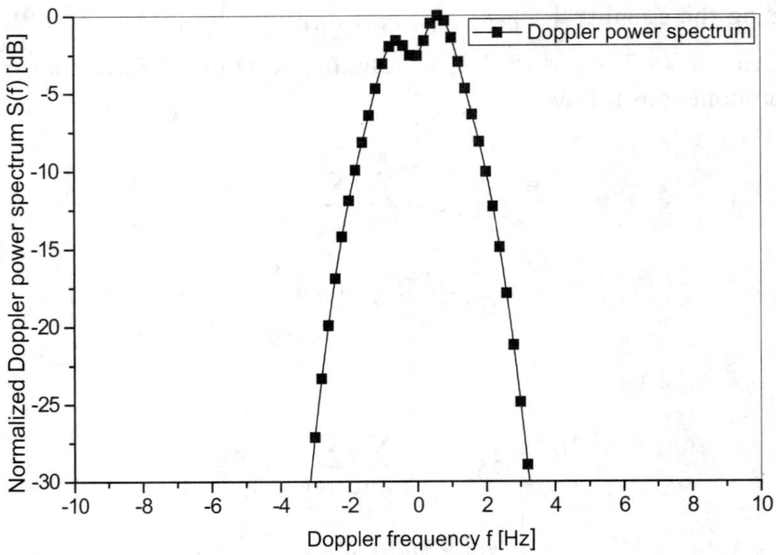

Figure 5.16 The normalized Doppler power spectrum.

Figure 5.16 shows an example of Doppler power spectrum. Isovelocity scattering environment with the carrier frequency f_c = 10 kHz and the speed of sound c = 1,500 m/s is assumed. The distance between the T_x and R_x is set to be R = 2 km. The water, T_x, and R_x depths are chosen to be h = 50m, h_T = 18m, and h_R = 18m, respectively. The T_x and R_x are moving with speeds $v_T = v_R$ = 2 m/s and have orientations $\gamma_T = \gamma_R = 40°$. The rest of the model parameters are chosen to be K = 0.6, $S = B = 1$, $\zeta_{\Delta Z_{DA10}} = \zeta_{\Delta Z_{DA11}} = \zeta_{\Delta Z_{UA10}} = \zeta_{\Delta Z_{UA11}} = 0.09, \mu_{DA10} = 178°, \mu_{DA11} = 177°,$ $\mu_{UA10} = 182°, \mu_{UA11} = 183°, \sigma_{DA10} = 10°, \sigma_{DA11} = 13.6°, \sigma_{UA10} = 10°, \sigma_{UA11} = 13.6°,$ $\alpha_{TR} = 180°$, $\eta_S = 0.4$, and $\eta_B = 1 - \eta_S$. We can observe that the Doppler power spectrum of SWA channels differs from the Doppler power spectrum in F-to-M cellular, A-to-G, and V-to-V radio channels due to different propagation mechanisms.

5.3.3.3 Level Crossing Rate and Average Fade Duration for UV-to-UV Acoustic Channels

The level crossing rate (LCR), $L(R)$, is introduced in Section 2.3.5.1 and defined as the rate at which the signal envelope crosses level R in the positive direction. The LCR can be written as [25]

$$L(R) = \int_0^\infty \dot{\alpha} p_{\alpha,\dot{\alpha}}(R, \dot{\alpha}) d\dot{\alpha} \tag{5.79}$$

where $\alpha = |h(t)|$ is the envelope level, $\dot{\alpha} = |\dot{h}(t)|$ is the envelope slope, and $p_{\alpha,\dot{\alpha}}(R, \dot{\alpha})$ is the joint pdf of the envelope level and the envelope slope.

Here, we will deploy the model described in Section 5.3.2 to calculate the LCR. First, we observe that the envelope $|h(t)|$ of the model described in Section 5.3.2 has a Ricean distribution. Then the envelope slope $|\dot{h}(t)|$ has a Gaussian distribution. Furthermore, the joint pdf $p_{\alpha,\dot{\alpha}}(\alpha, \dot{\alpha})$ cannot be represented as a product of a Rice pdf $p_{|h(t)|}(\alpha)$ and a Gaussian pdf $p_{|\dot{h}(t)|}(\dot{\alpha})$, because the real and imaginary components of the complex faded envelope $h(t)$ are correlated. Hence, the LCR for LoS conditions (Ricean fading, correlated in-phase and quadrature components) can be written as [30]

$$L(R) = \frac{R\sqrt{2\chi}}{\pi^{3/2} b_0} e^{-(R^2+\delta^2)/(2b_0)} \int_0^{\pi/2} \cosh\left(\frac{R\delta}{b_0}\cos\theta\right) \times\left[e^{-(\xi\delta\sin\theta)^2} + \sqrt{\pi}\xi\delta\sin\theta \operatorname{erf}(\xi\delta\sin\theta)\right]d\theta \tag{5.80}$$

where cosh(·) is the hyperbolic cosine function and erf(·) is the error function. The rest of the parameters are $\delta = \sqrt{K/(K+1)}$, $\chi = b_2 - b_1^2 / b_0$, $\xi = (2\pi f^{LoS} - b_1 / b_0)/\sqrt{2\chi}$, and

$$b_0 \triangleq E[h_i(t)^2] = E[h_q(t)^2] \tag{5.81}$$

$$b_1 \triangleq E[h_i(t)\dot{h}_q(t)] = E[h_q(t)\dot{h}_i(t)] \tag{5.82}$$

$$b_2 \triangleq E[\dot{h}_i(t)^2] = E[\dot{h}_q(t)^2] \tag{5.83}$$

where $h_i(t)$ and $h_q(t)$ denote the real and imaginary part of $h(t)$, respectively, and $\dot{h}_i(t)$ and $\dot{h}_q(t)$ are first derivatives of $h_i(t)$ and $h_q(t)$ with respect to time t.

When SWA scattering is characterized using the probability density functions in (5.48)–(5.51), closed-form expressions for b_0, b_1, and b_2 can be obtained by substituting (5.35) into (5.80). Parameter b_0 then becomes

$$2b_0 = 2b_0^{UA} + 2b_0^{DA} = \frac{1}{K+1} \tag{5.84}$$

where b_0^{UA} and b_0^{DA} are, respectively,

$$\begin{aligned} 2b_0^{UA} &= \Re\{R^{UA}(\tau)|_{\tau=0}\} \\ &= \frac{\eta_B}{(K+1)} \end{aligned} \tag{5.85}$$

$$\begin{aligned} 2b_0^{DA} &= \Re\{R^{DA}(\tau)| \, \tau = 0) \\ &= \frac{\eta_S}{(K+1)} \end{aligned} \tag{5.86}$$

and where $\Re\{\cdot\}$ denotes the real part operation. Similarly, parameters b_1 and b_2 can be obtained as follows:

$$b_n = b_n^{UA} + b_n^{DA} \tag{5.87}$$

where $n \in \{1, 2\}$ and b_n^{UA} and b_n^{DA} can be written as

$$b_1^{UA} = \Im\left\{\frac{1}{2}\frac{dR^{UA}(\tau)}{d\tau}|_{\tau=0}\right\}$$
$$= \frac{b_0}{2B}\sum_{b=1}^{B}\sum_{\hat{s}=b-1}^{b} 2\pi\left[f_{\text{Tmax}}\cos(\mu_{UAb\hat{s}} + \gamma_T) - f_{\text{Rmax}}\cos(\mu_{UAb\hat{s}} - \gamma_R)\right] \tag{5.88}$$

$$b_1^{DA} = \Im\left\{\frac{1}{2}\frac{dR^{DA}(\tau)}{d\tau}|_{\tau=0}\right\}$$
$$= \frac{b_0}{2S}\sum_{s=1}^{S}\sum_{\hat{b}=s-1}^{s} 2\pi\left[f_{\text{Tmax}}\cos(\mu_{DAs\hat{b}} + \gamma_T) - f_{\text{Rmax}}\cos(\mu_{DAs\hat{b}} - \gamma_R)\right] \tag{5.89}$$

$$b_2^{UA} = -\Re\left\{\frac{1}{2}\frac{d^2R^{UA}(\tau)}{d\tau^2}|_{\tau=0}\right\}$$
$$= \frac{b_0}{2B}\sum_{b=1}^{B}\sum_{\hat{s}=b-1}^{b} b_{UAb\hat{s}}^2 - a_{UAb\hat{s}}^2 + 2c_{UAb\hat{s}}^2 \tag{5.90}$$

$$b_2^{DA} = -\Re\left\{\frac{1}{2}\frac{d^2R^{DA}(\tau)}{d\tau^2}|_{\tau=0}\right\}$$
$$= \frac{b_0}{2S}\sum_{s=1}^{S}\sum_{\hat{b}=s-1}^{s} b_{DAs\hat{b}}^2 - a_{DAs\hat{b}}^2 + 2c_{DAs\hat{b}}^2 \tag{5.91}$$

and where $\Im\{\cdot\}$ denotes the imaginary part operation, and $a_{UAb\hat{s}}$, $b_{UAb\hat{s}}$, $c_{UAb\hat{s}}$, $a_{DAs\hat{b}}$, $b_{DAs\hat{b}}$, and $c_{DAs\hat{b}}$ are, respectively,

$$a_{UAb\hat{s}} = \frac{\zeta_{\Delta Z_{UAb\hat{s}}}^2}{2}\left[\frac{2\pi}{\lambda}\sin\mu_{UAb\hat{s}}\right]^2 \tag{5.92}$$

$$b_{UAb\hat{s}} = 2\pi\left[f_{\text{Rmax}}\cos(\mu_{UAb\hat{s}} - \gamma_R) - f_{\text{Tmax}}\cos(\mu_{UAb\hat{s}} + \gamma_T)\right] \tag{5.93}$$

$$c_{UAb\hat{s}} = \frac{\sigma_{UAb\hat{s}}^2}{2}\Bigg[2\pi\left[f_{\text{Rmax}}\sin(\mu_{UAb\hat{s}} - \gamma_R) - f_{\text{Tmax}}\sin(\mu_{UAb\hat{s}} + \gamma_T)\right]$$
$$+\zeta_{\Delta Z_{UAb\hat{s}}}^2\left[\frac{2\pi}{\lambda}\right]^2 \sin\mu_{UAb\hat{s}}\cos\mu_{UAb\hat{s}}\Bigg]^2 \tag{5.94}$$

$$a_{DAs\hat{b}} = \frac{\zeta_{\Delta Z_{DAs\hat{b}}}^2}{2}\left[\frac{2\pi}{\lambda}\sin\mu_{DAs\hat{b}}\right]^2 \tag{5.95}$$

$$b_{DAs\hat{b}} = 2\pi\left[f_{\text{Rmax}}\cos(\mu_{DAs\hat{b}} - \gamma_R) - f_{\text{Tmax}}\cos(\mu_{DAs\hat{b}} + \gamma_T)\right] \tag{5.96}$$

$$c_{DAs\hat{b}} = \frac{\sigma^2_{DAs\hat{b}}}{2}[2\pi[f_{R\max}\sin(\mu_{DAs\hat{b}} - \gamma_R) - f_{T\max}\sin(\mu_{DAs\hat{b}} + \gamma_T)] + \zeta^2_{\Delta Z_{DAs\hat{b}}}\left[\frac{2\pi}{\lambda}\right]^2 \sin\mu_{DAs\hat{b}}\cos\mu_{DAs\hat{b}}]^2 \tag{5.97}$$

The average fade duration (AFD) is the average time that the envelope spends below level R. When a LoS component is present, the AFD can be written as [25]

$$T(R) = \frac{1 - Q(\sqrt{2K}, \sqrt{2(K+1)R^2})}{L(R)} \tag{5.98}$$

where $Q(a, b)$ is the Marcum Q function.

In the presence of shadowing, the envelope $|h(t)|$ can be modeled using the extended Suzuki process (i.e., as a product of the Ricean process with cross-correlated components and the log-normal process). The LCR for LoS conditions with shadowing can thus be written as [30]

$$L(R) = \frac{R\sqrt{2\chi}}{\pi^{3/2}b_0}\int_0^\infty \frac{F(R,l)}{l}\frac{e^{-\frac{(\ln l)^2}{2\sigma^2}}}{\sqrt{2\pi}\sigma l}e^{-((R/l)^2+\delta^2)/(2b_0)}\int_C^{\pi/2}\cosh\left(\frac{R\delta}{lb_0}\cos\theta\right) \times\left[e^{-\left(\frac{\xi\delta\sin\theta}{F(R,l)}\right)^2} + \sqrt{\pi}\xi\delta\frac{\sin\theta}{F(R,l)}\operatorname{erf}\left(\xi\delta\frac{\sin\theta}{F(R,l)}\right)\right]d\theta dl \tag{5.99}$$

where $\cosh(\cdot)$ is the hyperbolic cosine function, $\operatorname{erf}(\cdot)$ is the error function, and σ is the variance of shadow process. The rest of the parameters are $\delta = \sqrt{K/(K+1)}$, $\chi = b_2 - b_1^2/b_0$, $\xi = (2\pi f^{LoS} - b_1/b_0)/\sqrt{2\chi}$, $F(R,l) = \sqrt{1+(\gamma/\chi)(\sigma R/l)^2}$, $\gamma = 2/L_c^2$, and L_c is the correlation distance.

To verify these models for LCR, we compare the theoretical results with those obtained from measured data. The experimental data was collected during the NRL ACOMMS09 experiment described in Section 5.2.2. The channel measurements are collected at f_c = 17 kHz and the speed of sound is c = 1440 m/s. The distance between the T_x and R_x is R = 1.5 km. The water, T_x, and R_x depths are h = 80m, h_T = 41m, and h_R = 45m, respectively. It is assumed that the T_x and R_x are relatively stationary—that is, slightly moving with waves (i.e., $\gamma_T = \gamma_R = 90°$).

Figures 5.17 and 5.18 compare the analytical LCRs with and without shadowing with the measured LCRs with and without shadowing, respectively.

The analytical curves in Figures 5.17 and 5.18 are obtained with the parameters K = 0.56, $S = B = 1$, $v_T = v_R$ = 0.001m/s, μ_{DA10} = 168.5°, μ_{DA11}= 167.5°, μ_{UA10} = 185.5°, μ_{UA11} = 186°, σ_{DA10} = 1.4°, σ_{DA11} = 1.9°, σ_{UA10} = 2.1°, σ_{UA11} = 2.6°, $\zeta_{\Delta Z_{DA10}} = \zeta_{\Delta Z_{DA11}}$ = 0.9, $\zeta_{\Delta Z_{UA10}} = \zeta_{\Delta Z_{UA11}}$ = 1.6, γ = 0.621, σ = 3.55, η_S = 0.58, and $\eta_B = 1 - \eta_S$. The results show good match between the measured and simulated results.

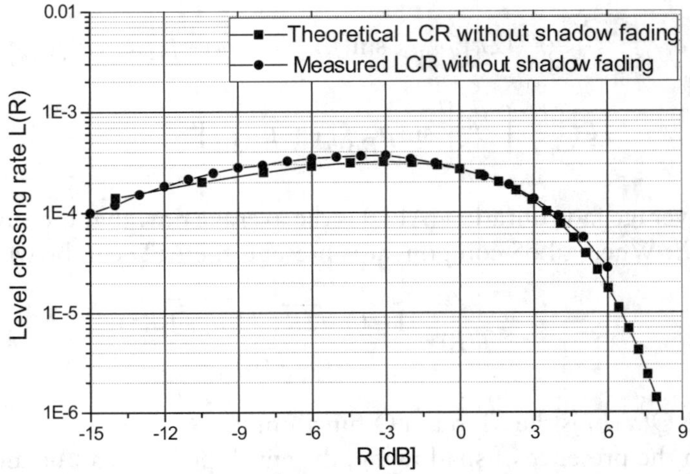

Figure 5.17 Measured and simulated level crossing rates *without* shadow fading.

5.3.4 Two-Dimensional Model for MIMO UV-to-UV Frequency-Flat Acoustic Fading Channels

The use of multiple hydrophones together with MIMO diversity techniques can improve the data rate or spectral efficiency of underwater acoustic communications. To model such a propagation environment, the model described in Section 5.3.2 is extended in [21] to include the effect of spatial diversity.

Here, we consider a medium- or long-range MIMO communication system with L_t transmit and L_r receive transducers. Like in Section 5.3.2, the propagation occurs in a shallow-water environment with constant sound speed (i.e., sound energy propagates along the plane waves) and the propagation is characterized with either LoS or NLoS conditions between the T_x and the R_x. The MIMO M-to-M SWA channel can be described by an $L_r \times L_t$ matrix $\mathbf{H}(t) = [h_{ij}(t)]_{L_r \times L_t}$ of channel impulse responses.

Figure 5.19 shows a SWA channel with $L_t = L_r = 2$ transducer elements. This elementary 2×2 transducer configuration will be used later to construct uniform linear arrays with an arbitrary number of transducer elements. As in Section 5.3.2,

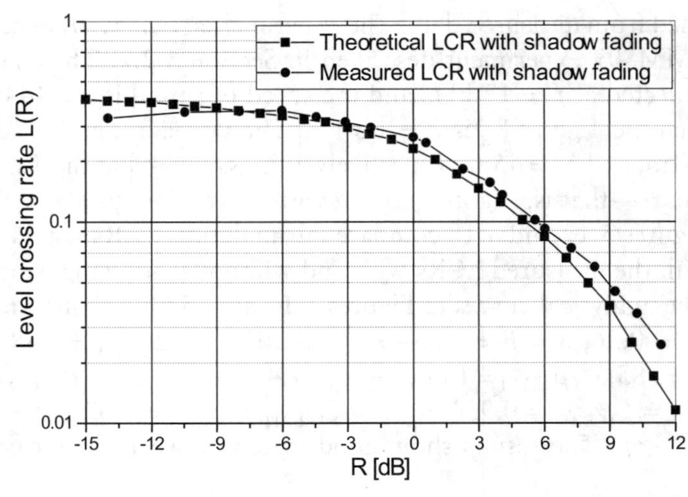

Figure 5.18 Measured and simulated level crossing rates *with* shadow fading.

the depths of T_x and R_x are denoted by h_T and h_R, respectively, while the orientations of T_x and R_x transducer arrays in the x–z plane are denoted by θ_T and θ_R, respectively, relative to the x-axis (as shown in Figure 5.20). The spacing between two adjacent transducer elements at the T_x and R_x is d_T and d_R, respectively. Finally, the T_x and R_x are moving with constant speeds v_T and v_R in directions described by angles γ_T and γ_R in the x-z plane (relative to the x-axis), respectively.

The symbols $\alpha^{(p)}_{Ts\hat{b}n}$ and $\alpha^{(p)}_{Tb\hat{s}m}$ are the angles of departure (AoDs) of micro-eigenrays that start from $T_x^{(p)}$ and impinge on scatterers $S^{(S)}_{s\hat{b}n}$ and $S^{(B)}_{b\hat{s}m}$, respectively, while $\alpha^{(q)}_{Rs\hat{b}n}$ and $\alpha^{(q)}_{Rb\hat{s}m}$ are the angles of arrival (AoAs) of micro-eigenrays scattered from $S^{(S)}_{s\hat{b}n}$ and $S^{(B)}_{b\hat{s}m}$, respectively, and arriving at $R_x^{(q)}$.

Assuming small angle spreads, $d_T << \min(h_T, h - h_T)$, and $d_R << \min(h_R, h - h_R)$, we can approximate the AoDs and AoAs in Figure 5.19 as $\alpha^{(p)}_{Ts\hat{b}n} \approx \alpha^{(\tilde{p})}_{Ts\hat{b}n} \approx \alpha_{Ts\hat{b}n}$, $\alpha^{(p)}_{Tb\hat{s}m} \approx \alpha^{(\tilde{p})}_{Tb\hat{s}m} \approx \alpha_{Tb\hat{s}m}$, $\alpha^{(q)}_{Rs\hat{b}n} \approx \alpha^{(\tilde{q})}_{Rs\hat{b}n} \approx \alpha_{Rs\hat{b}n}$, and $\alpha^{(q)}_{Rb\hat{s}m} \approx \alpha^{(\tilde{q})}_{Rb\hat{s}m} \approx \alpha_{Rb\hat{s}m}$, where $\alpha_{Ts\hat{b}n}$, $\alpha_{Tb\hat{s}m}$, $\alpha_{Rs\hat{b}n}$, and $\alpha_{Rb\hat{s}m}$ are defined in Figure 5.19. Finally, the symbol α_{TR} denotes the AoA of the LoS ray. For ease of reference, Figure 5.20 details the geometry of the LoS path, as well as the geometry of single-bounced surface and single-bounced bottom micro-eigenrays (i.e., $s = 1, \hat{b} = 0, b = 1, \hat{s} = 0$) scattered from the $S^{(S)}_{s=1,\hat{b}=0,n}$-th and $S^{(B)}_{b=1,\hat{s}=0,m}$-th micro-scatterers, respectively.

The horizontal spacing between the T_x and R_x is denoted by R, and the water depth is denoted by h. For medium- and long-range shallow-water communication, we can assume that the depths h, h_T, and h_R are much smaller than the distance R. Symbols $\varepsilon^{(p)}_{Ts\hat{b}n}$ and $\varepsilon^{(p)}_{Tb\hat{s}m}$ denote distances from $T_x^{(p)}$ to $S^{(S)}_{s\hat{b}n}$ and $S^{(B)}_{b\hat{s}m}$, respectively, where $S^{(S)}_{s\hat{b}n}$ and $S^{(B)}_{b\hat{s}m}$ denote the nth and mth micro-scatterers located around the macro-scatterers $S^{(S)}_{s\hat{b}}$ and $S^{(B)}_{b\hat{s}}$ at the surface and bottom, respectively, for $1 \le n \le N_{s\hat{b}}$

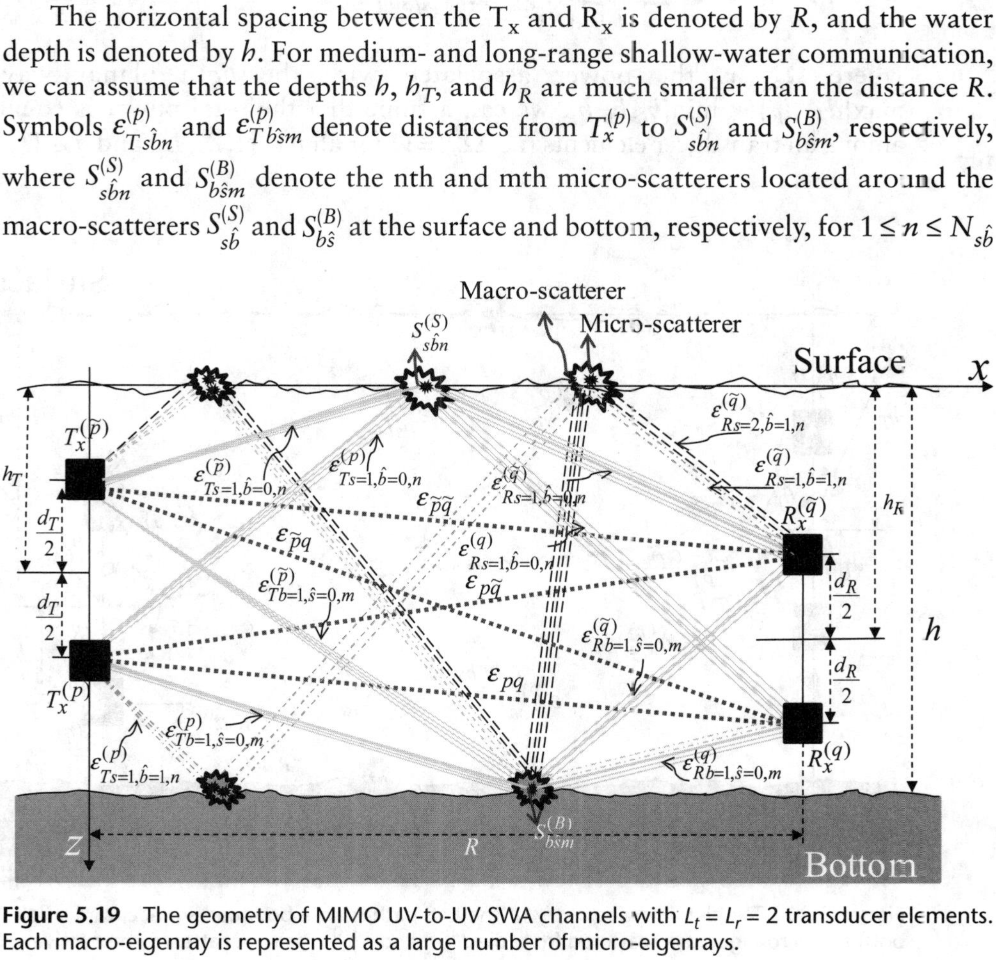

Figure 5.19 The geometry of MIMO UV-to-UV SWA channels with $L_t = L_r = 2$ transducer elements. Each macro-eigenray is represented as a large number of micro-eigenrays.

and $1 \le m \le M_{b\hat{s}}$, as shown in Figure 5.19. Similarly, symbols $\varepsilon_{Ts\hat{b}n}^{(\tilde{p})}$, $\varepsilon_{Tb\hat{s}m}^{(\tilde{p})}$, $\varepsilon_{Rs\hat{b}n}^{(q)}$, $\varepsilon_{Rb\hat{s}m}^{(q)}$, $\varepsilon_{Rs\hat{b}n}^{(\tilde{q})}$, $\varepsilon_{Rb\hat{s}m}^{(\tilde{q})}$, ε_{pq}, $\varepsilon_{\tilde{p}q}$, $\varepsilon_{p\tilde{q}}$, and $\varepsilon_{\tilde{p}\tilde{q}}$ denote distances $T_x^{(\tilde{p})}$–$S_{s\hat{b}n}^{(S)}$, $T_x^{(\tilde{p})}$–$S_{b\hat{s}m}^{(B)}$, $S_{s\hat{b}n}^{(S)}$–$R_x^{(q)}$, $S_{b\hat{s}m}^{(B)}$–$R_x^{(q)}$, $S_{s\hat{b}n}^{(S)}$–$R_x^{(\tilde{q})}$, $S_{b\hat{s}m}^{(B)}$–$R_x^{(\tilde{q})}$, $T_x^{(p)}$–$R_x^{(q)}$, $T_x^{(\tilde{p})}$–$R_x^{(q)}$, $T_x^{(p)}$–$R_x^{(\tilde{q})}$, and $T_x^{(\tilde{p})}$–$R_x^{(\tilde{q})}$, respectively. Finally, η_{TR} denotes the distance $T_x - R_x$.

Like in Section 5.3.2, the complex faded envelope under frequency-flat fading can be written as a superposition of LoS, upward-arriving (UA), and downward-arriving (DA) rays; that is,

$$h_{pq}(t) = h_{pq}^{LoS}(t) + h_{pq}^{UA}(t) + h_{pq}^{DA}(t) \tag{5.100}$$

but the respective components are now

$$h_{pq}^{LoS}(t) = \sqrt{\Omega_{pq}}\, a_{pq}^{LoS} e^{\left(j2\pi f_{pq}^{LoS} t - j\frac{2\pi}{\lambda} d_{pq}^{LoS}\right)} \tag{5.101}$$

$$h_{pq}^{UA}(t) = \sqrt{\Omega_{pq}} \sum_{b=1}^{B} \sum_{\hat{s}=b-1}^{b} \lim_{M_{b\hat{s}}\to\infty} \sum_{m=1}^{M_{b\hat{s}}} a_{pq,b\hat{s}m}^{UA} \cdot e^{j\left(2\pi f_{b\hat{s}m}^{UA} t - j\frac{2\pi}{\lambda} d_{pq,b\hat{s}m}^{UA} + \phi_{pq,b\hat{s}m}^{UA}\right)} \tag{5.102}$$

$$h_{pq}^{DA}(t) = \sqrt{\Omega_{pq}} \sum_{s=1}^{S} \sum_{\hat{b}=s-1}^{s} \lim_{N_{s\hat{b}}\to\infty} \sum_{n=1}^{N_{s\hat{b}}} a_{pq,s\hat{b}n}^{DA} \cdot e^{j\left(2\pi f_{s\hat{b}n}^{DA} t - j\frac{2\pi}{\lambda} d_{pq,s\hat{b}n}^{DA} + \phi_{pq,s\hat{b}n}^{DA}\right)} \tag{5.103}$$

where Ω_{pq} is the power associated with the pqth planar wave. For $\max\{d_T, d_R\} \ll \min\{h_t, h_r, h\}$, we can assume that the total power is equally split among all transducer elements (i.e, $\Omega_{pq} = \Omega$ for all $p \in \{1,\ldots, L_t\}$ and $q \in \{1,\ldots, L_r\}$).

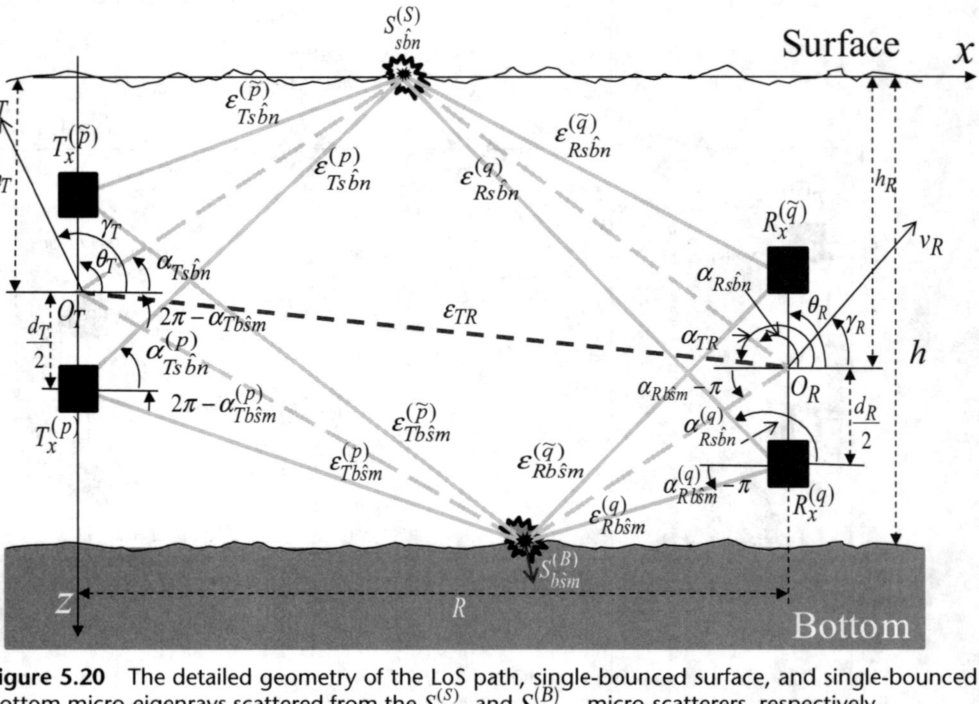

Figure 5.20 The detailed geometry of the LoS path, single-bounced surface, and single-bounced bottom micro-eigenrays scattered from the $S_{s,\hat{b},n}^{(S)}$ and $S_{b,\hat{s},m}^{(B)}$ micro-scatterers, respectively.

The channel gains, Doppler shift frequencies, phases, and path lengths in this model can be calculated as follows:

- *Channel gains.* The channel gains in this model are normalized such that the ensemble averages of the $a^{UA}_{pq,b\hat{s}m}$s and $a^{DA}_{pq,s\hat{b}n}$s are equal to 1. In addition, we introduce a set of parameters K_{pq} that describe how the total signal power between the pth transmit and the qth receive hydrophone is split between the LoS and NLoS components: between the pth transmit and qth receive hydrophone the power allocated to the LoS component is $K_{pq}/(1+K_{pq})$, while the total power allocated to the NLoS components is $1/(1+K_{pq})$. For $\max\{d_T, d_R\} \ll \min\{h_t, h_r, h\}$, we can assume that the Ricean fading factor is the same between all transducer elements (i.e., $K_{pq} = K$ for all $p \in \{1,\ldots,L_t\}$ and $q \in \{1,\ldots,L_r\}$). Furthermore, it is assumed that the relative power allocated to the UA and DA components is respectively η_B and η_S such that $\eta_B + \eta_S = 1$. These parameters have to be either set during simulations or estimated from measurements. Having all this in mind, the channel gains can be written as follows:

$$a^{LoS}_{pq} = \sqrt{\frac{K}{K+1}} \tag{5.104}$$

$$a^{UA}_{pq,b\hat{s}m} = \sqrt{\frac{1}{K+1}}\sqrt{\frac{\eta_B}{2B}}\sqrt{\frac{1}{M_{b\hat{s}}}} \tag{5.105}$$

$$a^{DA}_{pq,s\hat{b}n} = \sqrt{\frac{1}{K+1}}\sqrt{\frac{\eta_S}{2S}}\sqrt{\frac{1}{N_{s\hat{b}}}} \tag{5.106}$$

- *Doppler shifts.* As in Section 5.3.2, the Doppler shifts are

$$f^{LoS}_{pq} = f_{T\max}\cos(\alpha_{TR} + \pi - \gamma_T) + f_{R\max}\cos(\alpha_{TR} - \gamma_R) \tag{5.107}$$

$$f^{UA}_{b\hat{s}m} = \frac{1}{\lambda t}\Delta Z_{b\hat{s}m}(t)\sin\alpha_{Rb\hat{s}m} + f_{T\max}\cos(\alpha_{Tb\hat{s}m} - \gamma_T) + f_{R\max}\cos(\alpha_{Rb\hat{s}m} - \gamma_R) \tag{5.108}$$

$$f^{DA}_{s\hat{b}n} = \frac{1}{\lambda t}\Delta Z_{s\hat{b}n}(t)\sin\alpha_{Rs\hat{b}n} + f_{T\max}\cos(\alpha_{Ts\hat{b}n} - \gamma_T) + f_{R\max}\cos(\alpha_{Rs\hat{b}n} - \gamma_R) \tag{5.109}$$

- *Phases.* As in Section 5.3.2, we assume that the phases $\phi_{pq,s\hat{b}n}$ and $\phi_{pq,b\hat{s}m}$ are independent random variables uniformly distributed on the interval $[-\pi,\pi)$ and independent from any other random variable.
- *Path lengths.* From Figure 5.19, path lengths of the LoS, UA, and DA waves can be written as

$$d^{LoS}_{pq} = \varepsilon_{pq} \tag{5.110}$$

$$d^{UA}_{pq,b\hat{s}m} = \varepsilon^{(p)}_{Tb\hat{s}m} - \varepsilon^{UA}_{Tb\hat{s}} + \varepsilon^{(q)}_{Rb\hat{s}m} - \varepsilon^{UA}_{Rb\hat{s}} \tag{5.111}$$

$$d^{DA}_{pq,s\hat{b}n} = \varepsilon^{(p)}_{Ts\hat{b}n} - \varepsilon^{DA}_{Ts\hat{b}} + \varepsilon^{(q)}_{Rs\hat{b}n} - \varepsilon^{DA}_{Rs\hat{b}} \tag{5.112}$$

For shallow water acoustic propagation, we can assume that the water depths h_T, h_R, and h are much smaller than the distance R between the T_x and R_x—that is, $\max\{h_t, h_r, h\} \ll R$ (local scattering condition). Under this assumption, the distances $\varepsilon^{(p)}_{Ts\hat{b}n}$, $\varepsilon^{(p)}_{Tb\hat{s}m}$, $\varepsilon^{(q)}_{Rs\hat{b}n}$, $\varepsilon^{(q)}_{Rb\hat{s}m}$, and ε_{pq} can be approximated as follows [23]:

$$\varepsilon^{(p)}_{Ts\hat{b}n} \approx \frac{h_T}{\sin \alpha_{Rs\hat{b}n}} + \frac{L_t + 1 - 2p}{2} d_T \cos(\alpha_{Rs\hat{b}n} + \theta_T) \tag{5.113}$$

$$\varepsilon^{(p)}_{Tb\hat{s}m} \approx -\frac{h - h_T}{\sin \alpha_{Rb\hat{s}m}} + \frac{L_t + 1 - 2p}{2} d_T \cos(\alpha_{Rb\hat{s}m} + \theta_T) \tag{5.114}$$

$$\varepsilon^{(q)}_{Rs\hat{b}n} \approx \frac{h_R}{\sin \alpha_{Rs\hat{b}n}} - \frac{L_r + 1 - 2q}{2} d_R \cos(\alpha_{Rs\hat{b}n} - \theta_R) \tag{5.115}$$

$$\varepsilon^{(q)}_{Rb\hat{s}m} \approx -\frac{h - h_R}{\sin \alpha_{Rb\hat{s}m}} - \frac{L_r + 1 - 2q}{2} d_R \cos(\alpha_{Rb\hat{s}m} - \theta_R) \tag{5.116}$$

$$\begin{aligned} \varepsilon_{pq} \approx &\frac{L_t + 1 - 2p}{2} d_T \cos(\alpha_{TR} - \theta_T) \\ &- \frac{L_r + 1 - 2p}{2} d_R \cos(\alpha_{TR} - \theta_R) \end{aligned} \tag{5.117}$$

where p and q take values from the sets $p \in \{1,\ldots,L_t\}$ and $q \in \{1,\ldots,L_r\}$, respectively. Similarly, $\varepsilon^{DA}_{Ts\hat{b}}$, $\varepsilon^{DA}_{Rs\hat{b}}$, $\varepsilon^{UA}_{Tb\hat{s}}$, and $\varepsilon^{UA}_{Rb\hat{s}}$ can be written as

$$\varepsilon^{DA}_{Ts\hat{b}} = \frac{h_T}{\sin \mu_{Das\hat{b}}} \tag{5.118}$$

$$\varepsilon^{DA}_{Rs\hat{b}} = \frac{h_R}{\sin \mu_{DAs\hat{b}}} \tag{5.119}$$

$$\varepsilon^{UA}_{Tb\hat{s}} = -\frac{h - h_T}{\sin \mu_{UAb\hat{s}}} \tag{5.120}$$

$$\varepsilon^{UA}_{Rb\hat{s}} = -\frac{h - h_R}{\sin \mu_{UAb\hat{s}}} \tag{5.121}$$

where the mean AoA angles $\mu_{DAs\hat{b}}$ and $\mu_{UAb\hat{s}}$ are as depicted in Figure 5.12.

5.3.5 Spatial-Temporal Fading Characteristics of UV-to-UV Frequency-Flat MIMO Acoustic Fading Channels

Using the model described in Section 5.3.4, we can now derive the key temporal and spatial characteristics of UV-to-UV frequency-flat MIMO acoustic fading channels.

5.3.5.1 Space-Time Correlation Function

Assuming a shallow water acoustic scattering environment characterized by Gaussian pdfs in (5.48)–(5.51), we will now derive the space-time correlation function (ST-CF) of the complex faded envelope in (5.100).

The normalized space-time correlation function between two complex faded envelopes $h_{pq}(t)$ and $h_{\tilde{p}\tilde{q}}(t)$ is defined as

$$R_{pq,\tilde{p}\tilde{q}}(d_T,d_R,\tau)=\frac{E[h_{pq}(t)h_{\tilde{p}\tilde{q}}(t+\tau)^*]}{\sqrt{\text{Var}[h_{pq}(t)]\text{Var}[h_{\tilde{p}\tilde{q}}(t+\tau)^*]}} \tag{5.122}$$

where $(\cdot)^*$ denotes complex conjugate operation, E[·] is the statistical expectation operator, and Var[·] is the statistical variance operator. Since $h_{pq}^{UA}(t)$ and $h_{pq}^{DA}(t)$ are independent complex Gaussian random processes with zero means, (5.122) can be simplified to

$$R_{pq,\tilde{p}\tilde{q}}(d_T,d_R,\tau)=R_{pq,\tilde{p}\tilde{q}}^{LoS}(d_T,d_R,\tau)+R_{pq,\tilde{p}\tilde{q}}^{UA}(d_T,d_R,\tau)+R_{pq,\tilde{p}\tilde{q}}^{DA}(d_T,d_R,\tau) \tag{5.123}$$

where $R_{pq,\tilde{p}\tilde{q}}^{LoS}(d_T,d_R,\tau)$, $R_{pq,\tilde{p}\tilde{q}}^{UA}(d_T,d_R,\tau)$, and $R_{pq,\tilde{p}\tilde{q}}^{DA}(d_T,d_R,\tau)$ denote the normalized ST-CFs of the LoS, upward-arriving, and downward-arriving components, respectively.

Following a line of reasoning similar to Section 5.3.3, the space-time correlation functions of the LoS, UA, and DA components can be obtained as follows:

$$\begin{aligned}R^{LoS}(d_T,d_R,\tau)&=\frac{E\left[h_{pq}^{LoS}(t)h_{pq}^{LoS}(t+\tau)^*\right]}{\sqrt{\text{Var}[h_{pq}^{LoS}(t)]\text{Var}[h_{pq}^{LoS}(t+\tau)^*]}}\\&=\frac{K}{K+1}e^{j2\pi\tau[f_{T\max}\cos(\alpha_{TR}-\gamma_T)-f_{R\max}\cos(\alpha_{TR}-\gamma_R)]}\\&\quad\times e^{j\frac{2\pi}{\lambda}[(p-\tilde{p})d_T\cos(\alpha_{TR}-\theta_T)-(q-\tilde{q})d_R\cos(\alpha_{TR}-\theta_R)]}\end{aligned} \tag{5.124}$$

$$\begin{aligned}R^{UA}(d_T,d_R,\tau)&=\frac{E[h_{pq}^{UA}(t)h_{pq}^{UA}(t+\tau)^*]}{\sqrt{\text{Var}[h_{pq}^{UA}(t)]\text{Var}[h_{pq}^{UA}(t+\tau)^*]}}\\&=\frac{\eta_B}{2B(K+1)}\sum_{b=1}^{B}\sum_{\hat{s}=b-1}^{b}\exp\left(-\frac{\tau\zeta_{\Delta Z_{UAb\hat{s}}}^2}{2}\left[\frac{2\pi}{\lambda}\sin\mu_{UAb\hat{s}}\right]^2\right)\\&\quad\times\exp\left(j\frac{2\pi}{\lambda}[(p-\tilde{p})d_T\cos(\mu_{UAb\hat{s}}+\theta_T)-(q-\tilde{q})d_R\cos(\mu_{UAb\hat{s}}-\theta_R)]\right)\\&\quad\times\exp\left(-j2\pi\tau[f_{R\max}\cos(\mu_{UAb\hat{s}}-\gamma_R)-f_{T\max}\cos(\mu_{UAb\hat{s}}+\gamma_T)]\right)\\&\quad\times\exp\left\{-\frac{\sigma_{UAb\hat{s}}^2}{2}\left[\frac{2\pi}{\lambda}[(q-\tilde{q})d_R\sin(\mu_{UAb\hat{s}}-\theta_R)-(p-\tilde{p})d_T\sin(\mu_{UAb\hat{s}}+\theta_T)]\right.\right.\\&\quad+2\pi\tau[f_{R\max}\sin(\mu_{UAb\hat{s}}-\gamma_R)-f_{T\max}\sin(\mu_{UAb\hat{s}}+\gamma_T)]\\&\quad\left.\left.+\tau\zeta_{\Delta Z_{UAb\hat{s}}}^2\left[\frac{2\pi}{\lambda}\right]^2\sin\mu_{UAb\hat{s}}\cos\mu_{UAb\hat{s}}\right]^2\right\}\end{aligned} \tag{5.125}$$

$$R^{DA}(d_T,d_R,\tau)=\frac{E[h_{pq}^{DA}(t)h_{pq}^{DA}(t+\tau)^*]}{\sqrt{\mathrm{Var}[h_{pq}^{DA}(t)]\mathrm{Var}[h_{pq}^{DA}(t+\tau)^*]}}$$

$$=\frac{\eta_S}{2S(K+1)}\sum_{s=1}^{S}\sum_{\hat{b}=s-1}^{s}\exp\left(-\frac{\tau\zeta_{\Delta Z_{DAs\hat{b}}}^2}{2}\left[\frac{2\pi}{\lambda}\sin\mu_{DAs\hat{b}}\right]^2\right)$$

$$\times\exp\left(j\frac{2\pi}{\lambda}\left[(q-\tilde{q})d_R\cos(\mu_{DAs\hat{b}}-\theta_R)-(p-\tilde{p})d_T\cos(\mu_{DAs\hat{b}}+\theta_T)\right]\right)$$

$$\times\exp\left(j2\pi\tau\left[f_{\mathrm{Tmax}}\cos(\mu_{DAs\hat{b}}+\gamma_T)-f_{\mathrm{Rmax}}\cos(\mu_{DAs\hat{b}}-\gamma_R)\right]\right)$$

$$\times\exp\left\{-\frac{\sigma_{DAs\hat{b}}^2}{2}\left[\frac{2\pi}{\lambda}\left[(q-\tilde{q})d_R\sin(\mu_{DAs\hat{b}}-\theta_R)-(p-\tilde{p})d_T\sin(\mu_{DAs\hat{b}}+\theta_T)\right]\right.\right.$$

$$+2\pi\tau\left[f_{\mathrm{Rmax}}\sin(\mu_{DAs\hat{b}}-\gamma_R)-f_{\mathrm{Rmax}}\sin(\mu_{DAs\hat{b}}+\gamma_T)\right]$$

$$\left.\left.+\tau\zeta_{\Delta Z_{DAs\hat{b}}}^2\left[\frac{2\pi}{\lambda}\right]^2\sin\mu_{DAs\hat{b}}\cos\mu_{DAs\hat{b}}\right]^2\right\} \quad (5.126)$$

5.3.5.2 Space-Doppler Power Spectral Density

The Doppler power spectrum can be obtained by taking the Fourier transform of the ST-CF in (5.123) with respect to time lag τ, which essentially leads to a space-Doppler power spectral density (sD-psd). Applying (5.68) to (5.123) yields the following results [23]:

$$S_{pq,\tilde{p}\tilde{q}}(d_T,d_R,f)=S_{pq,\tilde{p}\tilde{q}}^{LoS}(d_T,d_R,f)+S_{pq,\tilde{p}\tilde{q}}^{UA}(d_T,d_R,f)+S_{pq,\tilde{p}\tilde{q}}^{DA}(d_T,d_R,f) \quad (5.127)$$

where $S_{pq,\tilde{p}\tilde{q}}^{LoS}(d_T,d_R,f)$, $S_{pq,\tilde{p}\tilde{q}}^{UA}(d_T,d_R,f)$, and $S_{pq,\tilde{p}\tilde{q}}^{DA}(d_T,d_R,f)$ denote the normalized sD-psds of the LoS, upward-arriving, and downward-arriving components, respectively.

By calculating the Fourier transformation of the space-time correlation function in (5.124), we obtain the space-Doppler power spectral density of the LoS component:

$$S_{pq,\tilde{p}\tilde{q}}^{LoS}(d_T,d_R,f)=\frac{K}{K+1}$$

$$\exp\left(j\frac{2\pi}{\lambda}\left[(p-\tilde{p})d_T\cos(\alpha_{TR}-\theta_T)-(q-\tilde{q})d_R\cos(\alpha_{TR}-\theta_R)\right]\right)$$

$$\delta(f-f_{\mathrm{Rmax}}\cos(\alpha_{TR}-\gamma_R)+f_{\mathrm{Tmax}}\cos(\alpha_{TR}-\gamma_T)) \quad (5.128)$$

Similarly, the D-psds of the UA and DA components are, respectively,

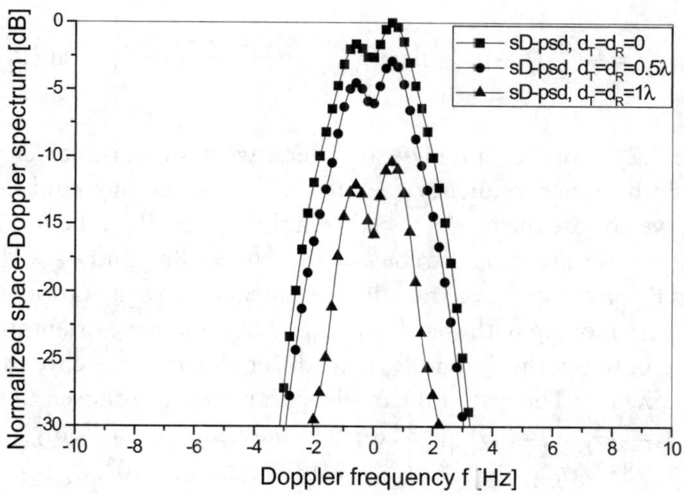

Figure 5.21 The normalized MIMO space-Doppler power spectral density for $d_T = d_R = [0, 0.5\lambda, 1\lambda]$.

$$S^{UA}_{pq,\tilde{p}\tilde{q}}(d_T, d_R, f) = \frac{\eta_B}{2B(K+1)} \sum_{b=1}^{B} \sum_{\hat{s}=b-1}^{b} \exp\left(-\frac{\sigma^2_{UAb\hat{s}}}{2} B^2_{UAb\hat{s}}\right)$$

$$\exp\left(\frac{\left(j2\pi f + C_{UAb\hat{s}} + 0.5\sigma^2_{UAb\hat{s}} A_{UAb\hat{s}} B_{UAb\hat{s}}\right)^2}{\sigma^4_{UAb\hat{s}} A^2_{UAb\hat{s}}}\right) \frac{\sqrt{\pi}}{0.5\sigma^2_{UAb\hat{s}} A_{UAb\hat{s}}}$$

$$\exp\left(j\frac{2\pi}{\lambda}\left[(p-\tilde{p})d_T \cos(\mu_{UAb\hat{s}} + \theta_T) - (q-\tilde{q})d_R \cos(\mu_{UAb\hat{s}} - \theta_R)\right]\right) \tag{5.129}$$

$$S^{DA}_{pq,\tilde{p}\tilde{q}}(d_T, d_R, f) = \frac{\eta_S}{2S(K+1)} \sum_{s=1}^{S} \sum_{\hat{b}=s-1}^{s} \exp\left(-\frac{\sigma^2_{DAs\hat{b}}}{2} B^2_{DAs\hat{b}}\right)$$

$$\exp\left(\frac{\left(j2\pi f + C_{DAs\hat{b}} + 0.5\sigma^2_{DAs\hat{b}} A_{DAs\hat{b}} B_{DAs\hat{b}}\right)^2}{\sigma^4_{DAs\hat{b}} A^2_{DAs\hat{b}}}\right) \frac{\sqrt{\pi}}{0.5\sigma^2_{DAs\hat{b}} A_{DAs\hat{b}}}$$

$$\exp\left(j\frac{2\pi}{\lambda}\left[(p-\tilde{p})d_T \cos(\mu_{DAs\hat{b}} + \theta_T) - (q-\tilde{q})d_R \cos(\mu_{DAs\hat{b}} - \theta_R)\right]\right) \tag{5.130}$$

where $A_{UAb\hat{s}}$, $C_{UAb\hat{s}}$, $A_{DAs\hat{b}}$, and $C_{DAs\hat{b}}$ are defined in (5.75)–(5.78), and $B_{UAb\hat{s}}$ and $B_{DAs\hat{b}}$ are, respectively

$$B_{UAb\hat{s}} = -\frac{2\pi}{\lambda}(p-\tilde{p})d_T \sin(\mu_{UAb\hat{s}} + \theta_T) + \frac{2\pi}{\lambda}(q-\tilde{q})d_R \sin(\mu_{UAb\hat{s}} - \theta_R) \tag{5.131}$$

$$B_{DAs\hat{b}} = -\frac{2\pi}{\lambda}(p-\tilde{p})d_T \sin(\mu_{DAs\hat{b}} + \theta_T) + \frac{2\pi}{\lambda}(q-\tilde{q})d_R \sin(\mu_{DAs\hat{b}} - \theta_R) \quad (5.132)$$

Figure 5.21 plots several sD-psds. Here, we assume isovelocity scattering environment with carrier frequency f_c = 10 kHz and speed of sound c = 1500 m/s. Furthermore, we choose the distance between the T_x and R_x to be R = 2 km. The water, T_x, and R_x depths are chosen to be h = 50m, h_T = 18m, and h_R = 18m, respectively. The T_x and R_x have $L_t = L_r = 2$ transducer elements with orientations $\theta_T = \theta_R = 90°$. The T_x and R_x are moving with speeds $v_T = v_R = 2$ m/s and have orientations $\gamma_T = \gamma_R = 40°$. The spacing between the T_x and R_x transducer elements are chosen from the set $d_T = d_R \in [0, 0.5\lambda, 1\lambda]$. The rest of the model parameters are chosen to be $K = 0.6$, $S = B = 1$, $\zeta_{\Delta Z_{DA10}} = \zeta_{\Delta Z_{DA11}} = \zeta_{\Delta Z_{UA10}} = \zeta_{\Delta Z_{UA11}} = 0.09$, $\mu_{DA10} = 178°$, $\mu_{DA11} = 177°$, $\mu_{UA10} = 182°$, $\mu_{UA11} = 183°$, $\sigma_{DA10} = 10°$, $\sigma_{DA11} = 13.6°$, $\sigma_{UA10} = 10°$, $\sigma_{UA11} = 13.6°$, $\alpha_{TR} = 180°$, $\eta_S = 0.4$ and $\eta_B = 1-\eta_S$. From Figure 5.21, we can observe that the sD-psd magnitude and the Doppler spread decrease as the spacings between the T_x and R_x elements increase, while the Doppler shift remains unchanged.

5.3.6 Modeling of Underwater Acoustic Particle Velocity Channels

A vector sensor package (VSP) generally consists of three orthogonally oriented sensors and an omnidirectional hydrophone, all packed in a compact housing. Because of its small size, the VSP might be more useful than a hydrophone array. In contrast to the hydrophone array, which has a nonnegligible vertical aperture and is difficult to deploy on an autonomous underwater vehicle AUV, the VSP can be easily deployed on the AUV and used for underwater acoustic communications.

A vector sensor package measures important nonscalar components of the acoustic field such as particle velocity and acceleration, which cannot be measured by a single scalar pressure sensor [31]. They have been used for sonar and target localization, to accurately estimate the azimuth and elevation of a source, to avoid the left-right ambiguity in linear towed arrays of scalar sensors, and to reduce acoustic noise due to their directive beam pattern [32–35]. Application of vector sensors as multichannel underwater equalizers was recently studied in [36, 37].

In general, there are two types of vector sensors: inertial and gradient. Inertial sensors truly measure the velocity or acceleration by responding to the acoustic particle motion, whereas gradient sensors employ a finite-difference approximation to estimate the gradients of the acoustic field such as velocity and acceleration. Each sensor type has its own merits and limitations. Depending on the application, system cost, required precision, and so on, one can choose the proper sensor type and technology.

Here, we consider a vector sensor system implemented in shallow water, as shown in Figure 5.22. In the two-dimensional x-z plane, there is one pressure transmitter and one vector sensor that measures in a single point, the pressure, as well as the x and z components of the particle velocity. In Figure 5.22, the pressure channel is represented by a straight line, whereas the pressure-equivalent velocity channels are shown by curved dashed lines. To define *pressure-equivalent velocity channels* $p^x(t)$ and $p^z(t)$, we first define the velocity channels v^x and v^z in m/s. According to

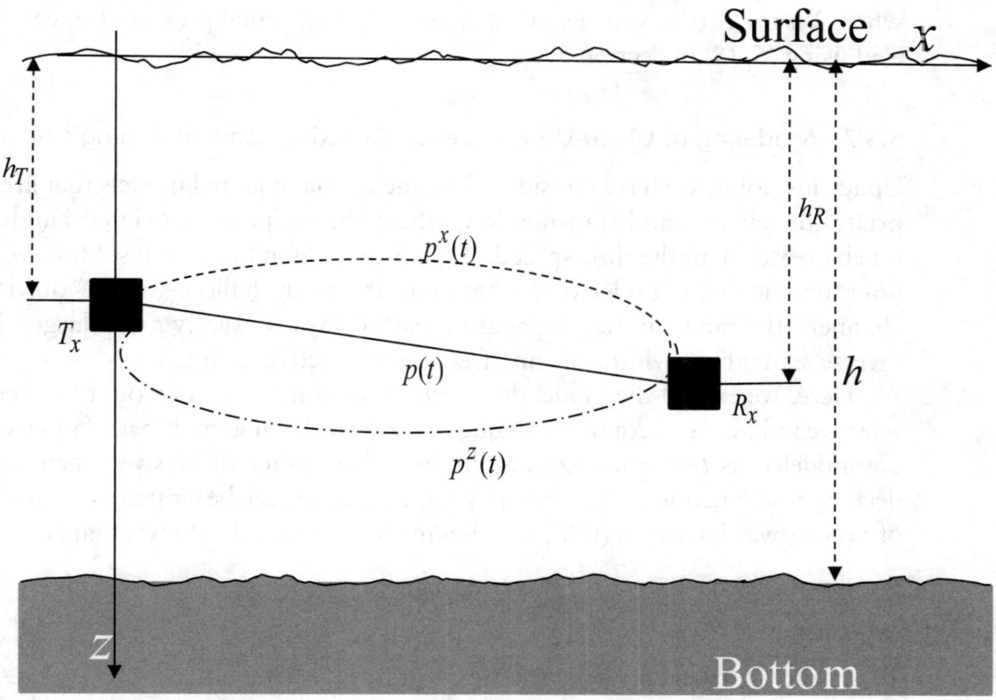

Figure 5.22 A system with one pressure transmitter and one vector sensor that measures the pressure, and the *x* and *z* components of the acoustic particle velocity.

the linearized momentum equation [38], the x and z components of the velocity at the frequency f_c at the receive side can be written as

$$v^x(t) = -\frac{1}{j\rho_0\omega_c}\frac{dp(t)}{dx} \tag{5.133}$$

$$v^z(t) = -\frac{1}{j\rho_0\omega_c}\frac{dp(t)}{dz} \tag{5.134}$$

where p_0 is the density of the fluid in kg/m^3 and $w_c = 2\pi f_c$ is the angular carrier frequency in rad/s. These equations imply that the velocity in a certain direction is proportional to the spatial pressure gradient in that direction [38]. Then, the associated pressure-equivalent velocity channels can be defined as

$$\begin{aligned} p^x(t) &= -\rho_0 c v^x(t) \\ &= \frac{\lambda}{j2\pi}\frac{dp(t)}{dx} \end{aligned} \tag{5.135}$$

$$\begin{aligned} p^z(t) &= -\rho_0 c v^z(t) \\ &= \frac{\lambda}{j2\pi}\frac{dp(t)}{dz} \end{aligned} \tag{5.136}$$

where λ is the carrier wavelength. Finally, note that signal pressure $p(t)$ can be modeled using (5.35) in Section 5.3.2.

5.3.7 Modeling of UV-to-UV Frequency-Selective Acoustic Fading Channels

Up to this point, we have considered frequency-flat channel models that are appropriate for narrowband transmission, where the reciprocal of signal bandwidth is much greater than the time spread of the propagation path delays. However, this is not often the case in underwater communications. In shallow-water acoustic (SWA) channels, the range in the propagation path delays is usually much larger than the inverse signal bandwidth, causing frequency-selective fading.

Here, we extend the model described in Section 5.3.4 to account for frequency-selective fading. To account for frequency-selective fading, multipath-fading channels are modeled as *time-varying* in addition to being time-dispersive. Then, the input delay-spread function under frequency-selective fading can be written as a superposition of LoS, upward-arriving (UA), and downward-arriving (DA) rays; that is,

$$h_{pq}(t,\tau) = h_{pq}^{LoS}(t,\tau) + h_{pq}^{UA}(t,\tau) + h_{pq}^{DA}(t,\tau) \tag{5.137}$$

To simplify further analysis, the time-variant transfer function is used instead of the input delay-spread function. The time-variant transfer function is the Fourier transform of the input delay-spread function and can be written as [23]

$$T_{pq}(t,f) = T_{pq}^{LoS}(t,f) + T_{pq}^{UA}(t,f) + T_{pq}^{DA}(t,f) \tag{5.138}$$

where the the respective components are

$$T_{pq}^{LoS}(t,f) = a_{pq}^{LoS} e^{\left(j2\pi f_{pq}^{LoS} t - j2\pi f \tau_{pq}^{LoS} - j\frac{2\pi}{\lambda} d_{pq}^{LoS}\right)} \tag{5.139}$$

$$T_{pq}^{UA}(t,f) = \sum_{b=1}^{B} \sum_{\hat{s}=b-1}^{b} \lim_{M_{b\hat{s}}\to\infty} \sum_{m=1}^{M_{b\hat{s}}} a_{pq,b\hat{s}m}^{UA} \cdot e^{j\left(2\pi f_{b\hat{s}m}^{UA} t - j2\pi f \tau_{b\hat{s}m}^{UA} - j\frac{2\pi}{\lambda} d_{pq,b\hat{s}m}^{UA} + \phi_{pq,b\hat{s}m}^{UA}\right)} \tag{5.140}$$

$$T_{pq}^{DA}(t,f) = \sum_{s=1}^{S} \sum_{\hat{b}=s-1}^{s} \lim_{N_{s\hat{b}}\to\infty} \sum_{n=1}^{N_{s\hat{b}}} a_{pq,s\hat{b}n}^{DA} \cdot e^{j\left(2\pi f_{s\hat{b}n}^{DA} t - j2\pi f \tau_{s\hat{b}n}^{DA} - j\frac{2\pi}{\lambda} d_{pq,s\hat{b}n}^{DA} + \phi_{pq,s\hat{b}n}^{DA}\right)} \tag{5.141}$$

The Doppler shift frequencies, phases, and path lengths are defined as in Section 5.3.4, while the amplitudes and time delays in this model can be calculated as follows:

- *Amplitudes*. The total signal power between the pth transmit and the qth receive hydrophone is split between the LoS and NLoS components: between the pth transmit and qth receive hydrophone the power allocated to the LoS component is $K_{pq}/(1 + K_{pq})$, while the total power allocated to the NLoS components is $1/(1 + K_{pq})$. For $\{d_T, d_R\} \ll \{h, h_T, h_R\}$, we can assume that the Ricean fading factor is the same between all hydrophone elements, (i.e., $K_{pq} = K$ for all $p \in \{1,\dots,L_t\}$ and $q \in \{1,\dots,L_r\}$). Furthermore, it is assumed that the relative power allocated to the UA and DA components is, respectively, η_B and η_S such that $\eta_B + \eta_S = 1$. We also assume that a power P_{pq} is channeled from transmit hydrophone p to receive hydrophone q. These parameters have to be either set during simulations or estimated from measurements. Having all this in mind, the amplitudes can be written as follows:

$$a_{pq}^{LoS} = \sqrt{\frac{K}{K+1}} L_S(\varepsilon_{TR}) L_A(\varepsilon_{TR}) \tag{5.142}$$

$$a_{pq,b\hat{s}m}^{UA} = \sqrt{\frac{1}{K+1}} \sqrt{\frac{\eta_B}{2B}} L_S(D_{b\hat{s}}^{UA}) L_A(D_{b\hat{s}}^{UA}) L_B(\theta_{b\hat{s}}^{UA})^b \sqrt{\frac{1}{M_{b\hat{s}}}} \tag{5.143}$$

$$a_{pq,s\hat{b}n}^{DA} = \sqrt{\frac{1}{K+1}} \sqrt{\frac{\eta_S}{2S}} L_S(D_{s\hat{b}}^{DA}) L_A(D_{s\hat{b}}^{DA}) L_B(\theta_{s\hat{b}}^{DA})^{\hat{b}} \sqrt{\frac{1}{N_{s\hat{b}}}} \tag{5.144}$$

where $L_S(\varepsilon_{TR})$ and $L_A(\varepsilon_{TR})$ are the LoS losses due to spherical spreading and absorption, respectively, $L_S(D_{b\hat{s}}^{UA})$ and $L_S(D_{s\hat{b}}^{DA})$ are the macro-eigenray propagation losses caused by spherical spreading, $L_A(D_{b\hat{s}}^{UA})$ and $L_A(D_{s\hat{b}}^{DA})$ are the macro-eigenray propagation losses caused by absorption, and $D_{b\hat{s}}^{UA}$ and $D_{s\hat{b}}^{DA}$ are the total distances traveled by UA and DA macro-eigenrays with $\hat{s}$, b and s, $\hat{b}$ surface-bottom interactions, respectively. Finally, $L_B(\theta_{s\hat{b}}^{DA})$ and $L_B(\theta_{b\hat{s}}^{UA})$ represent the impedance mismatch between the seawater and seabed.

The propagation loss caused by the spherical spreading is calculated as $L_S(D)=1/D$, while the absorbtion loss $L_A(D)$ is calculated using (5.29). Distances $D_{b\hat{s}}^{UA}$ and $D_{s\hat{b}}^{DA}$ are obtained using the method of images and can be written as [5]

$$D_{b\hat{s}}^{UA} = \sqrt{R^2 + (2\hat{s}h - h_R + (-1)^{(b-\hat{s})} h_T)^2} \tag{5.145}$$

$$D_{s\hat{b}}^{DA} = \sqrt{R^2 + (2\hat{b}h + h_R - (-1)^{(s-\hat{b})} h_T)^2} \tag{5.146}$$

Finally, the impedance mismatch between the sea water and seabed is calculated [5]

$$L_B(\theta) = \left| \frac{x\cos\theta - \sqrt{y^2 - \sin^2\theta}}{x\cos\theta + \sqrt{y^2 - \sin^2\theta}} \right| \tag{5.147}$$

where $x = \rho_1/\rho$, $y = c/c_1$, ρ and c denote the density and sound speed, respectively, in sea water, whereas ρ_1 and c_1 denote the density and sound speed, respectively, in the seabed, and incidence angles $\theta_{b\hat{s}}^{UA}$ and $\theta_{s\hat{b}}^{DA}$ are obtained as follows:

$$\theta_{b\hat{s}}^{UA} = \tan^{-1}\left(\frac{R}{2\hat{s}h - h_R + (-1)^{(b-\hat{s})} h_T}\right) \tag{5.148}$$

$$\theta_{s\hat{b}}^{DA} = \tan^{-1}\left(\frac{R}{2\hat{b}h + h_R - (-1)^{(s-\hat{b})} h_T}\right) \tag{5.149}$$

- *Multipath time delays.* The time delays of the multipath components can be inferred from the traversed distances and are given by the following expressions:

$$\tau_{pq}^{LoS} = \frac{\varepsilon_{pq}}{c_0} \tag{5.150}$$

$$\tau_{b\hat{s}m}^{(p,q)}(t) = \frac{D_{b\hat{s}}^{UA}}{c} \tag{5.151}$$

$$\tau_{s\hat{b}n}^{(p,q)}(t) = \frac{D_{s\hat{b}}^{DA}}{c} \tag{5.152}$$

where c is the speed of sound.

To simplify further analysis, we normalize all time delays with respect to the LoS arrival time. This normalization is reflected in the distances $D_{s\hat{b}}^{DA}$ and $D_{b\hat{s}}^{UA}$, and they can be rewritten as

$$\tilde{D}_{s\hat{b}}^{DA} = \sqrt{R^2 + (2\hat{b}h + h_R - (-1)^{(s-\hat{b})} h_T)^2} - \sqrt{R^2 + (h_T - h_R)^2} \tag{5.153}$$

$$\tilde{D}_{b\hat{s}}^{UA} = \sqrt{R^2 + (2\hat{s}h - h_R + (-1)^{(b-\hat{s})} h_T)^2} - \sqrt{R^2 + (h_T - h_R)^2} \tag{5.154}$$

Since the depths h, h_T, and h_R are much smaller than the distance R, distances $\tilde{D}_{s\hat{b}}^{DA}$ and $\tilde{D}_{b\hat{s}}^{UA}$ in (5.153) and (5.154) can be approximated, using $\sqrt{1+x} \approx 1 + x/2$ for small x, as follows:

$$\tilde{D}_{s\hat{b}}^{DA} \approx 2\left[\hat{b}^2 h^2 + \hat{b}hh_R - (-1)^{(s-\hat{b})}\hat{b}hh_T + (s-\hat{b})h_T h_R\right]/R \tag{5.155}$$

$$\tilde{D}_{b\hat{s}}^{UA} \approx 2\left[\hat{s}^2 h^2 - \hat{s}hh_R + (-1)^{(b-\hat{s})}\hat{s}hh_T + (b-\hat{s})h_T h_R\right]/R \tag{5.156}$$

5.3.8 Spectral-Spatial-Temporal Fading Characteristics of UV-to-UV Frequency-Selective Acoustic Fading Channels

Using the model described in Section 5.3.7, we can now derive the key temporal, spatial, and spectral characteristics of UV-to-UV frequency-selective acoustic fading channels.

5.3.8.1 Space-Time-Frequency Correlation Function and Space-Doppler Power Spectral Density

Assuming a shallow water acoustic scattering environment characterized by Gaussian pdfs in (5.48)–(5.51), we derive the space-time-frequency correlation function (STF-CF) of the complex faded envelope in (5.137).

The normalized STF-CF between two time-variant transfer functions $T_{pq}(t_1,f_1)$ and $T_{\tilde{p}\tilde{q}}(t_2,f_2)$, is defined as

$$R_{pq,\tilde{p}\tilde{q}}(t_1,f_1,t_2,f_2)=\frac{\mathrm{E}\left[T_{pq}(t_1,f_1)^{*}T_{\tilde{p}\tilde{q}}(t_2,f_2)\right]}{\sqrt{\mathrm{Var}[|T_{pq}(t_1,f_1)^{*}|^2]\mathrm{Var}[|T_{\tilde{p}\tilde{q}}(t_2,f_2)|^2]}} \tag{5.157}$$

where $(\cdot)^*$ denotes complex conjugate operation, E[·] is the statistical expectation operator, Var[·] is the statistical variance operator, $p,\tilde{p}\in\{1,\ldots,L_t\}$, and $q,\tilde{q}\in\{1,\ldots,L_r\}$. Since $T_{pq}^{UA}(t,f)$ and $T_{pq}^{DA}(t,f)$ are independent zero-mean random processes, (5.157) can be simplified to

$$R_{pq,\tilde{p}\tilde{q}}(t_1,f_1,t_2,f_2)=R_{pq,\tilde{p}\tilde{q}}^{LoS}(t_1,f_1,t_2,f_2)+R_{pq,\tilde{p}\tilde{q}}^{UA}(t_1,f_1,t_2,f_2)+R_{pq,\tilde{p}\tilde{q}}^{DA}(t_1,f_1,t_2,f_2) \tag{5.158}$$

where $R_{pq,\tilde{p}\tilde{q}}^{LoS}(t_1,f_1,t_2,f_2)$, $R_{pq,\tilde{p}\tilde{q}}^{UA}(t_1,f_1,t_2,f_2)$ and $R_{pq,\tilde{p}\tilde{q}}^{DA}(t_1,f_1,t_2,f_2)$ are the normalized STF-CFs of the LoS, UA, and DA components, respectively.

Following reasoning similar to Section 5.3.3, the space-time-frequency correlation functions of the LoS, UA, and DA components can be obtained as follows [23]:

$$\begin{aligned}
&R_{pq,\tilde{p}\tilde{q}}^{LoS}(t_1,f_1,t_2,f_2)\\
&=\frac{\mathrm{E}[T_{pq}^{LoS}(t_1,f_1)T_{\tilde{p}\tilde{q}}^{LoS}(t_2,f_2)^{*}]}{\sqrt{\mathrm{Var}[T_{pq}^{LoS}(t_1,f_1)]\mathrm{Var}[T_{\tilde{p}\tilde{q}}^{LoS}(t_2,f_2)^{*}]}}\\
&=\frac{K}{K+1}\exp\left(j\frac{2\pi f_c}{c}[\upsilon_T(t_2-t_1)\cos(\alpha_{TR}-\gamma_T)-\upsilon_R(t_2-t_1)\cos(\alpha_{TR}-\gamma_R)]\right)\\
&\exp\left(j\frac{2\pi f_1}{c}[\upsilon_T(t_2-t_1)\cos(\alpha_{TR}-\gamma_T)-\upsilon_R(t_2-t_1)\cos(\alpha_{TR}-\gamma_R)]\right)\\
&\exp\left(j\frac{2\pi(f_2-f_1)}{c}[\upsilon_T t_2\cos(\alpha_{TR}-\gamma_T)-\upsilon_R t_2\cos(\alpha_{TR}-\gamma_R)]\right)\\
&\exp\left(j\frac{2\pi f_c}{c}[(p-\tilde{p})d_T\cos(\alpha_{TR}-\theta_T)-(q-\tilde{q})d_R\cos(\alpha_{TR}-\theta_R)]\right)
\end{aligned} \tag{5.159}$$

$$
\begin{aligned}
&R^{UA}_{pq,\tilde{p}\tilde{q}}(t_1,f_1,t_2,f_2)\\
&=\frac{\eta_B}{2B(K+1)}\sum_{b=1}^{B}\sum_{\hat{s}=b-1}^{b}\exp\left(-j\frac{2\pi(f_2-f_1)}{c}\tilde{D}^{UA}_{b\hat{s}}\right)\\
&\exp\left(j\frac{2\pi f_c}{c}\left[(p-\tilde{p})d_T\cos(\mu_{UAb\hat{s}}+\theta_T)-(q-\tilde{q})d_R\cos(\mu_{UAb\hat{s}}-\theta_R)\right]\right)\\
&\exp\left(j\frac{2\pi(f_c+f_1)}{c}\left[v_T(t_2-t_1)\cos(\mu_{UAb\hat{s}}+\gamma_T)-v_R(t_2-t_1)\cos(\mu_{UAb\hat{s}}-\gamma_R)\right]\right)\\
&\exp\left(j\frac{2\pi(f_2-f_1)}{c}\left[v_Tt_2\cos(\mu_{UAb\hat{s}}+\gamma_T)-v_Rt_2\cos(\mu_{UAb\hat{s}}-\gamma_R)\right]\right)\\
&\exp\left(-\frac{(t_2-t_1)\zeta^2_{\Delta Z_{UAb\hat{s}}}}{2}\left[\frac{2\pi(f_c+f_1)}{c}\sin\mu_{UAb\hat{s}}\right]^2\right)\\
&\exp\left(-\frac{t_2\zeta^2_{\Delta Z_{UAb\hat{s}}}}{2}\left[\frac{2\pi(f_2-f_1)}{c}\sin\mu_{UAb\hat{s}}\right]^2\right)\\
&\exp\left\{-\frac{\sigma^2_{UAb\hat{s}}}{2}\left[\frac{2\pi(f_2-f_1)}{c}\frac{h_R+h_T-2h}{\sin\mu_{UAb\hat{s}}\tan\mu_{UAb\hat{s}}}+\frac{2\pi f_c}{c}(q-\tilde{q})d_R\sin(\mu_{UAb\hat{s}}-\theta_R)\right.\right.\\
&-\frac{2\pi f_c}{c}(p-\tilde{p})d_T\sin(\mu_{UAb\hat{s}}+\theta_T)\\
&+\frac{2\pi(f_2-f_1)}{c}\left[v_Rt_2\sin(\mu_{UAb\hat{s}}-\gamma_R)-v_Tt_2\sin(\mu_{UAb\hat{s}}+\gamma_T)\right]\\
&+\frac{2\pi(f_c+f_1)}{c}\left[v_R(t_2-t_1)\sin(\mu_{UAb\hat{s}}-\gamma_R)-v_T(t_2-t_1)\sin(\mu_{UAb\hat{s}}+\gamma_T)\right]\\
&+(t_2-t_1)\zeta^2_{\Delta Z_{UAb\hat{s}}}\left[\frac{2\pi(f_c+f_1)}{c}\right]^2\sin\mu_{UAb\hat{s}}\cos\mu_{UAb\hat{s}}\\
&\left.\left.+t_2\zeta^2_{\Delta Z_{UAb\hat{s}}}\left[\frac{2\pi(f_2-f_1)}{c}\right]^2\sin\mu_{UAb\hat{s}}\cos\mu_{UAb\hat{s}}\right]^2\right\}
\end{aligned}
\tag{5.160}
$$

$$
\begin{aligned}
&R^{DA}_{pq,\tilde{p}\tilde{q}}(t_1,f_1,t_2,f_2)\\
&= \frac{\eta_S}{2S(K+1)} \sum_{s=1}^{S} \sum_{\hat{b}=s-1}^{s} \exp\left(-j\frac{2\pi(f_2-f_1)}{c}\tilde{D}^{DA}_{s\hat{b}}\right)\\
&\exp\left(j\frac{2\pi f_c}{c}\left[(p-\tilde{p})d_T\cos(\mu_{DAs\hat{b}}+\theta_T)-(q-\tilde{q})d_R\cos(\mu_{DAs\hat{b}}-\theta_R)\right]\right)\\
&\exp\left(j\frac{2\pi(f_c+f_1)}{c}\left[\upsilon_T(t_2-t_1)\cos(\mu_{DAs\hat{b}}+\gamma_T)-\upsilon_R(t_2-t_1)\cos(\mu_{DAs\hat{b}}-\gamma_R)\right]\right)\\
&\exp\left(j\frac{2\pi(f_2-f_1)}{c}\left[\upsilon_T t_2\cos(\mu_{DAs\hat{b}}+\gamma_T)-\upsilon_R t_2\cos(\mu_{DAs\hat{b}}-\gamma_R)\right]\right)\\
&\exp\left(-\frac{(t_2-t_1)\zeta^2_{\Delta Z_{DAs\hat{b}}}}{2}\left[\frac{2\pi(f_c+f_1)}{c}\sin\mu_{DAs\hat{b}}\right]^2\right)\\
&\exp\left(-\frac{t_2\zeta^2_{\Delta Z_{DAs\hat{b}}}}{2}\left[\frac{2\pi(f_2-f_1)}{c}\sin\mu_{DAs\hat{b}}\right]^2\right)\\
&\exp\left\{-\frac{\sigma^2_{DAs\hat{b}}}{2}\left[\frac{2\pi(f_2-f_1)}{c}\frac{h_R+h_T}{\sin\mu_{DAs\hat{b}}\tan\mu_{DAs\hat{b}}}+\frac{2\pi f_c}{c}(q-\tilde{q})d_R\sin(\mu_{DAs\hat{b}}-\theta_R)\right.\right.\\
&-\frac{2\pi f_c}{c}(p-\tilde{p})d_T\sin(\mu_{DAs\hat{b}}+\theta_T)\\
&+\frac{2\pi(f_c+f_1)}{c}\left[\upsilon_R(t_2-t_1)\sin(\mu_{DAs\hat{b}}-\gamma_R)-\upsilon_T(t_2-t_1)\sin(\mu_{DAs\hat{b}}+\gamma_T)\right]\\
&+\frac{2\pi(f_2-f_1)}{c}\left[\upsilon_R t_2\sin(\mu_{DAs\hat{b}}-\gamma_R)-\upsilon_T t_2\sin(\mu_{DAs\hat{b}}+\gamma_T)\right]\\
&+(t_2-t_1)\zeta^2_{\Delta Z_{DAs\hat{b}}}\left[\frac{2\pi(f_c+f_1)}{c}\right]^2\sin\mu_{DAs\hat{b}}\cos\mu_{DAs\hat{b}}\\
&\left.\left.+t_2\zeta^2_{\Delta Z_{DAs\hat{b}}}\left[\frac{2\pi(f_2-f_1)}{c}\right]^2\sin\mu_{DAs\hat{b}}\cos\mu_{DAs\hat{b}}\right]^2\right\} \qquad (5.161)
\end{aligned}
$$

Under wide sense stationary uncorrelated scattering (WSSUS) conditions, the space-time-frequency correlation functions of the LoS, UA, and DA components simplify to [23]

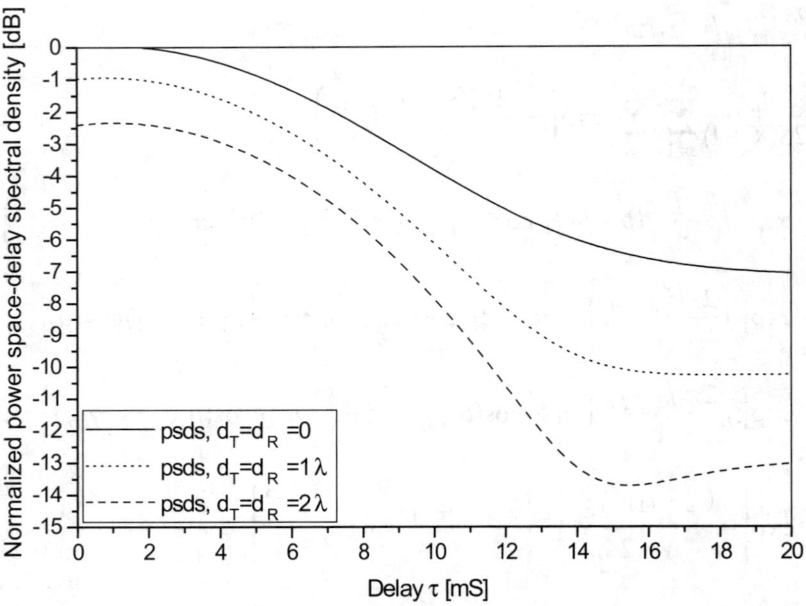

Figure 5.23 Normalized power space-delay spectral density (psds) for $d_T = d_R = [0, 1\lambda, 2\lambda]$.

$$
\begin{aligned}
&R_{pq,\tilde{p}\tilde{q}}^{LoS}(\Delta t, \Delta f) \\
&= \frac{K}{K+1}\exp\left(j\frac{2\pi f_c}{c}[\upsilon_T \Delta t\cos(\alpha_{TR}-\gamma_T) - \upsilon_R \Delta t \cos(\alpha_{TR}-\gamma_T)]\right) \\
&\quad \exp\left(j\frac{2\pi f_c}{c}[(p-\tilde{p})d_T\cos(\alpha_{TR}-\theta_T) - (q-\tilde{q})d_R\cos(\alpha_{TR}-\theta_R)]\right)
\end{aligned} \tag{5.162}
$$

$$
\begin{aligned}
&R_{pq,\tilde{p}\tilde{q}}^{UA}(\Delta t, \Delta f) \\
&= \frac{\eta_B}{2B(K+1)}\sum_{b=1}^{B}\sum_{\hat{s}=b-1}^{b}\exp\left(-\frac{\Delta t \zeta_{\Delta Z_{UAb\hat{s}}}^2}{2}\left[\frac{2\pi}{\lambda}\sin\mu_{UAb\hat{s}}\right]^2\right) \\
&\quad \exp\left(j\frac{2\pi}{\lambda}(p-\tilde{p})d_T\cos(\mu_{UAb\hat{s}}+\theta_T) - j\frac{2\pi\Delta f}{c}\tilde{D}_{b\hat{s}}^{UA} - j\frac{2\pi}{\lambda}\upsilon_R\Delta t\cos(\mu_{UAb\hat{s}}-\gamma_R)\right) \\
&\quad \exp\left(j\frac{2\pi}{\lambda}\left[-(q-\tilde{q})d_R\cos(\mu_{UAb\hat{s}}-\theta_R) + \upsilon_T\Delta t\cos(\mu_{UAb\hat{s}}+\gamma_T)\right]\right) \\
&\quad \exp\left\{-\frac{\sigma_{UAb\hat{s}}^2}{2}\left[\frac{2\pi\Delta f}{c}\frac{h_R+h_T-2h}{\sin\mu_{UAb\hat{s}}\tan\mu_{UAb\hat{s}}}\right.\right. \\
&+\frac{2\pi}{\lambda}[q-\tilde{q})d_R\sin((\mu_{UAb\hat{s}}-\theta_R) - (p-\tilde{p})d_T\sin(\mu_{UAb\hat{s}}+\theta_T)] \\
&+\frac{2\pi}{\lambda}[\upsilon_R\Delta t\sin(\mu_{UAb\hat{s}}-\gamma_R) - \upsilon_T\Delta t\sin(\mu_{UAb\hat{s}}+\gamma_T)] \\
&\left.\left.+\Delta t\zeta_{\Delta Z_{UAb\hat{s}}}^2\left[\frac{2\pi}{\lambda}\right]^2\sin\mu_{UAb\hat{s}}\cos\mu_{UAb\hat{s}}\right]^2\right\}
\end{aligned} \tag{5.163}
$$

$$
\begin{aligned}
&R^{DA}_{pq,\tilde{p}\tilde{q}}(\Delta t, \Delta f) \\
&= \frac{\eta_S}{2S(K+1)} \sum_{s=1}^{S} \sum_{\hat{b}=s-1}^{s} \exp\left(-\frac{\Delta t \zeta^2_{\Delta Z_{DAs\hat{b}}}}{2} \left[\frac{2\pi}{\lambda} \sin \mu_{DAs\hat{b}} \right]^2 \right) \\
&\exp\left(j\frac{2\pi}{\lambda}(q-\tilde{q})d_R \cos\left(\mu_{DAs\hat{b}} - \theta_R\right) - j\frac{2\pi\Delta f}{c} \tilde{D}^{DA}_{s\hat{b}} - v_R \Delta t \cos\left(\mu_{DAs\hat{b}} - \gamma_R\right)\right) \\
&\exp\left(j\frac{2\pi}{\lambda}\left[-(p-\tilde{p})d_T \cos\left(\mu_{DAs\hat{b}} + \theta_T\right) + v_T \Delta t \cos\left(\mu_{DAs\hat{b}} + \gamma_T\right)\right]\right) \\
&\exp\left\{ -\frac{\sigma^2_{DAs\hat{b}}}{2} \left[\frac{2\pi\Delta f}{c} \frac{h_R + h_T}{\sin \mu_{DAs\hat{b}} \tan \mu_{DAs\hat{b}}} \right.\right. \\
&+ \frac{2\pi}{\lambda}\left[(q-\tilde{q})d_R \sin\left(\mu_{DAs\hat{b}} - \theta_R\right) - (p-\tilde{p})d_T \sin\left(\mu_{DAs\hat{b}} + \theta_T\right)\right] \\
&+ \frac{2\pi}{\lambda}\left[v_R \Delta t \sin\left(\mu_{DAs\hat{b}} - \gamma_R\right) - v_T \Delta t \sin\left(\mu_{DAs\hat{b}} + \gamma_T\right)\right] \\
&\left.\left. + \Delta t \zeta^2_{\Delta Z_{DAs\hat{b}}} \left[\frac{2\pi f_c}{c} \right]^2 \sin \mu_{DAs\hat{b}} \cos \mu_{DAs\hat{b}} \right]^2 \right\}
\end{aligned} \tag{5.164}
$$

where λ denotes the carrier wavelength.

Here we note that the space-Doppler power spectral density of the time-variant transfer function is the Fourier transform of the space-time-frequency correlation function $R_{pq,\tilde{p}\tilde{q}}(\Delta t, \Delta f = 0)$ with respect to Δt, and is the same as the space-Doppler power spectral density in (5.127) of Section 5.3.5.2.

5.3.8.2 Power Space-Delay Spectral Density

The power space-delay spectral density (psds) of the time-variant transfer function is the inverse Fourier transform of the space-time-frequency correlation function $R_{pq,\tilde{p}\tilde{q}}(\Delta t = 0, \Delta f)$ with respect to Δf [25]. From (5.158), it follows that the psds is a summation of the psds functions of the LoS, UA, and DA components; that is,

$$
P_{pq,\tilde{p}\tilde{q}}(\tau) = P^{LoS}_{pq,\tilde{p}\tilde{q}}(\tau) + P^{UA}_{pq,\tilde{p}\tilde{q}}(\tau) + P^{DA}_{pq,\tilde{p}\tilde{q}}(\tau) \tag{5.165}
$$

By calculating the inverse Fourier transform of the STF-CF in (5.162), we obtain the psds of the LoS component as follows [23]:

$$
\begin{aligned}
P^{LoS}_{pq,\tilde{p}\tilde{q}}(\tau) &= \mathcal{F}^{-1}_{\Delta f}\{R^{LoS}_{pq,\tilde{p}\tilde{q}}(\Delta t = 0, \Delta f)\} = \frac{K}{K+1} \\
&\exp\left(j\frac{2\pi}{\lambda}[(p-\tilde{p})d_T \cos(\alpha_{TR} - \theta_T) - (q-\tilde{q})d_R \cos(\alpha_{TR} - \theta_R)]\right)\delta(\tau)
\end{aligned} \tag{5.166}
$$

Similarly, psds of the UA and DA components are, respectively,

$$P^{UA}_{pq,\tilde{p}\tilde{q}}(\tau)=\frac{\eta_B}{2B(K+1)}\sum_{b=1}^{B}\sum_{\hat{s}=b-1}^{b}\exp\left(-\frac{\sigma^2_{UAb\hat{s}}}{2}B^2_{UAb\hat{s}}\right)$$
$$\exp\left(j\frac{2\pi}{\lambda}\left[(p-\tilde{p})d_T\cos(\mu_{UAb\hat{s}}+\theta_T)-(q-\tilde{q})d_R\cos(\mu_{UAb\hat{s}}-\theta_R)\right]\right)$$
$$\exp\left(\frac{\left(j2\pi\tau-j2\pi\tilde{D}^{UA}_{b\hat{s}}/c-0.5\sigma^2_{UAb\hat{s}}E_{UAb\hat{s}}B_{UAb\hat{s}}\right)^2}{\sigma^4_{UAb\hat{s}}E^2_{UAb\hat{s}}}\right)\frac{\sqrt{\pi}}{0.5\sigma^2_{UAb\hat{s}}E_{UAb\hat{s}}} \tag{5.167}$$

$$P^{DA}_{pq,\tilde{p}\tilde{q}}(\tau)=\frac{\eta_S}{2S(K+1)}\sum_{s=1}^{S}\sum_{\hat{b}=s-1}^{s}\exp\left(-\frac{\sigma^2_{DAs\hat{b}}}{2}B^2_{DAs\hat{b}}\right)$$
$$\exp\left(j\frac{2\pi}{\lambda}\left[(p-\tilde{p})d_T\cos(\mu_{DAs\hat{b}}+\theta_T)-(q-\tilde{q})d_R\cos(\mu_{DAs\hat{b}}-\theta_R)\right]\right)$$
$$\exp\left(\frac{\left(j2\pi\tau-j2\pi\tilde{D}^{DA}_{s\hat{b}}/c-0.5\sigma^2_{DAs\hat{b}}E_{DAs\hat{b}}B_{DAs\hat{b}}\right)^2}{\sigma^4_{DAs\hat{b}}E^2_{DAs\hat{b}}}\right)\frac{\sqrt{\pi}}{0.5\sigma^2_{DAs\hat{b}}E_{DAs\hat{b}}} \tag{5.168}$$

where $B_{UAb\hat{s}}$ and $B_{DAs\hat{b}}$ are defined in (5.131) and (5.132), respectively, and $E_{UAb\hat{s}}$ and $E_{DAs\hat{b}}$ are defined as

$$E_{UAb\hat{s}}=\frac{2\pi}{c}\frac{h_R+h_T-2h}{\sin\mu_{UAb\hat{s}}\tan\mu_{UAb\hat{s}}} \tag{5.169}$$

$$E_{DAs\hat{b}}=\frac{2\pi}{c}\frac{h_R+h_T}{\sin\mu_{DAs\hat{b}}\tan\mu_{DAs\hat{b}}} \tag{5.170}$$

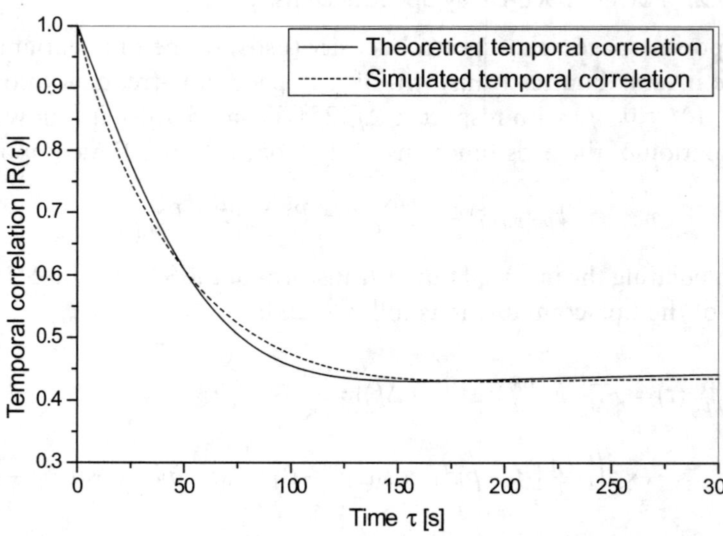

Figure 5.24 The magnitudes of simulated and theoretical temporal correlation functions.

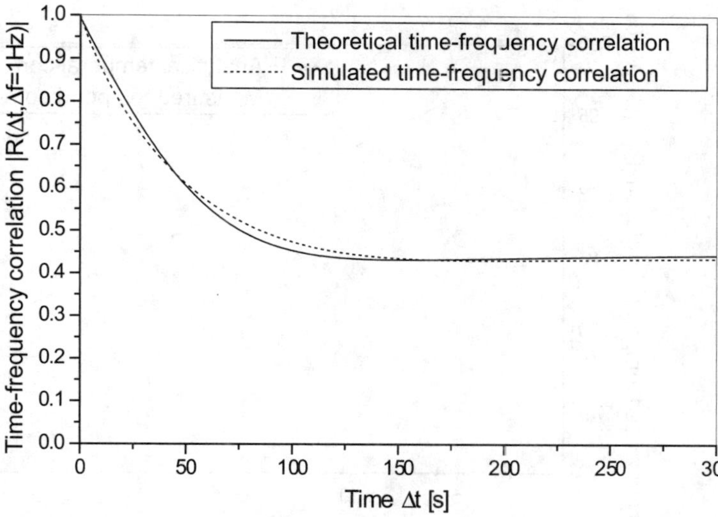

Figure 5.25 The magnitudes of simulated and theoretical time-frequency correlation functions.

Figure 5.23 plots several SISO and MIMO psds, assuming a WSSUS isovelocity scattering environment. We assume that the carrier frequency is $f_c = 17$ kHz and the speed of sound is $c = 1500$ m/s. The distance between the T_x and R_x is set to $R = 2$ km, while the water, T_x, and R_x depths are $h = 50$m, $h_T = 18$m, and $h_R = 18$m, respectively. The T_x and R_x have $L_t = L_r = 2$ vertically oriented elements (i.e., $\theta_T = \theta_R = 90°$). It is assumed that the T_x and R_x are stationary (i.e., $v_T = v_R = 0$ m/s). The spacing between the T_x and R_x transducer elements are chosen from the set $d_T = d_R \in [0, 1\lambda, 2\lambda]$. The rest of the model parameters are set to $K = 0$, $S = B = 1$, $\zeta_{\Delta Z_{DA10}} = \zeta_{\Delta Z_{DA11}} = \zeta_{\Delta Z_{UA10}} = \zeta_{\Delta Z_{UA11}} = 0.1$, $\mu_{DA10} = 160°$, $\mu_{DA11} = 157°$, $\mu_{UA10} = 190°$, $\mu_{UA11} = 191°$, $\sigma_{DA10} = 2.55°$, $\sigma_{DA11} = 3.1°$, $\sigma_{UA10} = 4°$, $\sigma_{UA11} = 6°$, $\alpha_{TR} = 180°$, $\eta_S = 0.8$, and $\eta_B = 1 - \eta_S$. From Figure 5.23, we can observe that the magnitude of the psds and the multipath spread decrease as the spacing between the T_x and R_x elements increases. In contrast to radio channels, where the later arriving multipaths carry less energy than the earlier arriving multipaths (i.e., the psds decays exponentially), in SWA channels, it is often the case that the later arriving multipaths are received as stronger signals than the earlier arriving multipaths. This phenomenon can be explained by observing how macro-eigenrays propagate through the waveguide. Note that macro-eigenrays cannot be received with equal strength at all vertical locations in water because their angles of departure and arrival are predetermined by the waveguide geometry. For example, if the hydrophone is placed near the center of waveguide (water column), most of the arriving macro-eigenrays can be detected and the psds decays exponentially ($d_T = d_R = 0$ in Figure 5.23). As the hydrophones move away from the center of waveguide, some macro-eigenrays, although arriving, cannot be recorded at that vertical location, causing the imbalance in the psds function ($d_T = d_R = 2\lambda$ in Figure 5.23).

5.4 Simulators for UV-to-UV Acoustic Channels

Simulation of UV-to-UV channels is a crucial part of link layer simulations and of the overall characterization of a UV-to-UV system performance. The channel models detailed in Section 5.3 are very useful mathematical tools for understanding

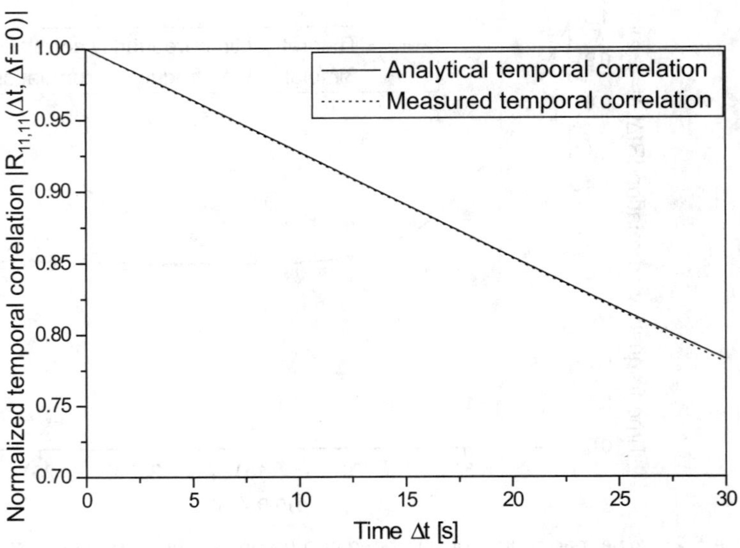

Figure 5.26 The magnitudes of analytical and measured temporal correlation functions.

propagation mechanisms and for statistical characterization of channels. However, these models assume an infinite number of scatterers, which makes them impractical for simulators. We now describe simulation models with finite numbers of scatterers that still match the statistical properties of the reference models in Section 5.3.

5.4.1 Simulators for UV-to-UV Frequency-Flat and Frequency-Selective Acoustic Fading Channels

Assuming an isovelocity SWA environment and using the reference model described in Section 5.3.2, the frequency-flat complex faded envelope can be written as

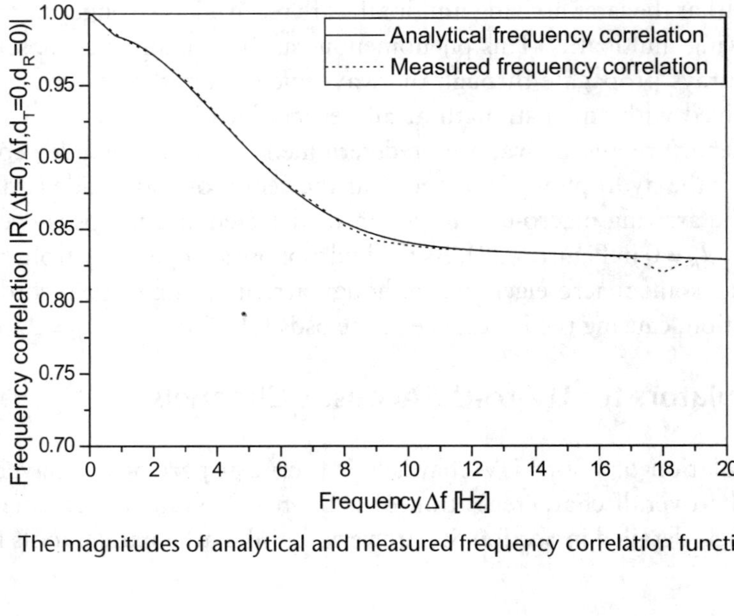

Figure 5.27 The magnitudes of analytical and measured frequency correlation functions.

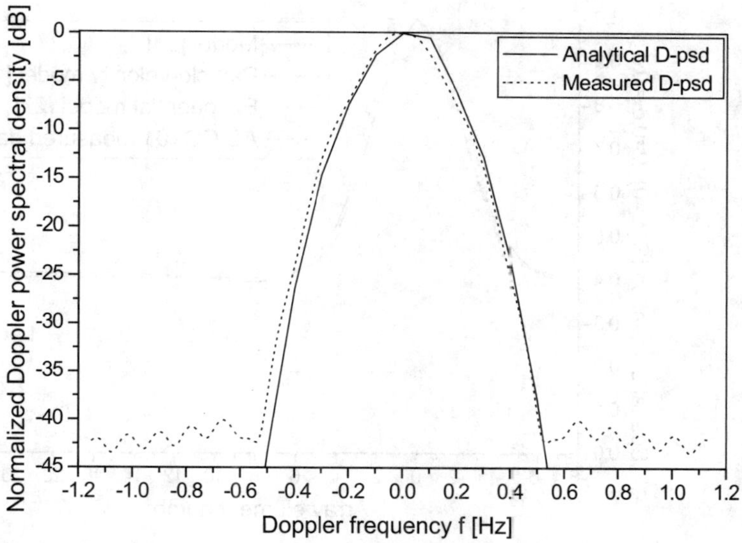

Figure 5.28 The normalized analytical and measured Doppler power spectral density.

$h(t) = h^{LoS}(t) + h^{UA}(t) + h^{DA}(t)$ where

$$h^{LoS}(t) = \sqrt{\frac{K}{K+1}} e^{j2\pi t[f_{T\max}\cos(\alpha_{TR}+\pi-\gamma_T)+f_{R\max}\cos(\alpha_{TR}-\gamma_R)]} \tag{5.171}$$

$$h^{UA}(t) = \sqrt{\frac{\eta_B}{2B(K+1)}} \sum_{b=1}^{B} \sum_{\hat{s}=b-1}^{b} \sqrt{\frac{1}{M_{b\hat{s}}}} \sum_{m=1}^{M_{b\hat{s}}} e^{j\phi_{b\hat{s}m}}$$
$$\times e^{j2\pi t\left(\frac{1}{\lambda t}\Delta Z_{b\hat{s}m}(t)\sin\alpha_{R\,b\hat{s}m}+f_{T\max}\cos(3\pi-\alpha_{Rb\hat{s}m}-\gamma_T)+f_{R\max}\cos(\alpha_{Rb\hat{s}m}-\gamma_R)\right)} \tag{5.172}$$

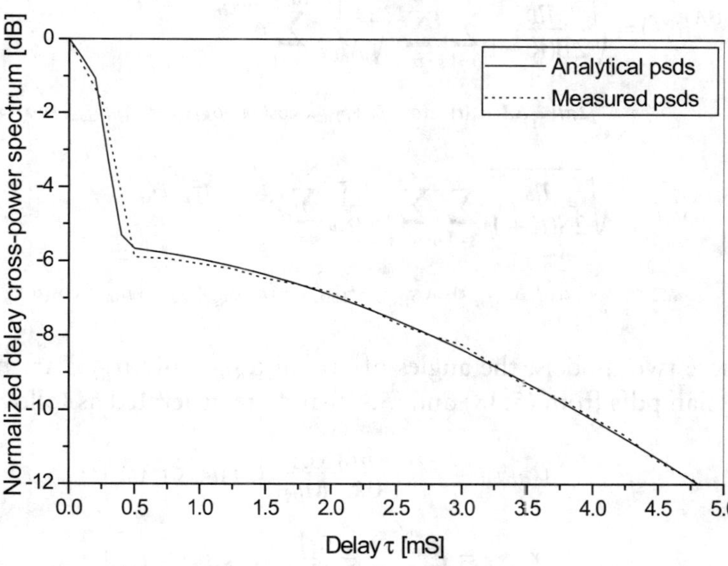

Figure 5.29 The normalized analytical and measured delay cross-power spectral density.

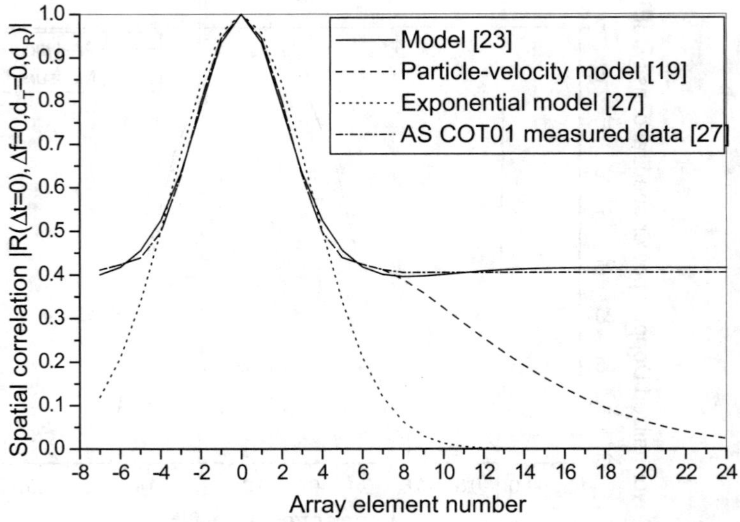

Figure 5.30 Comparison of the analytical spatial correlations in [19, 23, 27] with the measured spatial correlation in [27].

$$h^{DA}(t) = \sqrt{\frac{\eta_S}{2S(K+1)}} \sum_{s=1}^{S} \sum_{\hat{b}=s-1}^{s} \sqrt{\frac{1}{N_{s\hat{b}}}} \sum_{n=1}^{N_{s\hat{b}}} e^{j\phi_{s\hat{b}n}}$$
$$\times e^{j2\pi t\left(\frac{1}{\lambda t}\Delta Z_{s\hat{b}n}(t)\sin\alpha_{R s\hat{b}n} + f_{T\max}\cos(\pi-\alpha_{R s\hat{b}n}-\gamma_T)+f_{R\max}\cos(\alpha_{R s\hat{b}n}-\gamma_R)\right)} \tag{5.173}$$

Similarly, using the reference model described in Section 5.3.7, the frequency-selective input delay-spread function can be written as $T(t,f) = T^{LoS}(t,f) + T^{UA}(t,f) + T^{DA}(t,f)$, where the components are

$$T^{LoS}(t,f) = \sqrt{\frac{K}{K+1}} e^{j2\pi t[f_{T\max}\cos(\alpha_{TR}+\pi-\gamma_T)+f_{R\max}\cos(\alpha_{TR}-\gamma_R)]} \tag{5.174}$$

$$T^{UA}(t,f) = \sqrt{\frac{\eta_B}{2B(K+1)}} \sum_{b=1}^{B} \sum_{\hat{s}=b-1}^{b} \sqrt{\frac{1}{M_{b\hat{s}}}} \sum_{m=1}^{M_{b\hat{s}}} e^{j\phi_{b\hat{s}m} - j\frac{2\pi f}{c}\tilde{D}^{UA}_{b\hat{s}}}$$
$$\times e^{j2\pi t\left(\frac{1}{\lambda t}\Delta Z_{b\hat{s}m}(t)\sin\alpha_{Rb\hat{s}m} + f_{T\max}\cos(3\pi-\alpha_{Rb\hat{s}m}-\gamma_T)+f_{R\max}\cos(\alpha_{Rb\hat{s}m}-\gamma_R)\right)} \tag{5.175}$$

$$T^{DA}(t,f) = \sqrt{\frac{\eta_S}{2S(K+1)}} \sum_{s=1}^{S} \sum_{\hat{b}=s-1}^{s} \sqrt{\frac{1}{N_{s\hat{b}}}} \sum_{n=1}^{N_{s\hat{b}}} e^{j\phi_{s\hat{b}n} - j\frac{2\pi f}{c}\tilde{D}^{DA}_{s\hat{b}}}$$
$$\times e^{j2\pi t\left(\frac{1}{\lambda t}\Delta Z_{s\hat{b}n}(t)\sin\alpha_{Rs\hat{b}n} + f_{T\max}\cos(\pi-\alpha_{Rs\hat{b}n}-\gamma_T)+f_{R\max}\cos(\alpha_{Rs\hat{b}n}-\gamma_R)\right)} \tag{5.176}$$

In these two models, the angles of arrival $\alpha_{Rb\hat{s}m}$ and $\alpha_{Rb\hat{s}n}$ are modeled using the Gaussian pdfs from (5.48) and (5.49) and are generated as follows:

$$\alpha_{Rb\hat{s}m} = F^{-1}\left(\frac{m+0.1}{0.5\cdot M_{b\hat{s}}}, \mu_{UAb\hat{s}}, \sigma_{UAb\hat{s}}\right) \tag{5.177}$$

$$\alpha_{Rs\hat{b}n} = F^{-1}\left(\frac{n+0.1}{0.9\cdot N_{s\hat{b}}}, \mu_{DAs\hat{b}}, \sigma_{DAs\hat{b}}\right) \tag{5.178}$$

for m = 0,0.5,1,..., $M_{b\hat{s}}/2$, n = 0,0.5,1,...,$N_{s\hat{b}}/2$, where $F(\cdot)^{-1}$ denotes the inverse cumulative Gassian distribution function. The surface displacements $Z_{b\hat{s}m}(t)$ and $Z_{s\hat{b}n}(t)$ are modeled using Gaussian pdf from (5.50) and (5.51), and are generated as follows:

$$Z_{b\hat{s}m}(t) = F^{-1}\left(\frac{m+0.1}{1.8\cdot M_{b\hat{s}}}, 0, \zeta_{\Delta Z_{UAb\hat{s}}}\right) \quad (5.179)$$

$$Z_{s\hat{b}n}(t) = F^{-1}\left(\frac{n+0.1}{2.1\cdot N_{s\hat{b}}}, 0, \zeta_{\Delta Z_{DAs\hat{b}}}\right) \quad (5.180)$$

for m = 0,0.5,1,..., $M_{b\hat{s}}/2$, n = 0,0.5,1...,$N_{s\hat{b}}/2$, where $F(\cdot)^{-1}$ denotes the inverse cumulative Gassian distribution function. Finally, phases $\phi_{b\hat{s}m}$ and $\phi_{s\hat{b}n}$ are generated as independent uniform random variables on the interval $[-\pi,\pi)$.

To verify these models, we compare frequency-flat and frequency-selective simulation models with the theoretical models detailed in Sections 5.3.2 and 5.3.7, respectively. Figures 5.24 and 5.25 compare the simulated temporal and time-frequency correlation functions with the theoretical temporal and time-frequency correlation functions in (5.52) and (5.158), respectively. The theoretical and simulated curves in Figures 5.24 and 5.25 are obtained with the parameters f_c = 17 kHz, c = 1470 m/s, R = 5 km, h = 20m, h_T = 19m, h_R = 18m, K = 0.8, S = B = 1, v_T = 0.0006 m/s, v_R = 0.00025 m/s, μ_{DA10} = 166.5°, μ_{DA11} = 169°, μ_{UA10} = 194.9°, μ_{UA11} = 195.9°, σ_{DA10} = 3.25°, σ_{DA11} = 5°, σ_{UA10} = 2.7°, σ_{UA11} = 3.2°, $\zeta_{\Delta Z_{DA10}} = \zeta_{\Delta Z_{DA11}}$ = 0.0097, $\zeta_{\Delta Z_{UA10}} = \zeta_{\Delta Z_{UA11}}$ = 0.0087, η_B = 0.28, and $\eta_S = 1 - \eta_B$. It is assumed that the T_x and R_x are relatively stationary, or slightly moving with waves (i.e., $\gamma_T = \gamma_R$ = 90°). Finally, the number of scatterers in the simulators is set to $M_{b\hat{s}} = 10$ and $N_{s\hat{b}} = 10$. We observe a good match between theoretical and simulated curves. An even better match can be obtained by simulating with larger numbers of scatterers, at a cost of proportionally longer simulation times.

5.5 Underwater Acoustic Channel Measurement Campaigns and Modeling

A large number of papers have reported measurement results, and channel models based on these measurements, for underwater acoustic channels (e.g., [7–39]). The disadvantage of all channel models derived from measurements is that they are site specific (i.e., there is no guarantee that the model will be applicable to other environments). On the other hand, the theoretical models and simulators detailed in Sections 5.3 and 5.4 need to be validated against the measurements. Section 5.5.1 presents a comparison between the measured data and the model detailed in Section 5.3.7.

5.5.1 Validation of the Model for UV-to-UV Frequency-Selective Acoustic Fading Channels

In this section, we compare the theoretical results in Section 5.3.8 with the AUVFest07 measured data in [28], collected at calm sea. Furthermore, we compare the derived spatial correlation function $R_{pq,\tilde{p}\tilde{q}}(\Delta t = 0, \Delta f = 0)$ in (5.158) with the spatial correlation functions of the exponential model in [27], the particle-velocity model in [19], and the ASCOT01 measured data in [27].

The channel measurements in [28] are collected at $f_c = 17$ kHz, and the speed of sound is $c = 1470$ m/s. The distance between the T_x and R_x is $R = 5$ km. The water, T_x, and R_x depths are $h = 20$m, $h_T = 019$m, and $h_R = 18$m, respectively. The T_x and R_x are equipped with vertically oriented transducer elements (i.e., $\theta_T = \theta_R = 90°$), and are relatively stationary, or slightly moving with waves (i.e., $\gamma_T = \gamma_R = 90°$).

Figures 5.26 and 5.27 compare the analytical temporal and frequency correlation functions (i.e., $R_{11,11}(\Delta t,0)$) and $R_{11,11}(0,\Delta f)$ in (5.158), respectively with their measured counterparts, and Figures 5.28 and 5.29 show similar comparisons for sD-psds and psds, respectively.

The analytical curves in Figures 5.26–5.29 are obtained with the parameters $K=0.8$, $S=B=1$, $\upsilon_T=0.0006$ m/s, $\upsilon_R=0.00025$ m/s, $\mu_{DA10}=169°$, $\mu_{DA11}=167°$, $\mu_{UA10}=194°$, $\mu_{UA11}=195°$, $\sigma_{DA10}=3.25°$, $\sigma_{DA11}=5°$, $\sigma_{UA10}=2.7°$, $\sigma_{UA11}=3.2°$, $\zeta_{\Delta Z_{DA10}}=\zeta_{\Delta Z_{DA11}}=0.0097$, $\zeta_{\Delta Z_{UA10}}=\zeta_{\Delta Z_{UA11}}=0.0087$, $\eta_S=0.28$, and $\eta_B=1-\eta_S$. The Rice factor K is estimated using the moment-method in [41], and the parameters S and B are visually estimated from Figure 3c in [28]. The rest of the parameters (i.e., $[\upsilon_T, \upsilon_R, \zeta_{\Delta Z_{DAs\hat{b}}}, \zeta_{\Delta Z_{UAb\hat{s}}}, \mu_{DAs\hat{b}}, \mu_{UAb\hat{s}}, \sigma_{DAs\hat{b}}, \sigma_{UAb\hat{s}}, \eta_S]$ are estimated jointly using the maximum-likelihood approach from [26] and under the constraint $\eta_S + \eta_B = 1$. The maximum-likelihood function is optimized using the sequential quadratic programming algorithm [42], which is implemented in Matlab as a "fminsearch" function.

The results in Figures 5.26–5.29 show a good match between the measured and analytical data. This result illustrates the validity of the derived statistics and validates the estimation methods used to find the model parameters. Furthermore, the results in Figures 5.26 and 5.27 show that both temporal and frequency correlation functions decay exponentially with Δt and Δf, respectively. Finally, in contrast to radio channels where signals decorrelate relatively fast, in shallow-water acoustic (SWA) channels signals stay highly correlated. The reason for this behavior is the reduced amount of randomness in SWA channel. Note that macro-eigenrays are deterministic and the only randomness is introduced by micro-eigenrays that have small variations around macro-eigenrays.

Finally, Figure 5.30 compares the spatial correlation $R_{pq,\tilde{p}\tilde{q}}(0,0)$ in (5.158) (the model from [23]) with the analytical spatial correlation in [19] and the analytical and empirical spatial correlations in [27]. The ASCOT01 channel measurements are collected at the carrier frequency 1.2 kHz, and the speed of sound is $c = 1440$ m/s (i.e., $\lambda = 1.2$m) [27]. The distance between the T_x and R_x is $R = 10$ km. The water, T_x, and R_x depths are $h = 103$m, $h_T = 50$m, and $h_R = 50$m, respectively. The R_x is equipped with 33 vertically oriented transducer elements (i.e., $\theta_R = 90°$) with $L = d_R = 0.5$m element spacing, while the T_x is equipped with one vertical transducer (i.e., $\theta_T = 90°$). The vertical correlation is calculated with respect to the eighth element from the bottom of the 33-element array. The analytical spatial correlation in (5.158) is obtained with the estimated parameters $K = 0.71$, $S = B = 1$, $\mu_{DA11} = 173.8°$, $\mu_{UA11} = 173.4°$, $\mu_{UA10} = 183°$, $\mu_{UA11} = 184.5°$, $\sigma_{DA10} = 9.1°$, $\sigma_{DA11} = 10.2°$, $\sigma_{UA10} = 5.59°$, $\sigma_{UA11} = 8.89°$, $\eta_S = 0.44$, $\zeta_{\Delta Z_{DA10}} = \zeta_{\Delta Z_{DA11}} = 0.019$, $\zeta_{\Delta Z_{UA10}} = \zeta_{\Delta Z_{UA11}} = 0.015$, and $\eta_B = 1 - \eta_S$. Model parameters R, h, h_T, h_R, θ_T, θ_R, and $\alpha_{TR} \approx \pi$ are selected to match the measurement conditions described earlier. For the reference, Figure 5.30 also shows the spatial correlation functions of the exponential model in [27] (i.e, $\exp\{-L^2/(2\lambda)^2\}$) and the particle-velocity model in [19] (with parameters $\Lambda_b = 0.56$, $\mu_b = 183°$, $\mu_s = 173°$, $\sigma_b = 2.29°$, and $\sigma_s = 8°$). One can observe that the model from Section 5.3.7 provides a closer match with the experimental correlation than the models from [19] and [27]. Although the model from Section 5.3.7 has a large number of parameters that can be adjusted, they are not the reason for the improved fidelity to experimental data. The exponential model in [27] fits an exponential function with distance between hydrophones as a parameter into the measured data. On the other hand, the particle-velocity model in [19] models a channel impulse response in the SWA channel as a large number of random macro-eigenrays, which is an unrealistic assumption because the experimental results show that the number of different macro-eigenrays that arrive at the receiver rarely exceeds eight [17, 27, 28]. In contrast, the model in [23] (and detailed in Section 5.3.7) characterizes the SWA channel as a superposition of a LoS component and several macro-eigenrays, where each macro-eigenray is modeled as a large number of micro-eigenrays. This approach leads to a closer match with the experimental data.

The close agreement between the analytical and measured statistics in Figures 5.26–5.30 confirms the utility of the derived statistics.

5.6 Summary

This chapter first briefly described the propagation principles that govern signal propagation in acoustic underwater vehicle-to-underwater vehicle channels, then proceeded to describe important concepts of propagation modeling such as large-scale and small-scale fading effects. Then, statistical characterization of multipath propagation in UV-to-UV channels was presented, and several models for frequency-flat and frequency-selective channels and their statistics such as correlation function, Doppler spectrum, power delay profile, and level crossing rate, were detailed. Furthermore, simulators for frequency-flat and frequency-selective acoustic channels were presented, with a small number of propagation rays that match the statistical properties of the UV-to-UV channel. Finally, UV-to-UV measurement campaigns were reviewed, and validation of the frequency-selective model for UV-to-UV acoustic channels against the measurements was presented.

References

[1] Stojanović, M., "Acoustic (underwater) communications," *Encyclopedia of Telecommunications*, J. G. Proakis, Ed., New York: John Wiley and Sons, 2003.

[2] Stojanović, M., "Recent advances in high-speed underwater acoustic communications," *IEEE Journal of Oceanic Engineering*, Vol. 21, April 1996, pp. 125–136.

[3] Kilfoyle, D. B., and A. B. Baggeroer, "The state of the art in underwater acoustic telemetry," *IEEE Journal of Oceanic Engineering*, Vol. 25, January 2000, pp. 4–27.

[4] Catipović, J., "Performance limitations in underwater acoustic telemetry," *IEEE Journal of Oceanic Engineering*, Vol. 15, July 1990, pp. 205–216.

[5] Brekhovskikh, L. M., and Y. Lysanov, *Fundamentals of Ocean Acoustics,* 2nd ed., Berlin: Springer, 1991.

[6] Dahl, P. H., "High-frequency forward scattering from the sea surface: the characteristic scales of time and angle spreading," *IEEE Journal of Oceanic Engineering*, Vol. 26, January 2001, pp. 141–151.

[7] Farmer, D. M., G. B. Deane, and S. Vagle, "The influence of bubble clouds on acoustic propagation in the surf zone," *IEEE Journal of Oceanic Engineering*, Vol. 26, January 2001, pp. 113–124.

[8] Stojanović, M., and J. Preisig, "Underwater acoustic communication channels: propagation models and statistical characterization," *IEEE Communications Magazine*, Vol. 47, January 2009, pp. 84–87.

[9] Preisig, J., "Acoustic propagation considerations for underwater acoustic communications network development," *ACM SIGMOBILE Mobile Compomputing and Communications Review*, Vol. 11, Oct. 2007, pp. 2–10.

[10] Stojanović, M., J. Catipović, and J. G. Proakis, "Adaptive multichannel combining and equalization for underwater acoustic communications," *Journal of the Acoustical Society of America*, Vol. 94, Sept. 1993, pp. 1621–1631.

[11] Stojanović, M., "Low complexity OFDM detector for underwater channels,"*MTS/IEEE Proceedings of OCEANS'06*, Boston, MA, Sept. 2006, pp. 1–6.

[12] Li, W., and J. Preisig, "Estimation of rapidly time-varying sparse channels," *IEEE Journal of Oceanic Engineering*, Vol. 32, Oct. 2007, pp. 927–939.

[13] Owen, R. H., B. V. Smith, and R. F. W. Coates, "An experimental study of rough surface scattering and its effects on communication coherence," *MTS/IEEE Proceedings of OCEANS'94*, Vol. 3, Brest, France, Sept. 1994, pp. 483–488.

[14] Essebbar, A., G. Loubet, and F. Vial, "Underwater acoustic channel simulations for communication," *MTS/IEEE Proceedings of OCEANS'94*, Vol. 3, Brest, France, Sept. 1994, pp. 495–500.

[15] Zielinski, A., Y-H. Yoon, and L. Wu, "Performance analysis of digital acoustic communication in a shallow water channel," *IEEE Journal of Oceanic Engineering*, Vol. 20, Oct. 1995, pp. 293–299.

[16] Bjerrum-Niese, C., L. Bjorno, M. A. Pinto, and B. Quellec, "A simulation tool for high data-rate acoustic communication in a shallow-water, time-varying channel," *IEEE Journal of Oceanic Engineering*, Vol. 21, Apr. 1996, pp. 143–149.

[17] Chitre, M., "A high-frequency warm shallow water acoustic communications channel model and measurements," *Journal of the Acoustical Society of America*, Vol. 122, Nov. 2007, pp. 2580–2586.

[18] Yang, W. B., and T. C. Yang, "High-frequency channel characterization for M-ary frequency-shift-keying underwater acoustic communications," *Journal of the Acoustical Society of America*, Vol. 120, Nov. 2006, pp. 2615–2626.

[19] Abdi, A., and H. Guo, "Signal correlation modeling in acoustic vector sensor arrays," *IEEE Transactions on Signal Processing*, Vol. 57, Mar. 2009, pp. 892–903.

[20] Radosević A., J. G. Proakis, M. Stojanović, "Statistical characterization and capacity of shallow water acoustic channels," *MTS/IEEE Proceedings of OCEANS'09-EUROPE*, Bremen, Germany, May 2009, pp. 1–8.

[21] Zajić, A. G., "Statistical space-time-frequency characterization of MIMO shallow water acoustic channels," *MTS/IEEE Proceedings of OCEANS'09*, Biloxi, MS, Oct. 2009, pp. 1–6.

[22] Zajić, A. G., "Statistical modeling of underwater wireless channels," *IEEE Proceedings of Global Telecommunications Conference, GLOBECOM'10*, Miami, FL, Dec. 2010, pp. 1–5.

[23] Zajić, A. G., "Statistical modeling of MIMO mobile-to-mobile underwater channels," *IEEE Transactions on Vehicular Technology*, Vol. 60, April 2011, pp. 1337–1351.

[24] Jakes, W. C., *Microwave Mobile Communications*, 2nd ed. Piscataway, NJ: Wiley-IEEE Press, 1994.

[25] Stüber, G. L., *Principles of Mobile Communication*, 2nd ed., Norwell, MA: Kluwer, 2001.

[26] Zajić, A. G., G. L. Stüber, T. G. Pratt, and S. Nguyen, "Wideband MIMO mobile-to-mobile channels: geometry-based statistical modeling with experimental verification," *IEEE Transactions on Vehicular Technology*, Vol. 58, Feb. 2009, pp. 517–534.

[27] Yang, T. C., "A study of spatial processing gain in underwater acoustic communications," *IEEE Journal of Oceanic Engineering*, Vol. 32, July 2007, pp. 689–709.

[28] Yang, T. C., "Toward continuous underwater acoustic communications," *MTS/IEEE Proceedings of OCEANS'08*, Quebec City, Canada, Sept. 2008, pp. 1–6.

[29] Gradshteyn, I. S. and I. M. Ryzhik, *Table of Integrals, Series, and Products*, San Diego, CA, 5th ed., 1994.

[30] Pätzold, M., U. Killat, and F. Laue, "An extended Suzuki model for land mobile satellite channels and its statistical properties,"*IEEE Transactions on Vehicular Technology*, Vol. 47, May 1998, pp. 617–630.

[31] Nehorai, A., and E. Paldi, "Acoustic vector-sensor array processing," *IEEE Transactions on Signal Processing*, Vol. 42, Sept. 1994, pp. 2481–2491.

[32] Hochwald, B., and A. Nehorai, "Identifiably in array processing models with vector-sensor applications," *IEEE Transactions on Signal Processing*, Vol. 44, Jan. 1996, pp. 83–95.

[33] Zoltowski, M. D., and K. T. Wong, "Closed-form eigenstructure-based direction finding using arbitrary but identical subarrays on a sparse uniform Cartesian array grid," *IEEE Transactions on Signal Processing*, vol. 48, Aug. 2000, pp. 2205–2210.

[34] Hawkes, M., and A. Nehorai, "Wideband source localization using a distributed acoustic vectorsensor array," *IEEE Transactions on Signal Processing*, Vol. 51, June 2003, pp. 1479–1491.

[35] Hawkes, M., and A. Nehorai, "Acoustic vector-sensor beamforming and Capon direction estimation," *IEEE Transactions on Signal Processing*, Vol. 46, Sept. 1998, pp. 2291–2304.

[36] Abdi, A., and H. Guo, "A new compact multichannel receiver for underwater wireless communication networks," *IEEE Transactions on Wireless Communication*, Vol. 8, July 2009, pp. 3326–3329.

[37] Yang, T. C., and A. G. Zajić, "Using the vector sensors as receivers for underwater acoustic communications," *159th Acoustical Society of America Meeting*, Baltimore MD, April 2010.

[38] Cray, B. A., V. M. Evora, and A. H. Nuttall, "Highly directional acoustic receivers," *Journal of Acoustical Society of America*, Vol. 113, Mar. 2003, pp. 1526–1532.

[39] Aik, T. B., Q. S. Sen, and Z. Nan, "Characterization of multipath acoustic channels in very shallow waters for communications," *MTS/IEEE Proceedings of OCEANS'06 Asia-Pacific*, Singapore, May 2007, pp. 1–8.

[40] Badiey, M., Y. Mu, J. A. Simmen, and S. E. Forsythe, "Signal variability in shallow-water sound channels," *IEEE Journal of Oceanic Engineering*, Vol. 25, Oct. 2000, pp. 492–500.

[41] Greenstein, L., J., D. G. Michelson, and V. Erceg, "Moment-method estimation of the Ricean K-factor," *IEEE Communication Letters*, Vol. 3, June 1999, pp. 175–176.

[42] Fletcher, R., *Practical Methods of Optimization: Constrained Optimization*, 2nd ed., New York: John Wiley and Sons, 1981.

APPENDIX A

Probability and Random Processes

The theory of probability and random processes is an essential part in the design and performance analysis of communication systems. This appendix presents a brief review of the basic concepts of probability theory and random processes used in this book. A thorough treatment of this subject is available in a large number of textbooks such as [1, 2].

A.1 Random Variables and Probability

In many experiments, the outcomes of the experiment can be assigned numerical values. For instance, if we roll a die, each outcome has a value from 1 through 6. Another example is a midterm test score of a student in a class, which is again a number.

An experiment whose outcome is not known in advance is called a *random experiment*, and a value representing the outcome of a random experiment is called a *sample point* and will be denoted as s. A collection of possible outcomes of a random experiment is called an *event* and will be denoted as A. The event $A = \{s\}$ consisting of a single element s is called an *elementary event*. The set of all possible outcomes of a given random experiment is called the *sample space* of that experiment and will be denoted as Ω. Hence, a sample point is an element of the event (i.e., $s \in A$), and the event itself is a subset of the sample space (i.e., $A \subset \Omega$).

The sample space Ω is called the *certain event*, and the *empty set*, denoted by $\varnothing$, is the *impossible event*.

For example, if we assemble a deck of 52 playing cards with no jokers and draw a single card from the deck, then that card represents a sample point s; all 52 cards create a 52-element set called the sample space Ω; and an event A is any subset of the sample space Ω, including any singleton set (an elementary event), the empty set (an impossible event, with probability zero) and the sample space itself (a certain event, with probability one).

Random variables can be either discrete or continuous. Discrete random variables take integer values, usually as a result of counting. On the other hand, continuous random variables, in contrast, can take any value within a range of values.

Probability

A mapping $P: \mathcal{A} \rightarrow \mathbb{R}$ is called *probability* if the following conditions are fulfilled:

- (i) If $A \in \mathcal{A}$, then $0 \leq P(A) \leq 1$.
- (ii) $P(\Omega) = 1$.
- (iii) If $A_n \in \mathcal{A}$, $(n = 1, 2, \ldots)$ with $\cup_{n=1}^{\infty} A_n \in \mathcal{A}$ and $A_n \cap A_k = 0$ for any $n \neq k$, then also $P(\cup_{n=1}^{\infty} A_n) = \sum_{n=1}^{\infty} P(A_n)$.

Cumulative distribution function
The function F_η, defined as

$$F_\eta : \mathbb{R} \rightarrow [0,1], \quad x \rightarrow F_\eta(x) = P(\eta \le x) \tag{A.1}$$

is called the *cumulative distribution function* of random variable η. The cumulative distribution function $F_\eta(x)$ has the following properties: (a) $F_\eta(-\infty) = 0$; (b) $F_\eta(\infty) = 1$; and (c) $F_\eta(x)$ is nondecreasing (i.e., $F_\eta(x_1) \le F_\eta(x_2)$ if $x_1 \le x_2$).

Probability density function
Function p_η, defined as

$$p_\eta : \mathbb{R} \rightarrow \mathbb{R}, \quad x \rightarrow p_\eta(x) = \frac{dF_\eta(x)}{dx} \tag{A.2}$$

is called the *probability density function* of the random variable η, under the assumption that the cumulative distribution function $F_\eta(x)$ is differentiable with respect to x. The probability density function $p_\eta(x)$ has the following properties: (a) $p_\eta(x) \ge 0$ for all x; (b) $\int_{-\infty}^{\infty} p_\eta(x)dx = 1$; and (c) $F_n(x) = \int_{-\infty}^{x} p_\eta(x)dx$.

Joint cumulative distribution function
Function $F_{\eta1,\eta2}$, defined as

$$F_{\eta_1,\eta_2} : \mathbb{R}^2 \rightarrow [0,1], \quad (x_1,x_2) \rightarrow F_{\eta_1,\eta_2}(x_1,x_2) = P(\eta_1 \le x_1, \eta_2 \le x_2) \tag{A.3}$$

is called the *joint cumulative distribution function* of random variables η_1 and η_2.

Joint probability density function
The function $p_{\eta1,\eta2}$, defined as

$$p_{\eta_1,\eta_2} : \mathbb{R}^2 \rightarrow \mathbb{R}, \quad (x_1,x_2) \rightarrow p_{\eta_1,\eta_2}(x_1,x_2) = \frac{d^2F_{\eta_1,\eta_2}(x_1,x_2)}{dx_1x_2} \tag{A.4}$$

is called the *joint probability density function* of random variables η_1 and η_2, under the assumption that the joint cumulative distribution function $F_{\eta1,\eta2}(x_1, x_2)$ is partially differentiable with respect to x_1 and x_2.

Random variables η_1 and η_2 are said to be *statistically independent*, if the events $\{s \mid \eta_1(s) \le x_1\}$ and $\{s \mid \eta_2(s) \le x_2\}$ are independent for all $x_1, x_2 \in \mathbb{R}$. In this case, we can write $F_{\eta1,\eta2}(x_1, x_2) = F_{\eta1}(x_1)\, F_{\eta2}(x_2)$ and $p_{\eta1,\eta2}(x_1, x_2) = p_{\eta1}(x_1)\, p_{\eta2}(x_2)$. *Marginal probability density functions:* Given two random variables X and Y whose joint probability density function is known, the marginal probability density function of X is the probability distribution of X averaged over information about Y and is typically calculated by summing or integrating the joint probability distribution over Y. Hence, marginal probability density functions of the joint probability density function $p_{\eta1,\eta2}(x_1, x_2)$ can be obtained as

$$p_{\eta_1}(x_1) = \int_{-\infty}^{\infty} p_{\eta_1,\eta_2}(x_1,x_2)dx_2 \tag{A.5}$$

$$p_{\eta_2}(x_2) = \int_{-\infty}^{\infty} p_{\eta_1,\eta_2}(x_1, x_2)dx_1 \tag{A.6}$$

Expected value (mean value)
Value

$$\mathrm{E}[\eta] = \int_{-\infty}^{\infty} x p_n(x)dx \tag{A.7}$$

is called the *expected value* of the random variable η, and E [·] is called the *expected value operator.* The expected value operator E [·] is linear (i.e., $\mathrm{E}[\alpha\eta] = \alpha\mathrm{E}[\eta]$ for a constant $\alpha \in \mathbb{R}$, and $\mathrm{E}[\eta_1 + \eta_2] = \mathrm{E}[\eta_1] + \mathrm{E}[\eta_2]$). Furthermore, the expected value of a function $f(\eta)$ of random variable η is

$$\mathrm{E}[f(\eta)] = \int_{-\infty}^{\infty} f(\eta)p_\eta(x)dx \tag{A.8}$$

and a generalization to two random variables η_1 and η_2 leads to

$$\mathrm{E}[f(\eta_1,\eta_2)] = \int_{-\infty}^{\infty}\int_{-\infty}^{\infty} f(x_1,x_2)p_{\eta_1,\eta_2}(x_1,x_2)dx_1dx_2 \tag{A.9}$$

Variance
Value

$$\begin{aligned}\mathrm{Var}[\eta] &= \mathrm{E}[(\eta - \mathrm{E}[\eta])^2] \\ &= E[\eta^2] - (E[\eta])^2\end{aligned} \tag{A.10}$$

is the *variance* of random variable η, and Var[·] is the *variance operator.* The variance of a random variable η is a measure of how much η is concentrated near its expected value.

Covariance
Value

$$\begin{aligned}\mathrm{Cov}[\eta_1,\eta_2] &= \mathrm{E}[(\eta_1 - \mathrm{E}[\eta_1])(\eta_2 - \mathrm{E}[\eta_2])] \\ &= \mathrm{E}[\eta_1\eta_2] - \mathrm{E}[\eta_1]\mathrm{E}[\eta_2]\end{aligned} \tag{A.11}$$

is the *covariance* of two random variables η_1 and η_2. The covariance is a measure of how much random variables η_1 and η_2 change together.

Moments
Value

$$\mathrm{E}[\eta^k] = \int_{-\infty}^{\infty} x^k p_\eta(x)dx, \quad k = 0,1,\ldots \tag{A.12}$$

is the *kth moment* of a random variable η. Note that a knowledge of all the moments of η determines its distribution function p_η completely.

Characteristic function

The characteristic function of a random variable η is defined as the expected value

$$\begin{aligned}\Psi_\eta(\upsilon) &= \mathrm{E}[e^{j2\pi\upsilon\eta}] \\ &= \int_{-\infty}^{\infty} p_\eta(x) e^{j2\pi\upsilon x} dx\end{aligned} \tag{A.13}$$

where η is a real-valued valuable. Note that $\Psi_\eta(-\upsilon)$ is the Fourier transform of the probability density function $p_\eta(x)$. The characteristic function often provides a simple technique for determining the probability density function of a sum of statistically independent random variables.

Central limit theorem

Let η_n (n = 1,2, . . . , N) be statistically independent random variables with mean values $\mathrm{E}[\eta_n]$ and variances $\mathrm{Var}[\eta_n]$. Then, the random variable

$$\eta = \lim_{N\to\infty} \frac{1}{\sqrt{N}} \sum_{n=1}^{N} (\eta_n - \mathrm{E}[\eta_n]) \tag{A.14}$$

is normally distributed with the expected value $\mathrm{E}[\eta] = 0$ and variance

$$\mathrm{Var}[\eta] = \lim_{N\to\infty} \frac{1}{N} \sum_{n=1}^{N} \mathrm{Var}[\eta_n] \tag{A.15}$$

A.1.1 Important Probability Density Functions

This section presents a summary of several probability density functions that are often used in channel modeling. The corresponding statistical properties, such as the expected value and variance, are also discussed here.

Uniform distribution

Let $\theta \in [-\pi, \pi)$ be a real-valued random angle with the probability density function

$$p_\theta(x) = \frac{1}{2\pi}, \quad x \in [-\pi, \pi) \tag{A.16}$$

and zero otherwise. Then, $p_\theta(x)$ is called the *uniform distribution*. The expected value and variance of a uniformly distributed random variable θ are $\mathrm{E}[\theta] = 0$ and $\mathrm{Var}[\theta] = \pi^2/3$, respectively. This distribution is often used to characterize phases in channel impulse responses and/or angles of arrival/departure.

Gaussian distribution

Let μ be a real-valued random variable with the probability density function

$$p_\mu(x) = \frac{1}{\sqrt{2\pi}\sigma_\mu} e^{-\frac{(x-m_\mu)^2}{2\sigma_\mu^2}}, \quad x \in \mathbb{R} \tag{A.17}$$

Then $p_\mu(x)$ is called *Gaussian distribution.* In (A.17), the quantity $m_\mu \in \mathbb{R}$ denotes the expected value and $\sigma_\mu^2 \in (0,\infty)$ is the variance of μ; that is,

$$\mathrm{E}[\mu] = m_\mu \tag{A.18}$$

and

$$\begin{aligned}\mathrm{Var}[x] &= \mathrm{E}[\mu^2] - m_\mu^2 \\ &= \sigma_\mu^2\end{aligned} \tag{A.19}$$

To describe the distribution properties of a Gaussian distributed random variable μ, we often use the notation $\mu \sim N(m_\mu, \sigma_\mu^2)$ instead of giving the complete expression (A.17). For $m_\mu = 0$ and $\sigma_\mu = 1$, $N(0,1)$ is called the *standard normal distribution.*

This distribution is used to characterize angles of arrival/departure in underwater acoustic channels.

Multivariate Gaussian distribution

Let us consider n real-valued Gaussian distributed random variables $\mu_1, \mu_2, \ldots, \mu_n$ with expected values $m_{\mu i}$, $(i = 1, 2, \ldots, n)$ and variances $\sigma_{\mu i}$, $(i = 1, 2, \ldots, n)$. The *multivarite Gaussian distribution* of the Gaussian random variables $\mu_1, \mu_2, \ldots, \mu_n$ is defined by

$$p_{\mu_1\mu_2\ldots\mu_n}(x_1, x_2, \ldots, x_n) = \frac{1}{(\sqrt{2\pi})^n \sqrt{\det \mathbf{C}_\mu}} e^{-\frac{1}{2}(\mathbf{x}-\mathbf{m}_\mu)^T \mathbf{C}_\mu^{-1}(\mathbf{x}-\mathbf{m}_\mu)} \tag{A.20}$$

where $(\cdot)^{\mathrm{T}}$ denotes the transpose of a vector (or a matrix). In (A.20), $\mathbf{x}$ and $\mathbf{m}_\mu$ are column vectors given by

$$\mathbf{x} = \begin{bmatrix} x_1 \\ x_2 \\ \vdots \\ x_n \end{bmatrix} \in \mathbb{R}^{n\times 1} \tag{A.21}$$

and

$$\begin{aligned}\mathbf{m}_\mu &= \begin{bmatrix} \mathrm{E}[\mu_1] \\ \mathrm{E}[\mu_2] \\ \vdots \\ \mathrm{E}[\mu_n] \end{bmatrix} \\ &= \begin{bmatrix} m_{\mu_1} \\ m_{\mu_2} \\ \vdots \\ m_{\mu_n} \end{bmatrix} \in \mathbb{R}^{n\times 1}\end{aligned} \tag{A.22}$$

respectively, $\det \mathbf{C}_\mu$ is the determinant, and $(\mathbf{C}_\mu^{-1})$ is the inverse of the *covariance matrix*

$$\mathbf{C}_\mu = \begin{bmatrix} C_{\mu_1\mu_1} & \cdots & C_{\mu_1\mu_n} \\ C_{\mu_2\mu_1} & \cdots & C_{\mu_2\mu_n} \\ \vdots & \ddots & \vdots \\ C_{\mu_n\mu_1} & \cdots & C_{\mu_n\mu_n} \end{bmatrix} \in \mathbb{R}^{n\times n} \tag{A.23}$$

The elements of the covariance matrix $\mathbf{C}_\mu$ are given by

$$\begin{aligned} C_{\mu_i\mu_j} &= \mathrm{Cov}[\mu_i, \mu_j] \\ &= \mathrm{E}[(\mu_i - m_{\mu_i})(\mu_j - m_{\mu_j})], \quad i,j = 1,2,\ldots, n \end{aligned} \tag{A.24}$$

If the n random variables μ_i are normally distributed and each pair of these variables is uncorrelated, then the covariance matrix $\mathbf{C}_\mu$ is a diagonal matrix with diagonal entries $\sigma^2_{\mu_i}$. In this case, the joint probability density function (A.20) decomposes into a product of n Gaussian distributions, and the random variables μ_i are said to be statistically independent.

Rayleigh distribution

Let us consider two zero-mean statistically independent normally distributed random variables μ_1 and μ_2, each having a variance σ^2 [i.e., μ_1, $\mu_2 \sim N(0, \sigma^2)$]. Furthermore, let $\mu = \sqrt{\mu_1^2 + \mu_2^2}$ be a new random variable. Then, μ represents a *Rayleigh distributed* random variable. The probability density function $p_\mu(x)$ of Rayleigh distributed random variable μ is

$$p_\mu(x) = \begin{cases} \dfrac{x}{\sigma^2} e^{-\frac{x^2}{2\sigma^2}}, & x \geq 0, \\ 0, & x < 0, \end{cases} \tag{A.25}$$

A Rayleigh distributed random variable μ has an expected value

$$\mathrm{E}[\mu] = \sigma\sqrt{\frac{\pi}{2}} \tag{A.26}$$

and variance

$$\mathrm{Var}[\mu] = \sigma^2\left(2 - \frac{\pi}{2}\right) \tag{A.27}$$

This distribution is often used to characterize a received signal envelope under NLoS propagation conditions. Figure A.1 illustrates the probability density function $p_\mu(x)$ of Rayleigh distributed random variable x with variance $\sigma^2 = 2$.

Rice distribution

Let μ_1, $\mu_2 \sim N(0, \sigma^2)$ and $\rho \in \mathbb{R}$. Then, random variable $\mu = \sqrt{(\mu_1 + \rho)^2 + \mu_2^2}$ is a *Rice distributed* random variable. The probability density function $p_\mu(x)$ of Rice distributed random variable μ is

$$p_\mu(x) = \frac{x}{\sigma^2} e^{-\frac{x^2+\rho^2}{2\sigma^2}} I_0\left(\frac{x\rho}{\sigma^2}\right), \quad x \geq 0 \tag{A.28}$$

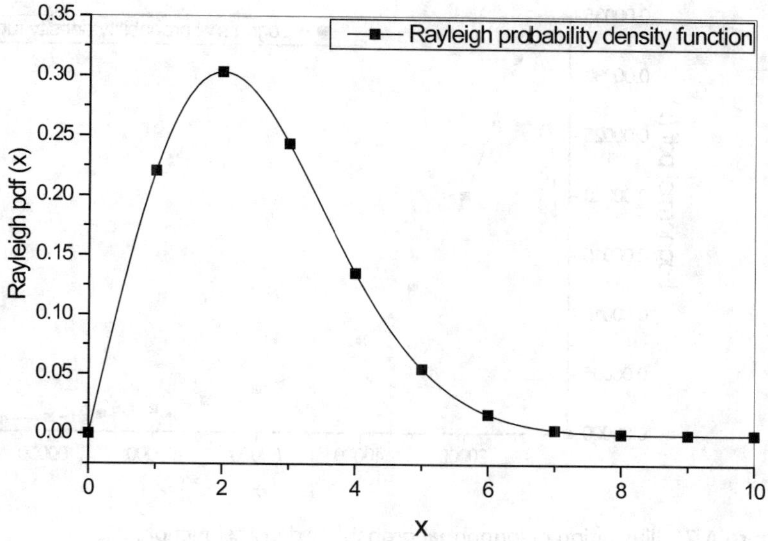

Figure A.1 Illustration of Rayleigh probability density function.

where $I_0(\cdot)$ denotes the modified Bessel function of zeroth order. For $\rho = 0$, the Rice distribution $p_\mu(x)$ results in the Rayleigh distribution described earlier. The mean and the variance of Rice distributed random variable μ are

$$\mathrm{E}[\mu] = \sigma\sqrt{\frac{\pi}{2}}e^{-\frac{\rho^2}{4\sigma^2}}\left\{\left(1+\frac{\rho^2}{2\sigma^2}\right)I_0\left(\frac{\rho^2}{4\sigma^2}\right)+\frac{\rho^2}{2\sigma^2}I_1\left(\frac{\rho^2}{4\sigma^2}\right)\right\} \tag{A.29}$$

$$\mathrm{Var}[\mu] = 2\sigma^2 + \rho^2 - \mathrm{E}[\mu]^2 \tag{A.30}$$

respectively, where $I_n(\cdot)$ denotes the modified Bessel function of nth order. This distribution is often used to characterize a received signal envelope under LoS propagation conditions.

Lognormal distribution

Let μ be a Gaussian distributed random variable with an expected value m_μ and variance σ_μ^2 (i.e., $\mu \sim N(m_\mu, \sigma_\mu^2)$). Then, random variable $\lambda = e^\mu$ is said to be *lognormally distributed*. The probability density function $p_\lambda(x)$ of a lognormally distributed random variable λ is

$$p_\lambda(x) = \begin{cases} \dfrac{1}{\sqrt{2\pi}\sigma_\mu x}e^{-\frac{(\ln x - m_\mu)^2}{2\sigma_\mu^2}}, & x \geq 0 \\ 0, & x < 0 \end{cases} \tag{A.31}$$

The expected value and variance of a lognormally distributed random variables λ are

$$\mathrm{E}[\lambda] = e^{m_\mu + \frac{\sigma_\mu^2}{2}} \tag{A.32}$$

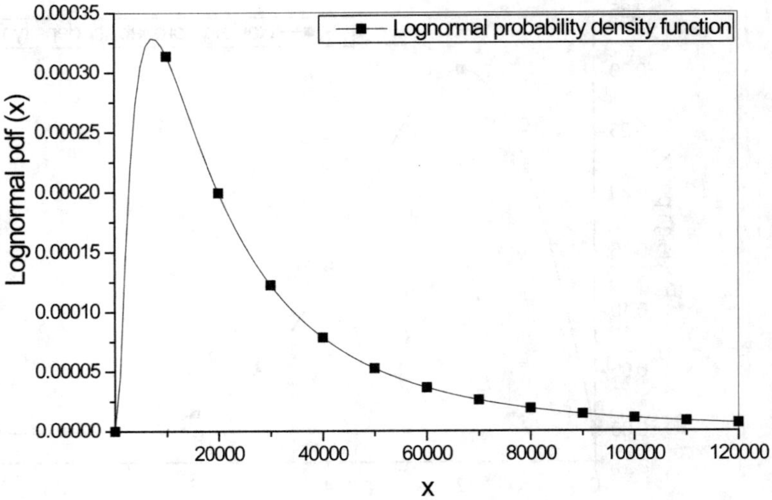

Figure A.2 Illustration of lognormal probability density function.

$$\mathrm{Var}[\lambda] = e^{2m_\mu+\sigma_\mu^2}\left(e^{\sigma_\mu^2}-1\right) \tag{A.33}$$

This distribution is often used to characterize shadowing in wireless channels. Figure A.2 illustrates the probability density function $p_\mu(x)$ of lognormal distributed random variable x with mean $m_\mu = log_{10}(20000)$ and variance $\sigma_\mu^2 = 1$.

Suzuki distribution

Let us assume that a random variable μ is Rayleigh distributed with the distribution defined in (A.25), that a random variable λ is lognormally distributed with the distribution defined in (A.31), and that μ and λ are statistically independent. Then, a random variable η defined by the product $\eta = \mu \cdot \lambda$ is said to be *Suzuki distributed.* The probability density function of a Suzuki distributed random variable η is

$$p_\eta(x) = \begin{cases} \dfrac{x}{\sqrt{2\pi\sigma^2}\sigma_\mu}\displaystyle\int_0^\infty \frac{1}{y^3}e^{-\frac{x^2}{2y^2\sigma^2}}.e^{-\frac{(\ln y-m_\mu)^2}{2\sigma_\mu}}dy, & x \ge 0 \\ 0, & x < 0 \end{cases} \tag{A.34}$$

A Suzuki distributed random variable η has an expected value and variance

$$\mathrm{E}[\eta] = \sigma\sqrt{\frac{\pi}{2}}e^{m_\mu+\frac{\sigma_\mu^2}{2}} \tag{A.35}$$

$$\mathrm{Var}[\eta] = \sigma^2 \cdot e^{2m_\mu+\sigma_\mu^2} \cdot \left(2\,e^{\sigma_\mu^2} - \frac{\pi}{2}\right) \tag{A.36}$$

This distribution is often used to characterize joint small-scale and large-scale effects in radio wireless channels.

Nakagami distribution

Let us consider a random variable ω distributed according to the probability density function

$$p_\omega(x) = \begin{cases} \dfrac{2m^m x^{2m-1} e^{-(m/\Omega)x^2}}{\Gamma(m)\Omega^m}, & m \geq 0.5, x \geq 0 \\ 0, & x < 0 \end{cases} \quad (A.37)$$

Then, ω is a *Nakagami distributed* random variable and the corresponding probability density function $p_\omega(x)$ is called the *Nakagami distribution*. Symbol $\Gamma(\cdot)$ in (A.37) denotes the Gamma function, Ω is the second moment of the random variable ω ($\Omega = \mathrm{E}[\omega^2]$), and m denotes the reciprocal value of the variance of ω^2 normalized to Ω^2 ($m = \Omega^2/\mathrm{E}[(\omega^2 - \Omega^2)]$). From the Nakagami distribution, we can obtain the one-sided Gaussian distribution and the Rayleigh distribution as special cases for $m = 0.5$ and $m = 1$, respectively.

Von Mises distribution

The von Mises distribution (also known as the circular normal distribution or Tikhonov distribution) is a continuous probability distribution defined on a circle. The von Mises probability density function for the angle $\alpha \in [-\pi, \pi)$ is given by [3]:

$$p_\alpha(x) = \frac{1}{2\pi I_0(k)} \exp[k\cos(x - \mu)] \quad (A.38)$$

where $I_0(\cdot)$ is the zeroth-order modified Bessel function of the first kind, $\mu \in [-\pi, \pi)$ is a measure of location (the distribution is clustered around μ), and k is a measure of concentration. When $k = 0$, the von Mises probability density function becomes a uniform probability density function on $[-\pi, \pi)$ (i.e., $p_\alpha(x) = 1/(2\pi)$). A von Mises distributed random variable α has the expected value and the variance given by

$$\mathrm{E}[\alpha] = \frac{I_1(k)}{I_0(k)} e^{j\mu} \quad (A.39)$$

$$\mathrm{Var}[\alpha] = 1 - \frac{I_1(k)}{I_0(k)} \quad (A.40)$$

This distribution is often used to characterize angles of arrival/departure in radio wireless channels.

A.1.2 Functions of Random Variables

In this book, we often use addition and multiplication of random variables. Hence, here we briefly review these mathematical principles.

Addition of two random variables

Let μ_1 and μ_2 be two random variables that are statistically characterized by the joint probability density function $p_{\mu_1\mu_2}(x_1,x_2)$. Then, the probability density function of the sum $\mu = \mu_1 + \mu_2$ is

$$\begin{aligned} p_\mu(y) &= \int_{-\infty}^{\infty} p_{\mu_1\mu_2}(x_1, y - x_1)dx_1 \\ &= \int_{-\infty}^{\infty} p_{\mu_1\mu_2}(x_2, y - x_2)dx_2 \end{aligned} \tag{A.41}$$

If the two random variables μ_1 and μ_2 are statistically independent, then the probability density function of μ is given by the convolution of the probability density functions of μ_1 and μ_2; that is,

$$\begin{aligned} p_\mu(y) &= p_{\mu_1}(y) * p_{\mu_2}(y) \\ &= \int_{-\infty}^{\infty} p_{\mu_1}(x_1)p_{\mu_2}(y - x_1)dx_1 \\ &= \int_{-\infty}^{\infty} p_{\mu_1}(y - x_2)p_{\mu_2}(x_2)dx_2 \end{aligned} \tag{A.42}$$

where $(\cdot)^*$ denotes the convolution operator.

Multiplication of two random variables

Let μ and λ be two random variables that are statistically described by the joint probability density function $p_{\mu\lambda}(x,y)$. Then, the probability density function of the random variable $\eta = \mu \cdot \lambda$ is

$$p_\eta(z) = \int_{-\infty}^{\infty} \frac{1}{|y|} p_{\mu\lambda}\left(\frac{z}{y}, y\right) dy \tag{A.43}$$

If the random variables μ and λ are statistically independent, the probability density function of the random variable η becomes

$$p_\eta(z) = \int_{-\infty}^{\infty} \frac{1}{|y|} p_\mu\left(\frac{z}{y}\right) \cdot p_\lambda(y)dy \tag{A.44}$$

A.2 Random Processes

A random process, also called a stochastic process, $X(t)$, is an ensemble of sample functions $\{X_1(t), X_2(t), \ldots, X_\eta(t)\}$ together with a probability rule that assigns a probability to any event associated with the observation of these functions. Then, the sample function $X_i(t)$ corresponds to the sample point s_i in the sample space and occurs with probability $P(s_i)$. The number of sample functions, η, in the ensemble may be finite or infinite. The function $X_i(t)$ is deterministic once the index i is known. Sample functions may be defined at discrete instants or be continuous in time, and their values at these time instants may also be discrete or continuous.

Suppose that we observe all the sample functions at some time instant t_1 and their values form the set of numbers $\{X_i(t_1)\}$, $i = 1, 2, \ldots, \eta$. Since $X_i(t_1)$ occurs with probability $P(s_i)$, the collection of numbers $\{X_i(t_1)\}$, $i = 1, 2, \ldots, \eta$ forms a random variable, denoted by $X(t_1)$. By observing the same set of functions at another time instant t_2 we obtain a different random variable $X(t_2)$. A collection of n such random variables, $X(t_1), X(t_2), \ldots, X(t_n)$ has the joint cdf

$$F_{X(t_1),\ldots X(t_n)}(x_1,\ldots,x_n) = P(X(t_1) < x_1,\ldots,X(t_n) < x_n) \tag{A.45}$$

A more compact notation can be obtained by defining vectors

$$\mathbf{x} = (x_1, x_2, \ldots, x_n) \tag{A.46}$$

$$\mathbf{X} = (X(t_1), X(t_2), \ldots, X(t_n)) \tag{A.47}$$

Then, the joint cdf and joint pdf are, respectively,

$$F_{\mathbf{x}(t)}(\mathbf{x}) = P(\mathrm{X}(t) \leq \mathrm{x}) \tag{A.48}$$

$$p_{\mathbf{x}(t)}(\mathbf{x}) = \frac{d^n F_{\mathbf{x}(t)}}{dx_1 dx_2 \cdots dx_n} \tag{A.49}$$

A random process is *strictly stationary* if and only if the joint density function $p_{\mathbf{x}(t)}(\mathbf{x})$ is invariant under shifts of the origin along the time axis; that is, when

$$p_{\mathbf{x}(t)}(\mathbf{x}) = p_{\mathbf{x}(t+r)}(\mathbf{x}) \tag{A.50}$$

holds for all sets of time instants $\{t_1, t_2, \ldots, t_n\}$ and all time shifts τ. Many important random processes that are encountered in practice are strictly stationary.

A.2.1 Moments and Correlation Functions

To describe the moments and correlation functions of a random process, it is useful to define the operator $E[\cdot]$ as an ensemble average and $< \cdot >$ as a time average.

The ensemble average of a random process at time t is

$$E[X(t)] = \int_{-\infty}^{\infty} x p_{X(t)}(x) dx \tag{A.51}$$

(i.e., it is the expected value of process $X(t)$ at time t).

The time average of a random process is

$$< X(t) > = \lim_{T \to \infty} \frac{1}{2T} \int_{-T}^{T} X(t) dt \tag{A.52}$$

(i.e., it is a function that represents an average of a particular sample function over all possible time instances). In general, the time average $< X(t) >$ is still a random variable, because it depends on the particular sample function that is selected for time averaging.

The *autocorrelation* of a random process $X(t)$ is defined as

$$R_{XX}(t_1,t_2) = \mathrm{E}[X(t_1)X(t_2)] \tag{A.53}$$

The *autocovariance* of a random process $X(t)$ is defined as

$$\begin{aligned} C_{XX}(t_1,t_2) &= \mathrm{E}[(X(t_1)-\mathrm{E}[X(t_1)])(X(t_2)-\mathrm{E}[X(t_2)])] \\ &= R_{XX}(t_1,t_2)-\mathrm{E}[X(t_1)]\mathrm{E}(X(t_2)] \end{aligned} \tag{A.54}$$

From (A.50) follows that a random process is strictly stationary only if

$$\mathrm{E}[X^n(t)] = \mathrm{E}[X^n] \quad \forall t,n \tag{A.55}$$

Then, for a strictly stationary random process, the following statements are also true:

$$\mathrm{E}[X(t)] = \mathrm{E}[X] \tag{A.56}$$

$$\mathrm{Var}[X(t)] = \mathrm{Var}[X] \tag{A.57}$$

$$R_{XX}(t_1,t_2) = R_{XX}(t_1-t_2) = R_{XX}(\tau) \tag{A.58}$$

$$C_{XX} = C_{XX}(t_1-t_2) = C_{XX}(\tau) \tag{A.59}$$

where $\tau = t_1 - t_2$.

If a random process satisfies the following conditions

$$\mathrm{E}[X(t)] = \mathrm{E}[X] \tag{A.60}$$

$$R_{XX}(t_1,t_2) = R_{XX}(\tau) \tag{A.61}$$

then it is said to be *wide sense stationary.* Note that a strictly stationary random process is always wide sense stationary, but the converse may not be true. However, a Gaussian random process is strictly stationary if it is wide sense stationary. The reason is that a joint Gaussian probability density function of the vector $\mathbf{X} = (X(t_1), X(t_2), \ldots, X(t_n))$ is completely described by the means and covariances of $X(t_i)$.

Properties of autocorrelation function

The autocorrelation function $R_{XX}(\tau)$ of a stationary random process satisfies the following properties:

$$\begin{aligned} R_{XX}(0) &= \mathrm{E}[X^2(t)] \\ R_{XX}(\tau) &= R_{XX}(-\tau) \\ |R_{XX}(\tau)| &\le R_{XX}(0) \\ R_{XX}(\infty) &= \mathrm{E}[X(t)]^2 \end{aligned} \tag{A.62}$$

Also, if $X(t) = X(t + \tau)$, then $R_{XX}(\tau) = R_{XX}(\tau + T)$ (i.e., if $X(t)$ is periodic, then $R_{XX}(\tau)$ is also periodic).

A random process is said to be *ergodic* if for all **X** and any function of **X**; that is, $g(\mathbf{X})$,

$$\begin{aligned} E[g(\mathbf{X})] &= \int_{-\infty}^{\infty} g(\mathbf{X}) p_{X(t)}(x)dx \\ &= \lim_{T\to\infty} \frac{1}{2T} \int_{-T}^{T} g[\mathbf{X}(t)]dt \\ &= \langle g[\mathbf{X}(t)] \rangle \end{aligned} \tag{A.63}$$

For a random process to be ergodic, it must be strictly stationary. However, not all strictly stationary random processes are ergodic. A random process is *ergodic in the mean* if

$$\langle X(t) \rangle = E[X] \tag{A.64}$$

and *ergodic in the autocorrelation* if

$$\langle X(t+\tau)X(t) \rangle = R_{XX}(\tau) \tag{A.65}$$

A.2.2 Crosscorrelation and Crosscovariance

Consider two random processes $X(t)$ and $Y(t)$. The *crosscorrelation* of $X(t)$ and $Y(t)$ is

$$R_{XY}(t_1,t_2) = E[X(t_1)Y(t_2)] \tag{A.66}$$

$$R_{YX}(t_1,t_2) = E[Y(t_1)X(t_2)] \tag{A.67}$$

The *correlation matrix* of $X(t)$ and $Y(t)$ is

$$\mathbf{R}(t_1,t_2) = \begin{bmatrix} R_{XX}(t_1,t_2) & R_{XY}(t_1,t_2) \\ R_{YX}(t_1,t_2) & R_{YY}(t_1,t_2) \end{bmatrix} \tag{A.68}$$

The *crosscovariance* of $X(t)$ and $Y(t)$ is

$$\begin{aligned} C_{XY}(t_1,t_2) &= E[(X(t_1) - E[X(t_1)])(Y(t_2) - E[Y(t_2)])] \\ &= R_{XY}(t_1,t_2) - E[X(t_1)]E[Y(t_2)] \end{aligned} \tag{A.69}$$

The *covariance matrix* of $X(t)$ and $Y(t)$ is

$$\mathbf{C}(t_1,t_2) = \begin{bmatrix} C_{XX}(t_1,t_2) & C_{XY}(t_1,t_2) \\ C_{YX}(t_1,t_2) & C_{YY}(t_1,t_2) \end{bmatrix} \tag{A.70}$$

If $X(t)$ and $Y(t)$ are each wide sense stationary and jointly wide sense stationary, then

$$\mathbf{R}(t_1,t_2) = \mathbf{R}(t_1 - t_2) = \mathbf{R}(\tau) \tag{A.71}$$

$$\mathbf{C}(t_1,t_2) = \mathbf{C}(t_1 - t_2) = \mathbf{C}(\tau) \tag{A.72}$$

where $\tau = t_1 - t_2$.

Properties of the crosscorrelation function

The crosscorrelation function $R_{XY}(\tau)$ has the following properties:

$$R_{XY}(\tau) = R_{XY}(-\tau)$$

$$|R_{XY}(\tau)| \le \frac{1}{2}[R_{XX}(0) + R_{YY}(0)]$$

$|R_{XY}(\tau)|^2 \le R_{XX}(0)R_{YY}(0)$, if X(t) and Y(t) have zero mean.

Classifications of random processes

Two random processes $X(t)$ and $Y(t)$ are said to be

- *uncorrelated* if and only if $C_{XY}(\tau) = 0$.
- *orthogonal* if and only if $R_{XY}(\tau) = 0$.
- *statistically independent* if and only if

$$p_{\mathbf{X}(t)\mathbf{Y}(t+\tau)}(\mathbf{x},\mathbf{y}) = p_{\mathbf{x}(t)}(\mathbf{x})p_{\mathbf{Y}(t+\tau)}(\mathbf{y}) \tag{A.73}$$

Note that if E[X] = 0 or E[Y] = 0, then the orthogonal random processes are uncorrelated and vice versa. Similarly, under the same conditions, the statistically independent random processes are also uncorrelated. However, the uncorrelated random processes are not necessarily statistically independent.

A.2.3 Complex-Valued Random Processes

A complex-valued random process is given by

$$Z(t) = X(t) + jY(t),$$

where $X(t)$ and $Y(t)$ are real random processes.

Autocorrelation function

The autocorrelation function of a complex-valued random process is

$$\begin{aligned} R_{ZZ}(t_1,t_2) &= \frac{1}{2}E[Z(t_1)Z^*(t_2)] \\ &= \frac{1}{2}E[(X(t_1)+jY(t_1))(X(t_2) - jY(t_2))] \\ &= \frac{1}{2}\{R_{XX}(t_1,t_2) + R_{YY}(t_1,t_2) \\ &= j[R_{YX}(t_1,t_2) - R_{XY}(t_1,t_2)]\} \end{aligned} \tag{A.74}$$

where $(\cdot)^*$ is the complex conjugate operation. If $Z(t)$ is wide sense stationary, then

$$R_{ZZ}(t_1, t_2) = R_{ZZ}(t_1 - t_2) = R_{ZZ}(\tau) \tag{A.75}$$

Crosscorrelation function

Consider two complex-valued random processes

$$Z(t) = X(t) + jY(t)$$

$$W(t) = U(t) + jV(t)$$

The crosscorrelation function of $Z(t)$ and $W(t)$ is

$$\begin{aligned} R_{ZW}(t_1, t_2) &= \frac{1}{2} \mathrm{E}[Z(t_1) W^*(t_2)] \\ &= \frac{1}{2} \{R_{XV}(t_1, t_2) + R_{YV}(t_1, t_2) \\ &\quad + j[R_{YU}(t_1, t_2) - R_{XV}(t_1, t_2)]\} \end{aligned} \tag{A.76}$$

If $X(t)$, $Y(t)$, $U(t)$, and $V(t)$ are pairwise wide sense stationary random processes, then

$$R_{ZW}(t_1, t_2) = R_{ZW}(t_1 - t_2) = R_{ZW}(\tau) \tag{A.77}$$

The crosscorrelation function of a complex wide sense stationary random process has the following properties

$$R^*_{ZW}(\tau) = R_{ZW}(-\tau) \tag{A.78}$$

$$R^*_{XX}(\tau) = R_{XX}(-\tau) \tag{A.79}$$

$$R_{XX}(\tau) = R^*_{XX}(-\tau) \tag{A.80}$$

A.2.4 Power Spectral Density

The power spectral density (psd) of a random process $X(t)$ is the Fourier transform of the autocorrelation function; that is,

$$S_{XX}(f) = \int_{-\infty}^{\infty} R_{XX}(\tau) e^{-j2\pi f \tau} d\tau \tag{A.81}$$

If $X(t)$ is real, then $R_{XX}(\tau)$ and $S_{XX}(f)$ are also real and even. If $X(t)$ is complex, then $R_{XX}(\tau) = R^*_{XX}(-\tau)$ and $S^*_{XX}(f) = S_{XX}(f)$, meaning that $S_{XX}(f)$ is real but not necessarily even.

The power of a random process $X(t)$ is

$$\begin{aligned} P &= \mathrm{E}[X^2(t)] \\ &= R_{XX}(0) \\ &= \int_{-\infty}^{\infty} S_{XX}(f)df \end{aligned} \tag{A.82}$$

This result is known as *Parseval's theorem.*
The *cross power spectral density* between two random processes $X(t)$ and $Y(t)$ is

$$S_{XY}(f) = \int_{-\infty}^{\infty} R_{XY}(\tau)e^{j2\pi f\tau}d\tau \tag{A.83}$$

If $X(t)$ and $Y(t)$ are both real random processes, then

$$R_{XY}(\tau) = R_{YX}(\tau) \tag{A.84}$$

$$S_{XY}(f) = S_{YX}(-f) \tag{A.85}$$

On the other hand, if $X(t)$ and $Y(t)$ are complex random processes, then

$$R^*_{XY}(\tau) = R_{YX}(-\tau) \tag{A.86}$$

$$S^*_{XY}(f) = S_{YX}(f) \tag{A.87}$$

A.2.5 Deterministic Continuous-Time Signals

A *deterministic (continuous-time) signal* is usually defined over a set of real numbers $\mathbb{R}$. The set $\mathbb{R}$ is the temporal space in which the variable t takes its values, (i.e., $t \in \mathbb{R}$). A deterministic signal is described as a function (mapping) $\hat{\mu}$, in which each value of t is assigned to a real-valued (or complex-valued) number $\hat{\mu}(t)$. (i.e., $\hat{\mu}$: $\mathbb{R} \to \mathbb{R}$ or ($\mathbb{C}$)). The following functions are of importance when operating with deterministic continuous-time signals.

Mean value
The mean value of a deterministic signal $\hat{\mu}(t)$ is defined as

$$\mathrm{E}[\hat{\mu}(t)] = \lim_{T\to\infty} \frac{1}{2T} \int_{-T}^{T} \hat{\mu}(t)dt \tag{A.88}$$

Mean power
The mean power of a deterministic signal $\hat{\mu}(t)$ is defined as

$$P_{\hat{\mu}} = \lim_{T\to\infty} \frac{1}{2T} \int_{-T}^{T} |\hat{\mu}(t)|^2 \, dt \tag{A.89}$$

Autocorrelation function
The autocorrelation function of a deterministic signal $\hat{\mu}(t)$ is defined as

$$R_{\hat{\mu}\hat{\mu}}(\tau) = \lim_{T\to\infty} \frac{1}{2T} \int_{-T}^{T} \hat{\mu}^*(t)\hat{\mu}(t+\tau)dt \tag{A.90}$$

Crosscorrelation function

The crosscorrelation function of deterministic signals $\hat{\mu}_1(t)$ and $\hat{\mu}_2(t)$ is defined as

$$R_{\hat{\mu}_1\hat{\mu}_2}(\tau) = \lim_{T\to\infty} \frac{1}{2T} \int_{-T}^{T} \hat{\mu}_1^*(t)\hat{\mu}_2(t+\tau)dt \tag{A.91}$$

Power spectral density

Power spectral density is defined as the Fourier transform of the autocorrelation function $R_{\hat{\mu}\hat{\mu}}(\tau)$; that is,

$$S_{\hat{\mu}\hat{\mu}}(f) = \int_{-\infty}^{\infty} R_{\hat{\mu}\hat{\mu}}(\tau)e^{-j2\pi f\tau}d\tau \tag{A.92}$$

Crosspower spectral density

Cross-power spectral density is denned as the Fourier transform of the crosscorrelation function $R_{\hat{\mu}_1\hat{\mu}_2}(\tau)$; that is,

$$S_{\hat{\mu}_1\hat{\mu}_2}(f) = \int_{-\infty}^{\infty} R_{\hat{\mu}_1\hat{\mu}_2}(\tau)e^{-j2\pi f\tau}d\tau \tag{A.93}$$

References

[1] Papoulis A., *Probability, Random Variables and Stochastic Processes,* New York: McGraw Hill, 1984.

[2] Leon-Garcia A., *Probability and Random Processes for Electrical Engineering,* Reading, MA: Addison-Wesley, 1989.

[3] Von Mises R. "Über die 'Ganzzahligkeit' der Atomgewicht und verwandte Fragen," *Physikal. Z.,* Vol. 19, 1918, pp. 490–500.

APPENDIX B

Description of Simulators for Fixed-to-Mobile Cellular, Vehicle-to-vehicle, Air-to-Ground, and Underwater Vehicle-to-Underwater Vehicle Channels

The purpose of a channel simulator is to reproduce the features of a physical channel that are critical for testing the performance of a communication system. These features include effects of large-scale fading (i.e., path loss and shadowing), effects of small-scale fading (i.e., multipath fading), and channel statistics such as correlation function, Doppler spectrum, and level crossing rate. To help readers better understand, use, and implement the propagation theory presented in Chapters 2–5, this book also provides MATLAB-based simulators for fixed-to-mobile (F-to-M) cellular, vehicle-to-vehicle (V-to-V), air-to-ground (A-to-G), and underwater vehicle-to-underwater vehicle (UV-to-UV) channels. This chapter explains how to use these simulators.

B.1 Simulator for Fixed-to-Mobile Cellular Channels

This section describes the MATLAB-based simulator for fixed-to-mobile (F-to-M) cellular channels, which can be found on the CD associated with this book, in the *F-to-M Channel* directory. This simulator is based on the 2-D channel models described in Sections 2.3.2, 2.3.3, 2.3.4, 2.3.5, and 2.4.2 and on the 3-D model described in [1].

The simulator is initiated by running the *F_to_M_channel_start.m* file. This starts the graphic user interface (GUI) shown in Figure B.l. The input parameters are divided into five categories: system, type of propagation, mobile movement, SISO/MIMO block, and antenna orientation parameters.

Among the system parameters are the carrier frequency (in GHz), sampling frequency (in kHz), number of samples for the baseband signal that is modulated onto the carrier signal, and the distance between the transmitter and the receiver (in meters).

The "type of propagation" parameters allow one to choose between Rayleigh fading and Ricean fading (i.e., between non-line-of-sight and line-of-sight propagation). If Ricean fading is selected, the value for parameter K should also be specified. For F-to-M cellular systems, the parameter K ranges between 1 and 10. The spread of scatterers and maximum elevation angle around the receiver implicitly select between isotropic and nonisotropic scattering and between 2-D and 3-D models, respectively. If the spread of scatterers around the receiver is zero, the simulator

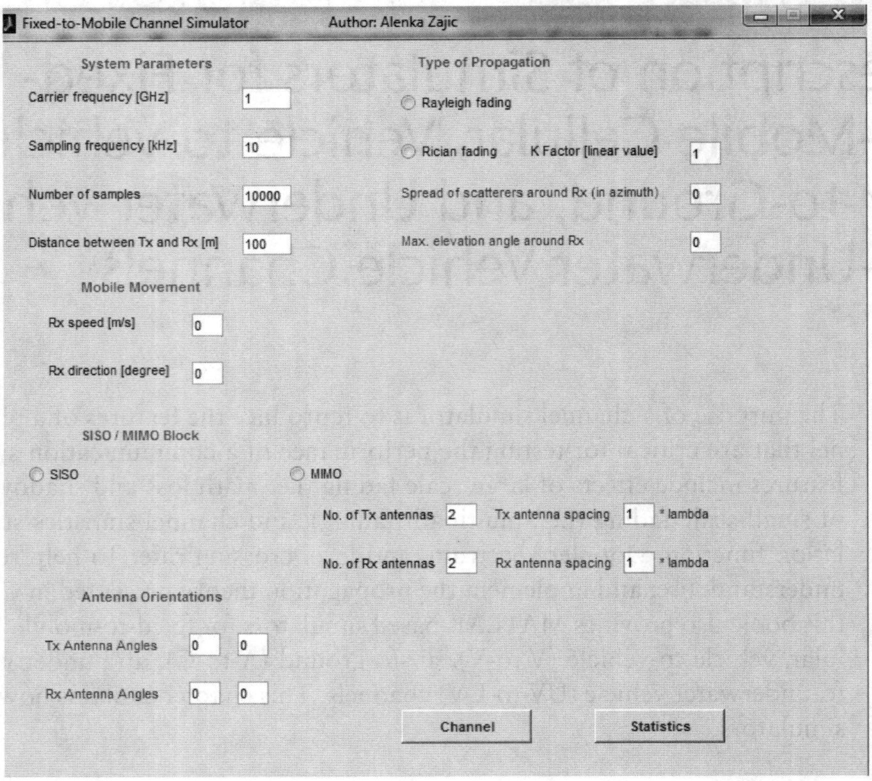

Figure B.1 GUI for fixed-to-mobile (cellular) channel simulator.

assumes isotropic scattering and uses the narrowband model described in Sections 2.3.2 and 2.3.3. On the other hand, when the spread of scatterers around the receiver is larger than zero, the simulator assumes nonisotropic scattering and uses the narrowband model described in Section 2.3.4. For F-to-M cellular systems, this value varies between 1 and 10. Similarly, if the maximum elevation angle around the receiver is zero, the simulator uses 2-D narrowband model described in Sections 2.3.2 and 2.3.4. If the maximum elevation angle around the receiver is greater than zero, the simulator uses the 3-D model from [1]. For F-to-M cellular systems, the elevation angle in 3-D models is typically between 10° and 80°.

In the mobile movement category, the user selects the velocity of the receiver (in m/s) and its direction of movement (in degrees).

In the SISO/MIMO block category, the user chooses between a single-input single-output (SISO) and a multiple-input multiple-output (MIMO) system. If the MIMO option is selected, the user needs to specify the number of transmit antennas and the spacing among them, as well as the number of receive antennas and the spacing among them.

Finally, in the antenna orientation category, the user specifies the azimuthal and elevation angles for antennas. Here, we assume that a MIMO system is using same antenna elements at the transmitter and at the receiver.

The simulator generates the channel impulse response as well as the channel statistics (the auto- and cross-correlation functions, the Doppler spectrum, and the level crossing rate). The channel impulse response is obtained when the "Channel" button is pressed, while the statistics are obtained when the "Statistics" button is pressed. The results for the channel impulse response are saved in a file named *channel.m*. This file contains the matrix *H* with channel impulse responses and the time array *t*. The results for the channel statistics are saved in a file named *channel_statistics.mat*. The file contains the correlation matrix called *SpaceTime_CorrelationFunction* and the corresponding time delay *tau*, the Doppler spectrum matrix called *sDoppler_psd* and the corresponding frequency array *f*, and the level crossing rate *LCR* and the average fade duration *AFD* with the corresponding crossing levels R_s.

To illustrate how this simulator works, here we present several examples.

Example 1: Varying the Receiver Velocity

To illustrate how changes in the receiver's velocity impact the channel impulse response, we select the following channel parameters: carrier frequency is 2 GHz, sampling frequency is 100 kHz, number of samples is 10,000, and distance between the transmitter and the receiver is 1000m. The 2-D, isotropic, and non-line-of-sight type of propagation is selected and the SISO block is chosen. The channel parameters are shown in Figure B.2. In the first simulation run, the receiver velocity is set to 10 m/s, and in the second to 30 m/s. The results obtained for the channel impulse response are shown in Figure B.3. The results show that, as the velocity of the receiver increases, the variability of the channel impulse response increases. To quantify this variability in the channel, we can use the level crossing rates generated by the simulator for these two channels. These level crossing rates are shown in Figure B.4. The results show that the channel impulse response with the receiver velocity υ_R = 30 m/s is varying three to four times more than the channel impulse response with the receiver velocity υ_R=10 m/s.

Example 2: Line-of-Sight vs. Non-Line-of-Sight Propagation

To illustrate how LoS propagation differs from NLoS propagation, we select the following channel parameters: carrier frequency is 1 GHz, sampling frequency is 10 kHz, number of samples is 10,000, and distance between the transmitter and the receiver is 500m. The 2-D and isotropic type of propagation is selected. The velocity and direction of the receiver are set to υ_R = 10 m/s and γ_R = 0°, respectively. We choose the SISO option and set the antenna directions to 0°. In the first simulation run we select the Rayleigh fading, and in the second simulation run we select the Rician fading with K = 5. The results shown in Figure B.5 illustrate the impact of direct path on channel correlation: when present, the LoS path increases the autocorrelation with respect to time.

Example 3: Isotropic vs. Nonisotropic Scattering

To illustrate how isotropic scattering differs from nonisotropic scattering, we select the same parameters as in Example 2 and choose NLoS propagation. In the first simulation run, we select the spread of scatterers around the receiver to be 0, and, in the second simulation run, we select it to be 5. The results shown in Figure B.6

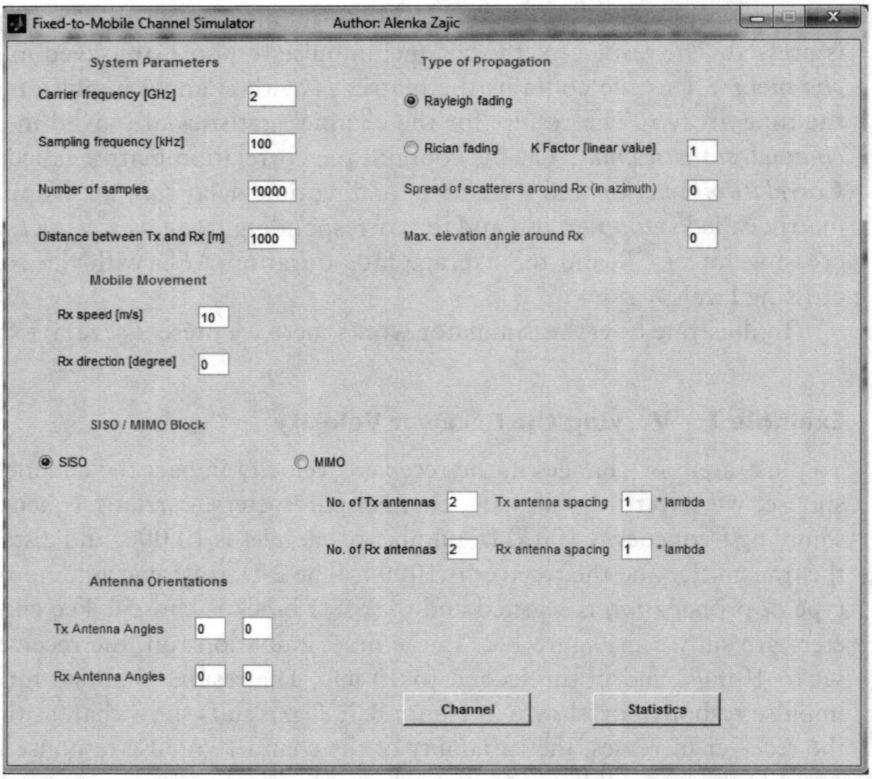

Figure B.2 Model parameters in Example 1: varying the receiver velocity.

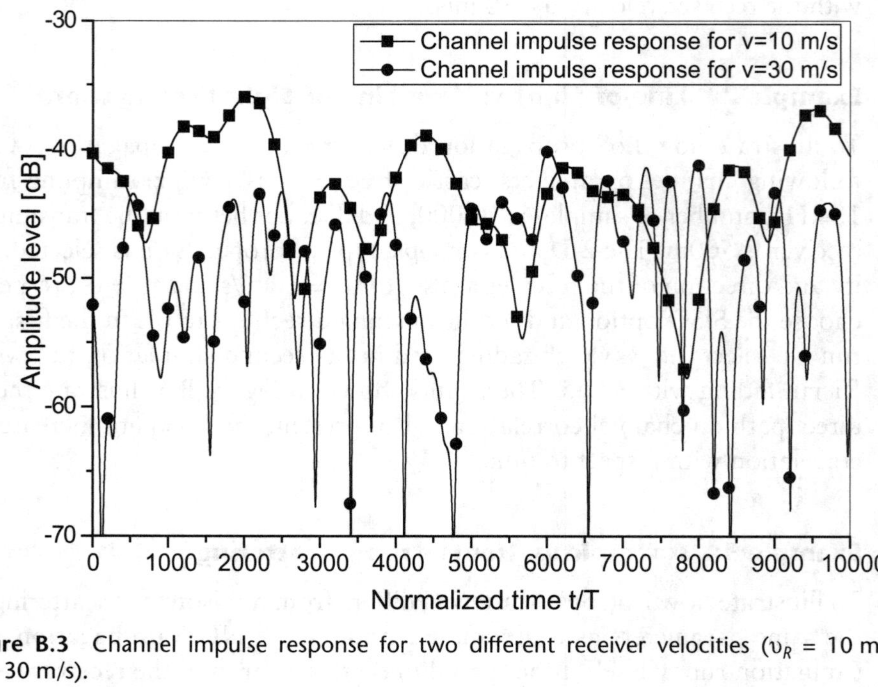

Figure B.3 Channel impulse response for two different receiver velocities (υ_R = 10 m/s and υ_R = 30 m/s).

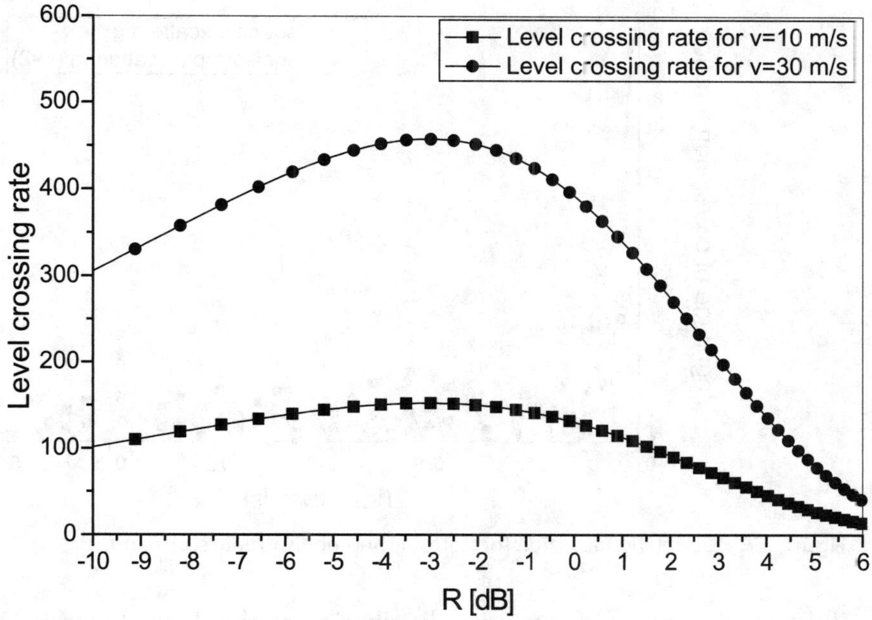

Figure B.4 Level crossing rate for two different receiver velocities (υ_R = 10 m/s and υ_R = 30 m/s).

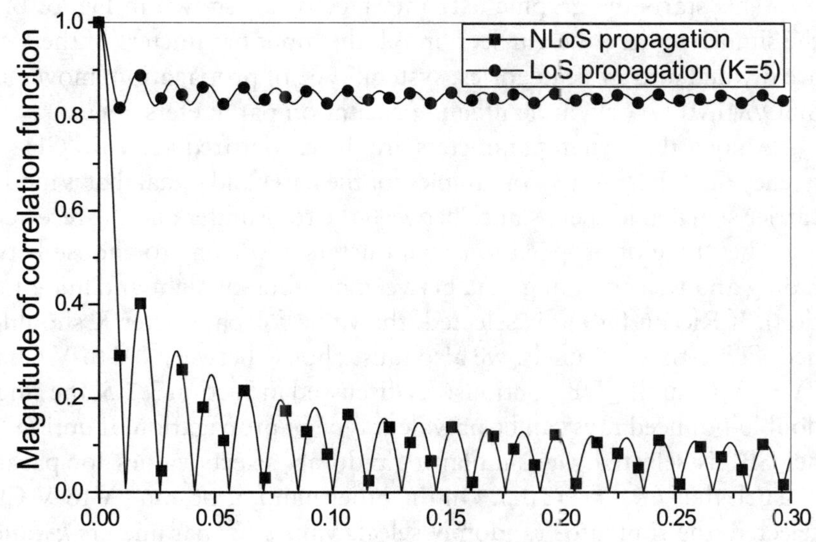

Figure B.5 Correlation function for NLoS and LoS propagation.

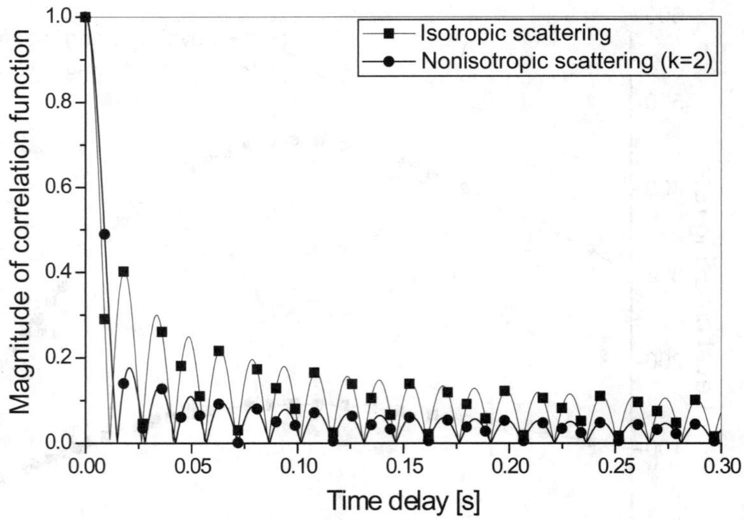

Figure B.6 Correlation function for isotropic and nonisotropic scattering.

illustrate the impact of scattering distribution on the channel correlation: the channel gets more decorrelated when nonisotropic scattering is present.

B.2 Simulator for Vehicle-to-Vehicle Channels

This section describes the MATLAB-based simulator for vehicle-to-vehicle (V-to-V) channels, which can be found on the CD associated with this book, in the *V-to-V Channel* directory. This simulator is based on the 2-D channel models described in Sections 3.1.3, 3.1.4, 3.1.5, and 3.1.6, and on the 3-D model described in Sections 3.1.7 and 3.1.8. The simulator is initiated by running the *V_to_V_channel_start.m* file. This starts the graphic user interface (GUI) shown in Figure B.7, Similarly to the simulator described in Section B.l, the input parameters of the V-to-V simulator are divided into five categories: system, type of propagation, movement of vehicles, SISO/MIMO block, and antenna orientation parameters.

Among the system parameters are the carrier frequency (in GHz), sampling frequency (in kHz), number of samples for the baseband signal that is modulated onto the carrier signal, and the distance between the transmitter and the receiver (in meters).

The "type of propagation" parameters allow one to choose between Rayleigh fading and Ricean fading (i.e., between non-line-of-sight and line-of-sight propagation). If Ricean fading is selected, the value for parameter K should also be specified. For V-to-V channels, we also must choose between "V-to-V Channel_SB" and "V-to-V Channel_DB" options. As discussed in Section 3.1.5, the single-bounced or double-bounced rays can be prevalent type of propagation. If option "V-to-V Channel_SB" is selected, the simulator randomly selects values for parameters k_T and k_R such that $k_R + k_T \geq 0.5$. On the other hand, if option "V-to-V Channel_DB" is selected, the simulator randomly selects values for parameters k_T and k_R such that $k_T + k_R \leq 0.5$. If the user wants to change the policy for selecting single-bounced versus double-bounced rays, this can be done in the file *Input_Parameters.m*. The spreads of scatterers and maximum elevation angles around the transmitter and

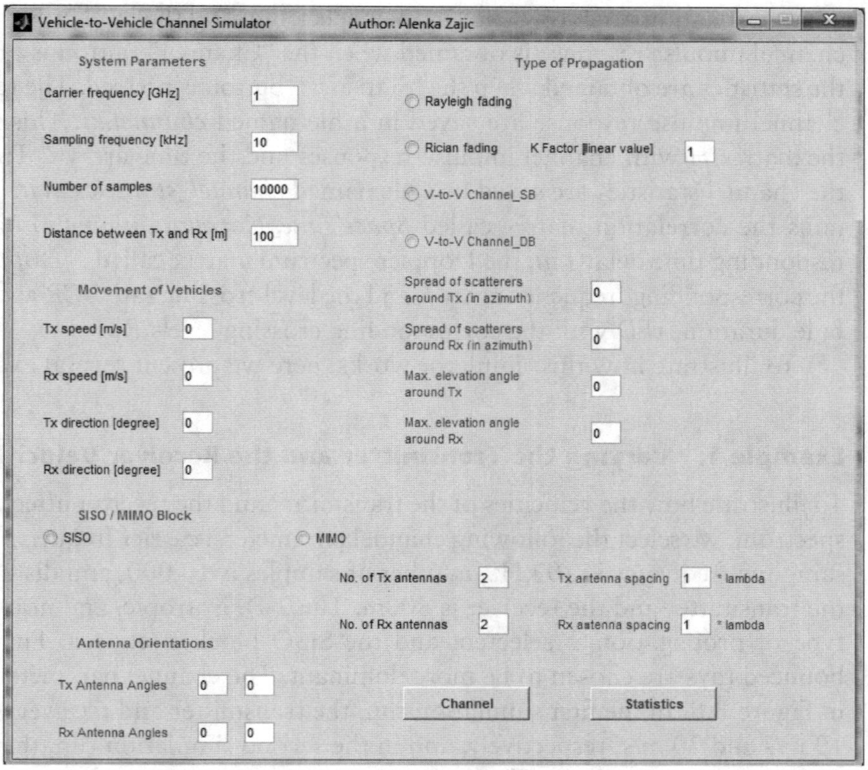

Figure B.7 GUI for vehicle-to-vehicle channel simulator.

receiver implicitly select between isotropic and nonisotropic scattering and between 2-D and 3-D models, respectively. If the spreads of scatterers around the transmitter and receiver are zero, the simulator assumes isotropic scattering, while when the spreads of scatterers around the transmitter and receiver are larger than zero, the simulator assumes nonisotropic scattering. Similarly, if the maximum elevation angles around the transmitter and receiver are zero, the simulator uses the 2-D narrowband model described in Sections 3.1.5 and 3.1.6. If the maximum elevation angles around the transmitter and receiver are greater than zero, the simulator uses the narrowband version of the 3-D model described in Sections 3.1.7 and 3.1.8. For V-to-V systems, the elevation angles typically vary between 0° and 20°.

In the movement of vehicles category, the user selects the velocities and directions of movement for both the transmitter and the receiver.

In the SISO/MIMO block category, the user chooses between a single-input single-output (SISO) and a multiple-input multiple-output (MIMO) system. If the MIMO option is selected, we must also specify the number of transmit antennas and the spacing among them, as well as the number of receive antennas and the spacing among them.

Finally, in the antenna orientation category, the user specifies the azimuthal and elevation angles for antennas. Here, we assume that a MIMO system is using same antenna elements at the transmitter and at the receiver.

Similar to the F-to-M channel simulator, the V-to-V channel simulator generates the channel impulse response as well as the channel statistics (the auto- and cross-correlation functions, the Doppler spectrum, and the level crossing rate). The channel impulse response is obtained when the "Channel" button is pressed, while the statistics are obtained when the "Statistics" button is pressed. The results for the channel impulse response are saved in a file named *channel.m*. This file contains the matrix H with channel impulse responses and the time array t. The results for the channel statistics are saved in a file named *channel_statistics.mat*. The file contains the correlation matrix called *SpaceTime_CorrelationFunction* and the corresponding time delay *tau*, the Doppler spectrum matrix called *s_Doppler_psd* and the corresponding frequency array f, and the level crossing rate *LCR* and the average fade duration *AFD* with the corresponding crossing levels R_s.

To illustrate how this simulator works, here we present several examples.

Example 1: Varying the Transmitter and the Receiver Velocities

To illustrate how the velocities of the transmitter and the receiver affect the Doppler spectrum, we select the following channel parameters: carrier frequency is 2.4 GHz, sampling frequency is 10 kHz, number of samples is 10,000, and distance between the transmitter and the receiver is 300m. The 2-D, isotropic, and non-line-of-sight type of propagation is selected, and the SISO block is chosen. Finally, double-bounced rays are chosen to be more dominant. The channel parameters are shown in Figure B.8. In the first simulation run, the transmitter and receiver velocities are 10 m/s and 20 m/s, respectively, and in the second simulation run, they are 30 m/s and 50 m/s, respectively. The obtained Doppler spectra are shown in Figure B.9. The results show that, as the velocities of the transmitter and the receiver increase, the Doppler spread increases and the Doppler amplitude decreases.

Example 2: Single-Bounced vs. Double-Bounced Rays

In contrast to F-to-M cellular channels, where only single-bounced rays can be dominant under NLoS conditions, in V-to-V channels double-bounced rays can also be dominant. For example, single-bounced rays are prevalent on interstate highways, while double-bounced rays are prevalent in surface street environments [2]. To illustrate how the channel differs when single-bounced or double-bounced rays are prevalent, we select the following channel parameters: carrier frequency is 1 GHz, sampling frequency is 10 kHz, number of samples is 10,000, and distance between the transmitter and the receiver is 500m. The 2-D and isotropic type of propagation is selected. The velocities of the transmitter and the receiver are $\upsilon_T = 10$ m/s and $\upsilon_R = 10$ m/s, respectively. The directions of the transmitter and the receiver are $\gamma_T = 0°$ and $\gamma_R = 0°$, respectively. We choose the SISO option and set the antenna directions to 0°. In the first simulation run, we select the "V-to-V Channel_SB" option, and, in the second simulation run, we select the "V-to-V Channel_DB" option.

Figure B.10 compares the autocorrelation functions of the channel impulse responses for these two simulation runs. The results show that the signals are more self-correlated when the single-bounced rays are dominant. This is an expected result because double-bounced signals experience more phase shifts, which decorrelates them faster.

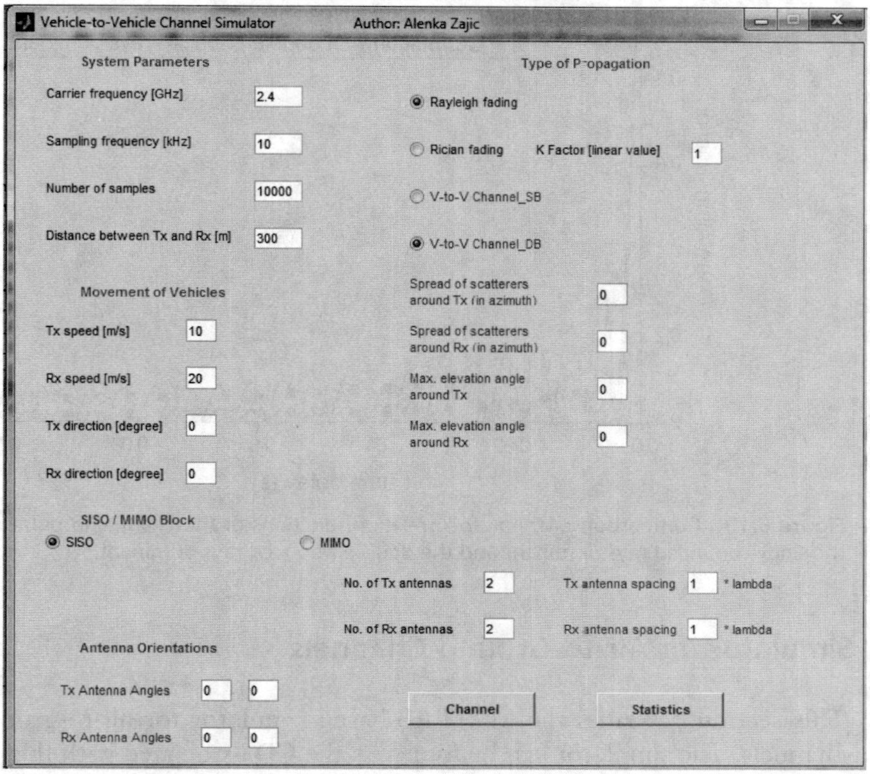

Figure B.8 Model parameters in example 1: varying the transmitter and the receiver velocities.

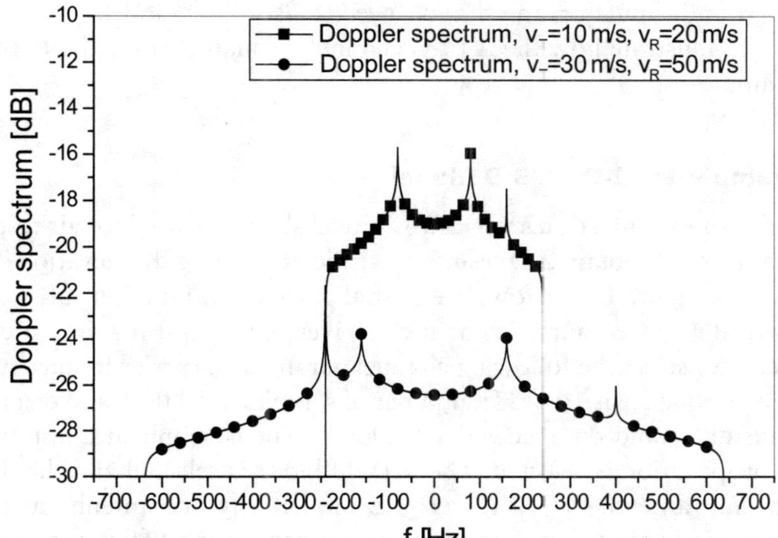

Figure B.9 The Doppler spectrum for two different sets of velocities: υ_T = 10 m/s, υ_R = 20 m/s and υ_T = 30 m/s, υ_T = 50 m/s.

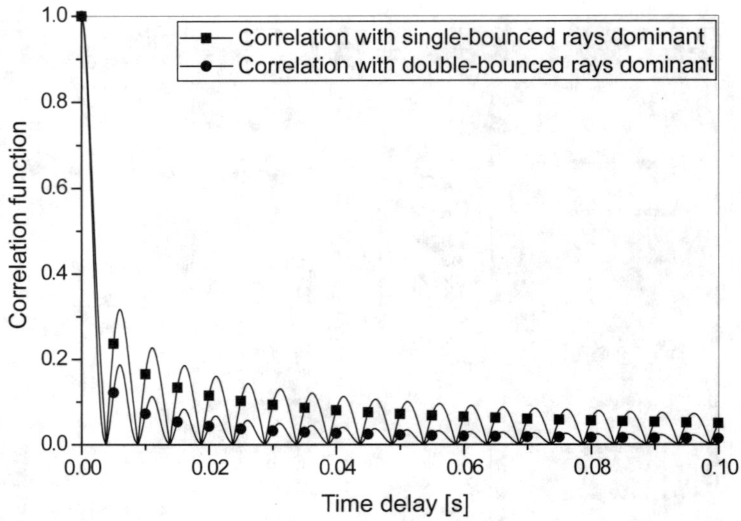

Figure B.10 Comparison of the autocorrelation functions of the channel impulse responses with the single-bounced rays dominant and the double-bounced rays dominant.

B.3 Simulator for Air-to-Ground Channels

This section describes the MATLAB-based simulator for air-to-ground (A-to-G) channels. The simulator can be found on the CD associated with this book, in the *A-to-G Channel* directory. This simulator is based on the 3-D model described in Sections 4.2.5, 4.2.6, and 4.3.1.

The simulator is initiated by running the *A_to_G_channel_start.m* file. This file starts graphic user interface (GUI) shown in Figure B.11. The model used for the air-to-ground simulator is very similar to the model for F-to-M cellular channels. The only differences are that both the transmitter and the receiver move in the air-to-ground simulator, and that we need to use a 3-D model.

To illustrate how the A-to-G channel simulator works, here we present several examples.

Example 1: 2-D vs. 3-D Model

This example illustrates why a 3-D model should be used for air-to-ground (A-to-G) channels. We obtain 2-D results by artificially setting the elevation in the 3-D model to zero. Figure B.12 shows the spatial correlation functions of two uniformly and vertically spaced antennas at the receiver for several maximum elevation angles. Here, we select the following channel parameters: carrier frequency is 2 GHz, sampling frequency is 100 kHz, number of samples is 10,000, and distance between the transmitter and the receiver is 10,000m. The isotropic and non-line-of-sight type of propagation is selected. The MIMO block is selected and the distance between transmit antennas is set to zero (to model only one antenna at the transmitter), while the distance between receive antennas is varied from zero to 5 λ. The transmitter velocity is $\upsilon_T = 100$ m/s, the receiver velocity is $\upsilon_R = 50$ m/s, and the orientations of the transmitter and the receiver are $\gamma_T = 0°$ and $\gamma_R = 0°$, respectively. We run

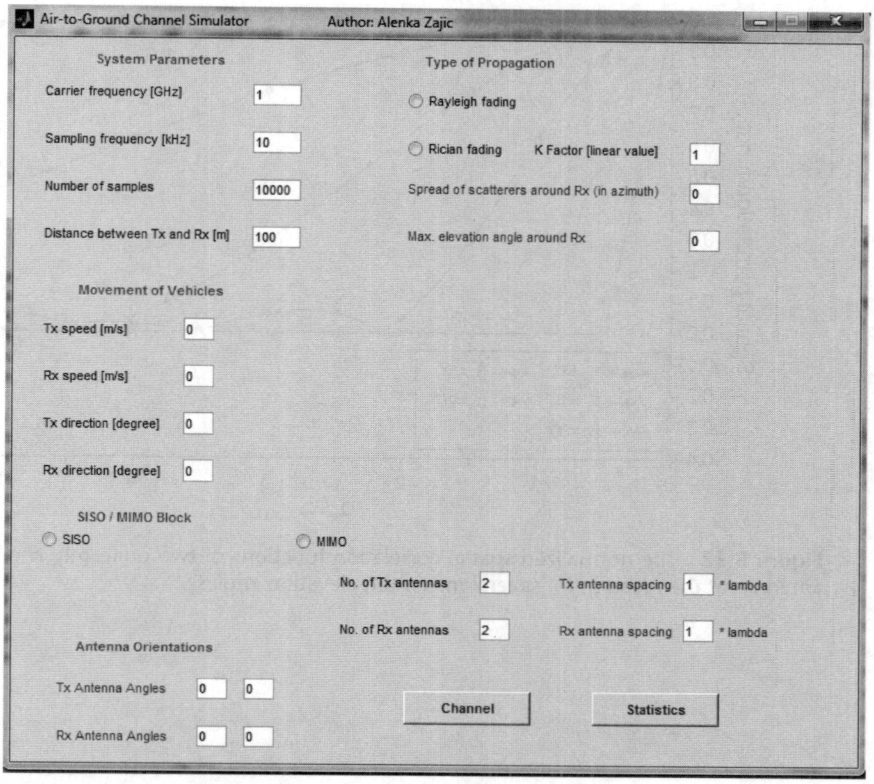

Figure B.11 GUI for air-to-ground channel simulator.

multiple simulations with different values for the "Rx Antenna Spacing" parameter (X-axis in Figure B.l2) and for the "Max. elevation angle around Rx" parameter (different lines in Figure B.12). The GUI with one such set of the channel parameters is shown in Figure B.13.

The results in Figure B.12 show that, as the maximum elevation angle increases from 1° to 80° and beyond, the correlation between the two receive antennas is dramatically reduced. This means that, if we ignore the 3-D aspect of the propagation, it will appear that no diversity gain is possible from using more than one antenna, whereas in reality significant diversity is present even if the elevation angles of the scatterers are modest (e.g., 20°).

Example 2: Varying the Receiver Velocity

To illustrate how the change in receiver velocity impacts the A-to-G channel impulse response, we select the following channel parameters: carrier frequency is 2.1 GHz, sampling frequency is 100 kHz, number of samples is 10,000, and distance between the transmitter and the receiver is 20,000m. The 3-D, nonisotropic, and non-line-of-sight type of propagation is selected, and the SISO block is chosen. The GUI with these parameters is shown in Figure B.14. In the first simulation run, the receiver velocity is set to 50 m/s, and in the second to 100 m/s. To quantify the variability in the channel, Figure B.15 compares the level crossing rates for the two different velocities of the receiver. The results show that the channel impulse response with the

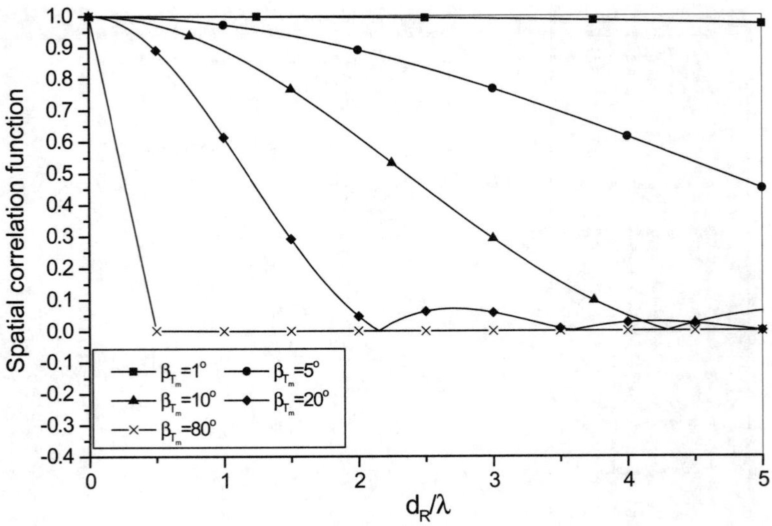

Figure B.12 The normalized spatial correlation functions of two uniformly and vertically spaced antennas at the receiver, for several maximum elevation angles.

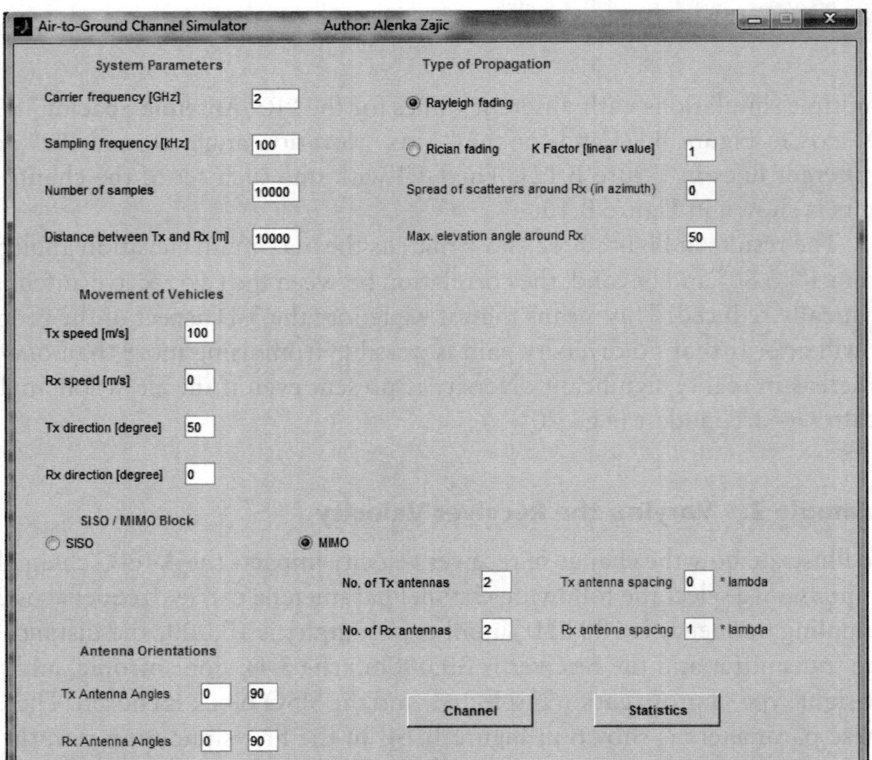

Figure B.13 The channel parameters used in Example 1: 2-D vs. 3-D model.

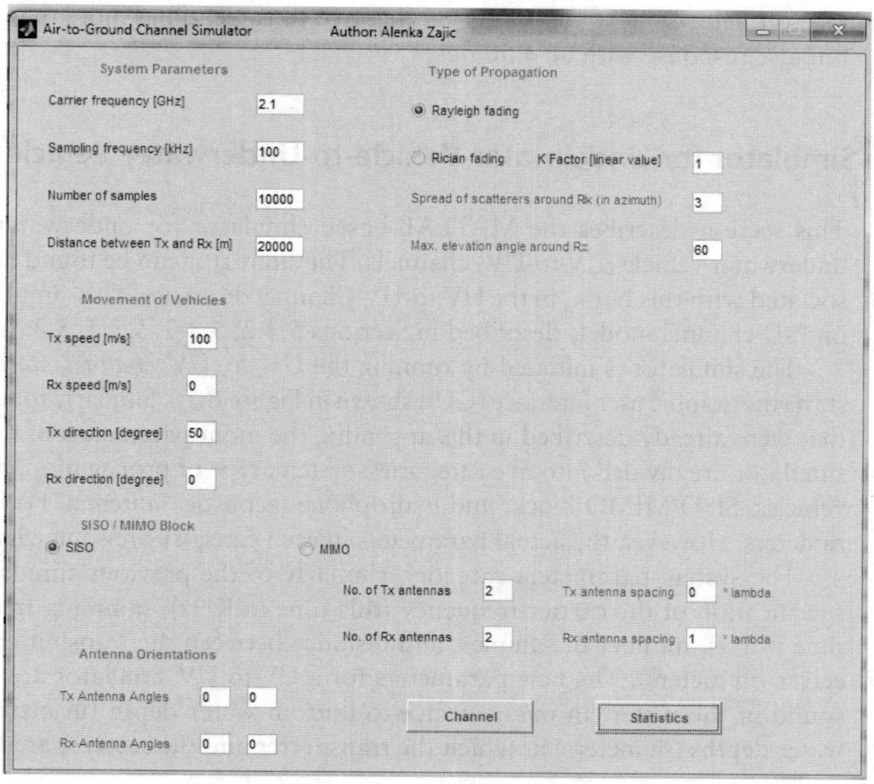

Figure B.14 Model parameters in Example 2: varying the receiver velocity.

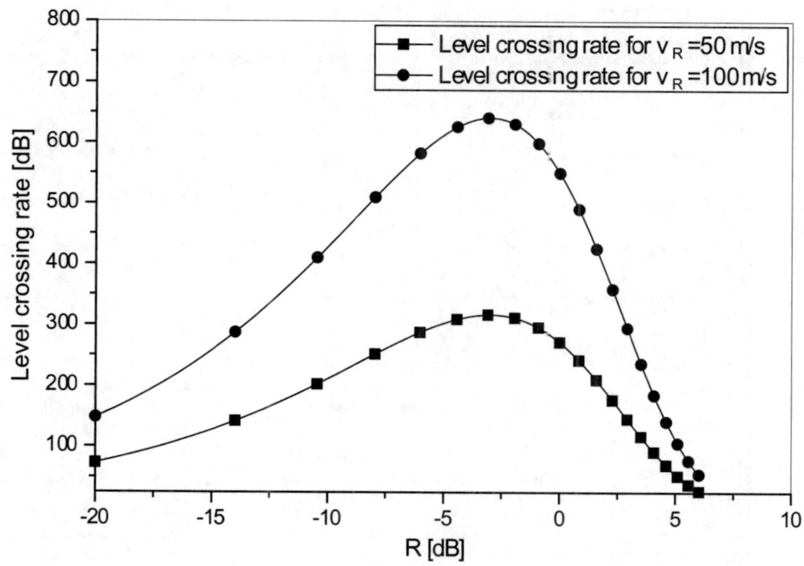

Figure B.15 Level crossing rate for receiver velocities (υ_R = 50 m/s and υ_R = 100 m/s).

receiver velocity υ_R = 100 m/s is varying two to three, times more than the channel impulse response with υ_R = 50 m/s.

B.4 Simulator for Underwater Vehicle-to-Underwater Vehicle Channels

This section describes the MATLAB-based simulator for underwater vehicle-to-underwater vehicle (UV-to-UV) channels. The simulator can be found on the CD associated with this book, in the UV-to-UV Channel directory. This simulator is based on 2-D channel models described in Sections 5.3.2, 5.3.3, 5.3.4, 5.3.5, and 5.4.1.

The simulator is initiated by running the *UV_to_UV_channel_start.m* file. This starts the graphic user interface (GUI) shown in Figure B.16. Similarly to the simulators that were already described in this appendix, the input parameters of the UV-to-UV simulator are divided into five categories: system, type of propagation, movement of vehicles, SISO/MIMO block, and hydrophone (acoustic "antenna") orientation parameters. However, the actual parameters in each category are somewhat different.

The system parameters category, similarly to the previous simulators, allows specification of the carrier frequency (this time in KHz), sampling frequency (this time in Hz), number of samples, and distance between the transmitter and the receiver (in meters). The new parameters for a UV-to-UV simulator are the speed of sound in the water (in m/s), surface-to-bottom water depth (in meters), and the water depths (in meters) at which the transmitter and the receiver are located.

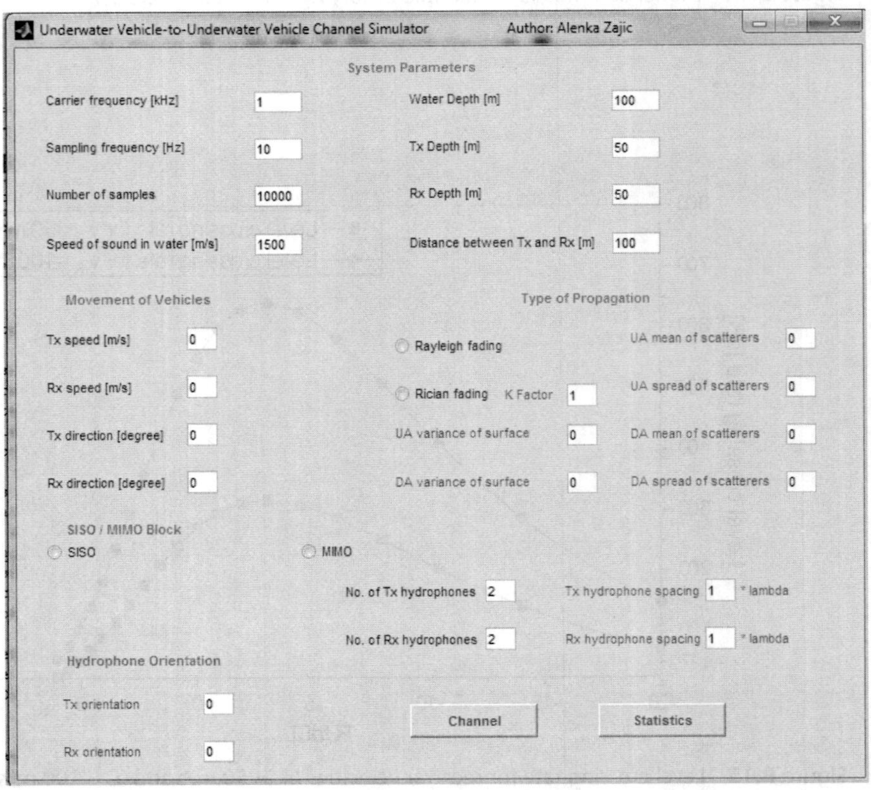

Figure B.16 GUI for underwater vehicle-to-underwater vehicle channel simulator.

Like in other models in this appendix, the "type of propagation" parameters allow us to choose between Rayleigh fading and Ricean fading (i.e., between non-line-of-sight and line-of-sight propagation). If Ricean fading is selected, the value for parameter K should also be specified. For UV-to-UV systems, the parameter K ranges between 0.2 and 5. Additional "type-of-propagation" parameters for UV-to-UV channels allow us to specify the impact of surface movement (variance of surface) on upward- and downward-arriving rays. These numbers vary between 0 and 1. Finally, we can specify the mean angles and the spread angles of scatterers that impact the upward- and downward-arriving rays, respectively. The mean angle of scatterers that impacts the upper-arriving rays varies between 0° and 180°, while the mean angle of scatterers that impacts the downward-arriving rays varies between 180° and 360°. The spread angle is typically between 0° and 10°. In the movement of vehicles category, the user selects the velocities of the transmitter and the receiver (in m/s) and their directions of movement (in degrees). In the SISO/MIMO block category, the user chooses between a single-input single-output (SISO) option and multiple-input multiple-output (MIMO) option. If the MIMO option is selected, user needs to specify the number of transmit hydrophones and the spacing among them, as well as the number of receive hydrophones and the spacing among them.

Finally, in the hydrophone orientation category, the user specifies the azimuthal angles for the hydrophones. Here, we assume that MIMO system is using the same hydrophone elements at the transmitter and at the receiver.

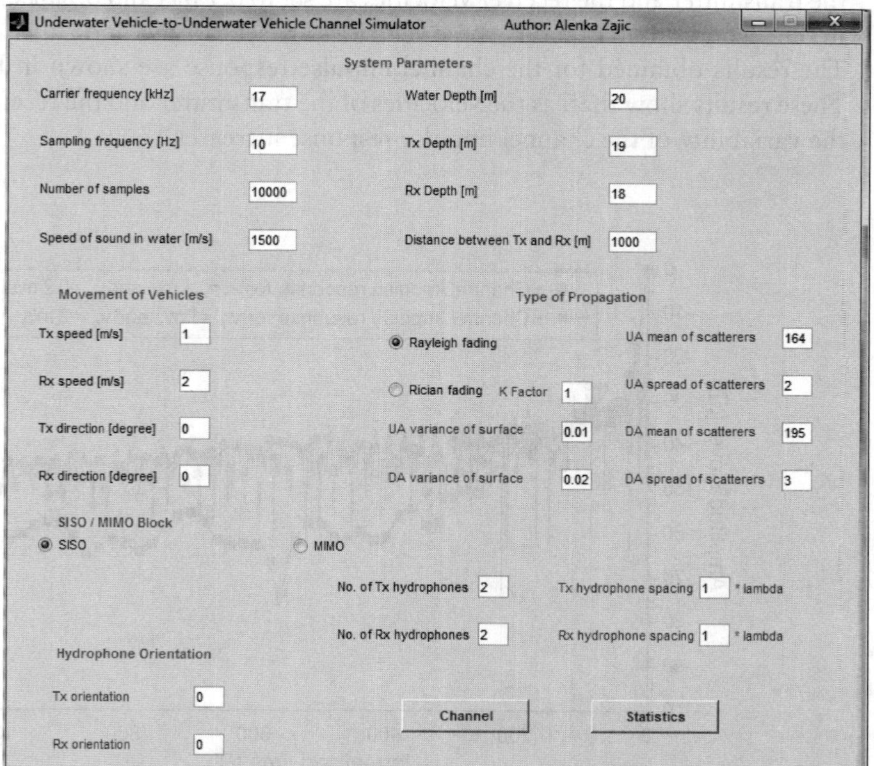

Figure B.17 Model parameters for example 1 of UV-to-UV channel simulator.

The simulator provides the channel impulse response as well as the channel statistics (the auto- and cross-correlation functions and the Doppler spectrum). The channel impulse response is obtained when the "Channel" button is pressed, while the statistics are obtained when the "Statistics" button is pressed. The results for the channel impulse response are saved in a file named *channel.m*. This file contains the matrix *h* with channel impulse responses and the time array *t*. The results for the channel statistics are saved in a file named *channel_statistics.mat*. The file contains the correlation matrix called *corr* and the corresponding time delay *tau*, along with the Doppler spectrum matrix called *Dopp* and the corresponding frequency array *f*. To illustrate how this simulator works, here we present several examples.

Example I: Varying Transmitter and Receiver Velocities

To illustrate how the change in receiver velocity impacts the channel impulse response, we select the following channel parameters: carrier frequency is 17 kHz, sampling frequency is 100 Hz, number of samples is 10,000, and distance between the transmitter and the receiver is 1000m. The speed of sound in water is 1500 m/s, water depth is 20m, transmitter depth is 19m, and receiver depth is 18m. The variance of surface is 0.01 for upward-arriving rays and 0.02 for downward-arriving rays, while the mean angle and the spread angle are 164° and 2°, respectively, for upward-arriving rays and 195° and 3°, respectively, for downward-arriving rays. Non-line-of-sight propagation is selected, and the SISO block is chosen. The GUI with these channel parameters is shown in Figure B.17. In the first simulation run, the transmitter and the receiver velocities are set to 0.1 m/s and 0.2 m/s, respectively. In the second simulation run, these velocities are set to 1 m/s and 2 m/s, respectively. The results obtained for the channel impulse response are shown in Figure B.18. These results show that, as the velocities of the transmitter and the receiver increase, the variability of the channel impulse response increases.

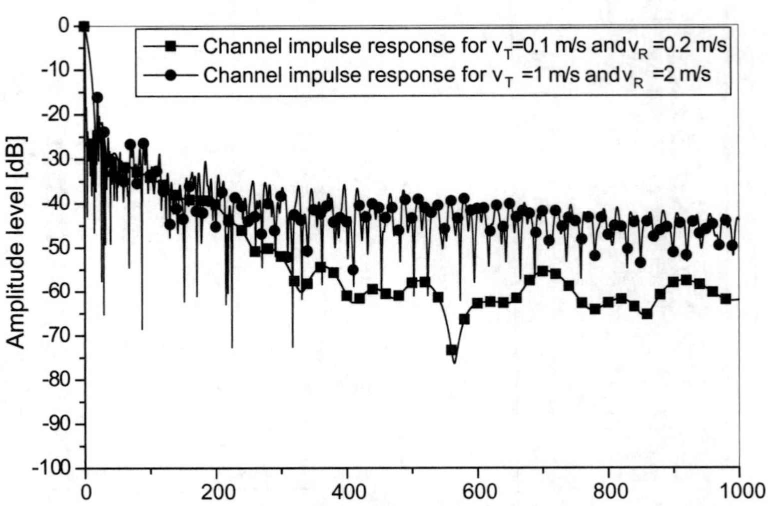

Figure B.18 Channel impulse response for two different transmitter and receiver velocities.

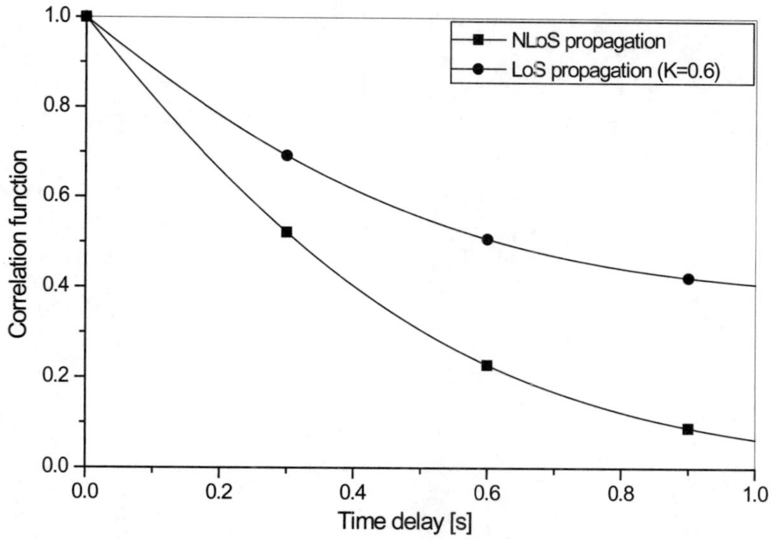

Figure B.19 Correlation function for NLoS and LoS propagation.

Example 2: Line-of-Sight vs. Non-Line-of-Sight

To illustrate how LoS propagation differs from NLoS propagation, we select the following channel parameters: carrier frequency is 20 kHz, sampling frequency is 10 Hz, number of samples is 10,000, and distance between the transmitter and the receiver is 2000m. The speed of sound in water is 1500 m/s, water depth is 100m, transmitter depth is 50m, and receiver depth is 50m. The variance of surface is 0.1 for both upward- and downward-arriving rays. The mean angle and the spread angle are 166° and 2°, respectively, for upward-arriving rays and 192° and 3°, respectively, for downward-arriving rays. The velocities and directions of both the transmitter and the receiver are $\upsilon_T = \upsilon_R = 0.05$ m/s and $\gamma_T = \gamma_R = 0°$, respectively. We choose the SISO option and set the hydrophone directions to 0°. In the first simulation run we select Rayleigh fading, while for the second run we select Rician fading with $K = 0.6$. The results for channel autocorrelation are shown in Figure B.19. These results illustrate the impact of a direct (LoS) path on channel autocorrelation: when present, the LoS path increases the autocorrelation with respect to time.

References

[1] Parsons J. D. and A. M. D. Turkmani, "Characterisation of mobile radio signals: model description," *IEE Proceedings I, Communications, Speech, and Vision*, Vol. 138, Dec. 1991, pp. 549–556.

[2] Zajić A. G., G. L. Stüber, T. G. Pratt, and S. Nguyen, "Wideband MIMO mobile-to-mobile channels: geometry-based statistical modeling with experimental verification," *IEEE Transactions on Vekicular Technology*, Vol. 58, Feb. 2009, pp. 517–34.

About the Author

Alenka G. Zajić was born in Belgrade, Serbia, on April 28, 1977. She received the B.Sc. and M.Sc. degrees from the School of Electrical Engineering, University of Belgrade, in 2001 and 2003, respectively. She received her Ph.D. degree in Electrical and Computer Engineering from the Georgia Institute of Technology in 2008. Currently, she is an Assistant Professor in the School of Electrical and Computer Engineering at the Georgia Institute of Technology. Prior to that, she was visiting faculty in the School of Computer Science at the Georgia Institute of Technology, a post-doctoral fellow in the Naval Research Laboratory, and a design engineer at Skyworks Solutions. Her research interests span areas of electromagnetics, wireless communications, signal processing, and computer engineering. Dr. Zajić received the Best Paper Award at ICT 2008, the Best Student Paper Award at WCNC 2007, and she was also the recipient of the Dan Noble Fellowship in 2004, awarded by Motorola and the IEEE Vehicular Technology Society for quality impact in the area of vehicular technology. In 2012, she received the Neal Shepherd Memorial Best Propagation Paper Award.

Index

N

O

P

Recent Titles in the Artech House Mobile Communications Series

John Walker, Series Editor

3G CDMA2000 Wireless System Engineering, Samuel C. Yang

3G Multimedia Network Services, Accounting, and User Profiles, Freddy Ghys, Marcel Mampaey, Michel Smouts, and Arto Vaaraniemi

802.11 WLANs and IP Networking: Security, QoS, and Mobility, Anand R. Prasad and Neeli R. Prasad

Achieving Interoperability in Critical IT and Communications Systems, Robert I. Desourdis, Peter J. Rosamilia, Christopher P. Jacobson, James E. Sinclair, and James R. McClure

Advances in 3G Enhanced Technologies for Wireless Communications, Jiangzhou Wang and Tung-Sang Ng, editors

Advances in Mobile Information Systems, John Walker, editor

Advances in Mobile Radio Access Networks, Y. Jay Guo

Applied Satellite Navigation Using GPS, GALILEO, and Augmentation Systems, Ramjee Prasad and Marina Ruggieri

Artificial Intelligence in Wireless Communications, Thomas W. Rondeau and Charles W. Bostian

Broadband Wireless Access and Local Network: Mobile WiMax and WiFi, Byeong Gi Lee and Sunghyun Choi

CDMA for Wireless Personal Communications, Ramjee Prasad

CDMA Mobile Radio Design, John B. Groe and Lawrence E. Larson

CDMA RF System Engineering, Samuel C. Yang

CDMA Systems Capacity Engineering, Kiseon Kim and Insoo Koo

CDMA Systems Engineering Handbook, Jhong S. Lee and Leonard E. Miller

Cell Planning for Wireless Communications, Manuel F. Cátedra and Jesús Pérez-Arriaga

Cellular Communications: Worldwide Market Development, Garry A. Garrard

Cellular Mobile Systems Engineering, Saleh Faruque

Cognitive Radio Techniques: Spectrum Sensing, Interference Mitigation, and Localization, Kandeepan Sithamparanathan and Andrea Giorgetti

The Complete Wireless Communications Professional: A Guide for Engineers and Managers, William Webb

EDGE for Mobile Internet, Emmanuel Seurre, Patrick Savelli and Pierre-Jean Pietri

Emerging Public Safety Wireless Communication Systems, Robert I. Desourdis, Jr., et al.

The Future of Wireless Communications, William Webb

Geographic Information Systems Demystified, Stephen R. Galati

GPRS for Mobile Internet, Emmanuel Seurre, Patrick Savelli, and Pierre-Jean Pietri

GPRS: Gateway to Third Generation Mobile Networks, Gunnar Heine and Holger Sagkob

GSM and Personal Communications Handbook, Siegmund M. Redl, Matthias K. Weber, and Malcolm W. Oliphant

GSM Networks: Protocols, Terminology, and Implementation, Gunnar Heine

GSM System Engineering, Asha Mehrotra

Handbook of Land-Mobile Radio System Coverage, Garry C. Hess

Handbook of Mobile Radio Networks, Sami Tabbane

High-Speed Wireless ATM and LANs, Benny Bing

Interference Analysis and Reduction for Wireless Systems, Peter Stavroulakis

Introduction to 3G Mobile Communications, Second Edition, Juha Korhonen

Introduction to Communication Systems Simulation, Maurice Schiff

Introduction to Digital Professional Mobile Radio, Hans-Peter A. Ketterling

Introduction to GPS: The Global Positioning System, Ahmed El-Rabbany

An Introduction to GSM, Siegmund M. Redl, Matthias K. Weber, and Malcolm W. Oliphant

Introduction to Mobile Communications Engineering, José M. Hernando and F. Pérez-Fontán

Introduction to Radio Propagation for Fixed and Mobile Communications, John Doble

Introduction to Wireless Local Loop, Broadband and Narrowband, Systems, Second Edition, William Webb

IS-136 TDMA Technology, Economics, and Services, Lawrence Harte, Adrian Smith, and Charles A. Jacobs

Location Management and Routing in Mobile Wireless Networks, Amitava Mukherjee, Somprakash Bandyopadhyay, and Debashis Saha

LTE Air Interface Protocols, Mohammad T. Kawser

Mobile Data Communications Systems, Peter Wong and David Britland

Mobile IP Technology for M-Business, Mark Norris

Mobile Satellite Communications, Shingo Ohmori, Hiromitsu Wakana, and Seiichiro Kawase

Mobile Telecommunications Standards: GSM, UMTS, TETRA, and ERMES, Rudi Bekkers

Mobile-to-Mobile Wireless Channels, Alenka Zajić

Mobile Telecommunications: Standards, Regulation, and Applications, Rudi Bekkers and Jan Smits

Multiantenna Digital Radio Transmission, Massimiliano Martone

Multiantenna Wireless Communications Systems, Sergio Barbarossa

Multi-Gigabit Microwave and Millimeter-Wave Wireless Communications, Jonathan Wells

Multipath Phenomena in Cellular Networks, Nathan Blaunstein and Jørgen Bach Andersen

Multiuser Detection in CDMA Mobile Terminals, Piero Castoldi

OFDMA for Broadband Wireless Access, Slawomir Pietrzyk

Personal Wireless Communication with DECT and PWT, John Phillips and Gerard MacNamee

Practical Wireless Data Modem Design, Jonathon Y. C. Cheah

Prime Codes with Applications to CDMA Optical and Wireless Networks, Guu-Chang Yang and Wing C. Kwong

Quantitative Analysis of Cognitive Radio and Network Performance, Preston Marshall

QoS in Integrated 3G Networks, Robert Lloyd-Evans

Radio Engineering for Wireless Communication and Sensor Applications, Antti V. Räisänen and Arto Lehto

Radio Propagation in Cellular Networks, Nathan Blaunstein

Radio Resource Management for Wireless Networks, Jens Zander and Seong-Lyun Kim

Radiowave Propagation and Antennas for Personal Communications, Third Edition, Kazimierz Siwiak and Yasaman Bahreini

RDS: The Radio Data System, Dietmar Kopitz and Bev Marks

Resource Allocation in Hierarchical Cellular Systems, Lauro Ortigoza-Guerrero and A. Hamid Aghvami

RF and Baseband Techniques for Software-Defined Radio, Peter B. Kenington

RF and Microwave Circuit Design for Wireless Communications, Lawrence E. Larson, editor

Sample Rate Conversion in Software Configurable Radios, Tim Hentschel

Signal Processing Applications in CDMA Communications, Hui Liu

Smart Antenna Engineering, Ahmed El Zooghby

Software Defined Radio for 3G, Paul Burns

Spread Spectrum CDMA Systems for Wireless Communications, Savo G. Glisic and Branka Vucetic

Technologies and Systems for Access and Transport Networks, Jan A. Audestad

Third-Generation and Wideband HF Radio Communications, Eric E. Johnson, Eric Koski, William N. Furman, Mark Jorgenson, and John Nieto

Third Generation Wireless Systems, Volume 1: Post-Shannon Signal Architectures, George M. Calhoun

Traffic Analysis and Design of Wireless IP Networks, Toni Janevski

Transmission Systems Design Handbook for Wireless Networks, Harvey Lehpamer

UMTS and Mobile Computing, Alexander Joseph Huber and Josef Franz Huber

Understanding Cellular Radio, William Webb

Understanding Digital PCS: The TDMA Standard, Cameron Kelly Coursey

Understanding GPS: Principles and Applications, Second Edtion, Elliott D. Kaplan and Christopher J. Hegarty, editors

Understanding WAP: Wireless Applications, Devices, and Services, Marcel van der Heijden and Marcus Taylor, editors

Universal Wireless Personal Communications, Ramjee Prasad

WCDMA: Towards IP Mobility and Mobile Internet, Tero Ojanperä and Ramjee Prasad, editors

Wireless Communications in Developing Countries: Cellular and Satellite Systems, Rachael E. Schwartz

Wireless Communications Evolution to 3G and Beyond, Saad Z. Asif

Wireless Intelligent Networking, Gerry Christensen, Paul G. Florack, and Robert Duncan

Wireless LAN Standards and Applications, Asunción Santamaría and Francisco J. López-Hernández, editors

Wireless Sensor and Ad Hoc Networks Under Diversified Network Scenarios, Subir Kumar Sarkar

Wireless Technician's Handbook, Second Edition, Andrew Miceli

For further information on these and other Artech House titles, including previously considered out-of-print books now available through our In-Print-Forever® (IPF®) program, contact:

Artech House
685 Canton Street
Norwood, MA 02062
Phone: 781-769-9750
Fax: 781-769-6334
e-mail: artech@artechhouse.com

Artech House
16 Sussex Street
London SW1V 4RW UK
Phone: +44 (0)20 7596-8750
Fax: +44 (0)20 7630-0166
e-mail: artech-uk@artechhouse.com

Find us on the World Wide Web at: www.artechhouse.com

